Elasticity for Engineers

Consulting Editor
Professor P. B. Morice
University of Southampton

Elasticity for Engineers

D. S. Dugdale

*John Brown Professor of Cutting Tool Technology,
University of Sheffield*

C. Ruiz

*University Lecturer in Engineering and
Fellow of Exeter College, University of Oxford.*

McGRAW-HILL · LONDON

New York · St Louis · San Francisco · Düsseldorf · Johannesburg
Kuala Lumpur · Mexico · Montreal · New Delhi · Panama
Rio de Janeiro · Singapore · Sydney · Toronto

Published by
McGraw-Hill Publishing Company Limited
MAIDENHEAD · BERKSHIRE · ENGLAND

07 094152 1

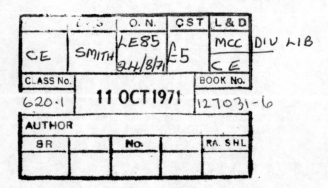

PRINTED AND BOUND IN GREAT BRITAIN

Preface

The subject of elasticity is regarded here as simply a means of calculating stresses. Although inelastic deformation must often be expected and allowed for, elasticity still provides a firm basis for determining the proportions of structures and machine parts.

It is assumed that the reader has already had some acquaintance with the theory of beams and shafts, as may be found in basic texts on the strength of materials. This is adequate for finding stress in members of uniform section. These results may be extended to practical situations by using safety factors based on experience. However, there has been a growing awareness for over half a century of the need for more accurate estimates of local stresses associated with geometrical irregularities. To take a couple of examples, an aircraft accident could be caused by the growth of a fatigue crack from a notch in an engine component, or a ship might be lost through brittle fracture from a cracked weld. Therefore, to attain reliability and avoid costly failures, the engineer must have a keen appreciation of stress concentrations.

Although formulas and tabulated data are available in books on design, the present book is intended to satisfy those who wish to gain the full understanding of stress analysis that comes from deriving values from first principles. The aim has been to give a closely reasoned account of methods of analysis, while leaving the reader in a position to deal with truly practical situations. As the variety of practical problems is unlimited, it was thought best to select half a dozen topics of basic relevance to mechanical and structural engineering and to develop them to a level that allows easy transition to specialized texts and papers.

Chapter 1 is intended to serve the remainder of the book by providing a few of the fundamental relationships. Chapter 2 deals chiefly with the complex variable method for plane elasticity, as the need was felt for this rather one-sided treatment. Real stress functions of the Airy type are indeed simpler to use in simple situations, but they have limitations, and have been well treated in other books. In chapter 3, the intention has been to develop thin plate theory to a practically useful stage without leaving behind any areas of inadequate physical reasoning.

From this point onwards, emphasis is placed on methods of dealing with engineering problems rather than on obtaining mathematical solutions. Chapter 4 is concerned with energy methods, including an easy introduction to the variational method of analysis. Chapter 5 deals with two methods of stress analysis that have become widely used, namely, the finite difference or relaxation method and the method of finite elements, both of which can be used in conjunction with digital computers. The calculation of stress in curved shells tends to require specialized techniques which are introduced in chapter 6, and which have obvious importance in the design of pressure vessels of all kinds. When theoretical methods are unsuitable for giving an answer of sufficient accuracy in a limited time, experimental methods can be used. Some of these are outlined in chapter 7. These methods may, of course, be used when working loads are unknown, and calculation is therefore impossible.

Regarding mathematical standards, the technology undergraduate need not fear that he will come up against any impassable barriers, though he may not have time to explore everything in the first reading. Naturally, the authors have tried to make easy any paths that appeared to be inadequately mapped out in existing texts. Where texts giving a fuller or more advanced treatment are mentioned, the author's name may be found listed alphabetically at the end of the book, with the full reference.

The book was written while the authors worked in the Department of Mechanical Engineering, University of Sheffield, and one of the authors (D.S.D.) wishes to acknowledge the support of Messrs. Firth Brown Tools and the John Brown Group.

D. S. Dugdale
C. Ruiz

Contents

1. Elements of Stress and Strain

The practical reason for wishing to know values of stresses acting within a load-carrying member is to predict whether the member will deform excessively or develop a sudden or progressive fracture. If tensile tests are carried out on two bars, one having twice the cross-sectional area of the other, the thicker bar would be expected to sustain twice the breaking load of the thinner one. As this is usually found to be so, the strength of the material may be reported as the breaking force divided by the area of section, that is, as a stress. With a knowledge of this stress, a designer can provide members of sufficient size to withstand the known working loads with safety.

Average stress may be defined as a force divided by the area of the surface through which it acts. When this surface is made smaller without limit, stress tends towards a definite value at a point. Stress will, in general, have different values at different points, and it will be one of the main purposes of this book to examine permissible distributions of stress within a solid body. As a starting point, it is necessary to have a precise means for describing forces which act in more than one direction, and the relative disposition of the elementary surfaces through which these forces act.

1.1 Stress Components

Rectangular coordinates may be set up at some point within a continuous solid. For the moment, it is unnecessary to enquire whether the body has an external boundary, or to fix the coordinate axes in any relationship with such a boundary. By referring to these three axes, the disposition of any elementary surface can be unambiguously specified by giving the direction of its normal. For example, consider an elementary block having its edges parallel to the coordinate axes. The faces of this block have areas that can be written δA_x, δA_y, and δA_z, the subscript showing the reference axis which is normal to each surface. Similarly, the directions of force vectors δF_x, δF_y, and δF_z are indicated by subscripts. Direct stress can now be defined as a force divided by a surface area, this surface being perpendicular to the direction of the force. As illustrated in Fig. 1.1(a), the direct stress in the x-direction is given by $\sigma_{xx} = \delta F_x/\delta A_x$. Two subscripts are used here, the first defining the surface on which the stress acts, and the second defining the direction in which the stress acts. Other direct stresses σ_{yy} and σ_{zz} may be defined in a similar way. For brevity, direct stresses are often written with a single subscript, e.g., σ_x.

Forces can also act tangentially on the faces of this elementary block. Such a force is marked δF_y in Fig. 1.1(a) and it acts in the y-direction on the face of area δA_x. This gives rise to a shear stress $\sigma_{xy} = \delta F_y / \delta A_x$. The subscripts refer respectively to the normal to the surface on which the stress acts and to the direction in which the stress acts. Each stress component acts equally on a

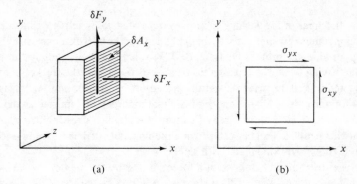

Fig. 1.1 Direct and shear stresses

pair of opposite faces of the elementary block, whether it is a direct stress or a shear stress. Stress is considered positive when it corresponds with a positively directed force acting on the face which is outermost with respect to the origin of positively directed coordinate axes. Positive shear stresses are shown acting on four faces of an element in Fig. 1.1(b). However, this scheme of shear stresses can be simplified by considering moments of forces applied to the element, these moments being taken about some axis perpendicular to the (x,y) plane. Equilibrium requires that the so-called complementary shear stress σ_{yx} should be numerically equal to σ_{xy}. This means that only three independent shear stresses may be present, these being σ_{xy}, σ_{yz}, and σ_{zx}. With the three components of direct stress, it is seen that a total of six stress components may act at any chosen point.

It is not necessary that stresses should always be considered as acting normally or tangentially to a surface, as the resultant of these components may be considered as acting obliquely on the surface. When a particular surface is specified, stress components multiplied by the area of this surface become equal to forces, and can be treated as force vectors. For example, considering an area δA_y situated in the (x,z) plane as shown in Fig. 1.2(a), the stress acting obliquely in the n-direction may be written $Y_n = \delta F_n / \delta A_y$. Direct and shear components referred to the coordinate axes may be found by resolving force vector δF_n in the appropriate directions. Let this vector be represented by line ON, as in Fig. 1.2(b). Point N can now be projected on to the x-axis to give point X. The

2

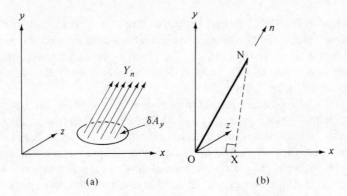

(a)   (b)

Fig. 1.2 Oblique stress

force component acting in the x-direction is then given by $\delta F_x = \delta F_n(\text{OX}/\text{ON})$. Hence the shear stress component acting in the x-direction is given by $\sigma_{yx} = Y_n(\text{OX}/\text{ON})$.

It should be noted that stresses can be treated as vectors only when they act on a specified surface. When reference axes are changed, both directions of forces and directions of normals to elementary surfaces will change simultaneously. To examine this situation, the geometry of rectangular coordinate systems must be discussed in more detail.

1.1.1 Direction cosines   A direction cosine is a property of two lines only, and is the cosine of the angle between them. In Fig. 1.3(a), a vector is

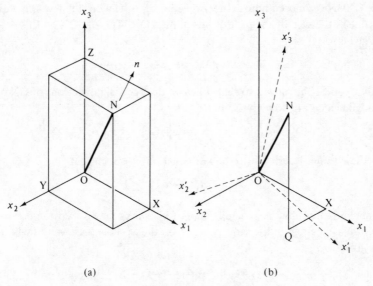

(a)   (b)

Fig. 1.3 Direction cosines

3

represented as line ON. This may be imagined to join opposite corners of a right-angled block. If ON is taken to be of unit length, its direction cosines with respect to the three coordinate axes x_1, x_2, and x_3 are numerically equal to the lengths of edges OX, OY, and OZ. These numbers specify direction n and may be written n_1, n_2, and n_3.

A further generalization is needed to express the collective directions of a new set of coordinate axes x'_1, x'_2, and x'_3. The general direction cosine may be written a_{ij}, where the two subscripts indicate the two particular lines to which the direction cosine refers. The first subscript specifies an axis of the second set, that is, either x'_1, x'_2, or x'_3. The second subscript specifies an axis of the original set, that is, either x_1, x_2, or x_3. This order of subscripts appears to be established by usage. To specify the directions of three new axes relative to three original axes, nine direction cosines are needed, each relating one of the new axes to one of the original axes, as shown by the following array.

$$
\begin{array}{c|ccc}
 & x_1 & x_2 & x_3 \\
\hline
x'_1 & a_{11} & a_{12} & a_{13} \\
x'_2 & a_{21} & a_{22} & a_{23} \\
x'_3 & a_{31} & a_{32} & a_{33} \\
\end{array}
$$

As the three new axes are required to be mutually at right angles, these nine values are not independent. In Fig. 1.3(b), line ON has projections OX, XQ, and QN on the original axes, and these distances may be written x_1, x_2, and x_3. Since OXQN is a closed figure, the projection x'_1 of ON on the new axis x'_1 is the same as the sum of the projections of lines OX, XQ, and QN on this new axis, giving

$$x'_1 = a_{11} x_1 + a_{12} x_2 + a_{13} x_3$$

Now let point N be moved so that it lies on the x'_1-axis. Projections of ON on the original axes are now given by

$$x_1 = a_{11}\text{ON} \qquad x_2 = a_{12}\text{ON} \qquad x_3 = a_{13}\text{ON}.$$

With these values inserted into the last equation, it is seen that

$$a_{11}{}^2 + a_{12}{}^2 + a_{13}{}^2 = 1$$

Further equations for the general values of x'_2 and x'_3 may be written down, but if point N is retained in the same position on the x'_1-axis, these values will be zero, giving

$$a_{21} a_{11} + a_{22} a_{12} + a_{23} a_{13} = 0$$

$$a_{31} a_{11} + a_{32} a_{12} + a_{33} a_{13} = 0$$

These orthogonality relations may be generalized and collectively expressed by using the summation convention,

$$a_{ik} a_{jk} = \delta_{ij} \tag{1.1}$$

Either of the subscripts i and j may be replaced by 1, 2, or 3. On the left, the subscript k occurs twice, so this subscript is put equal to 1, 2, and 3 successively, and the three terms are added together. Such a subscript indicating summation is called a dummy subscript. The delta symbol has value unity when i is the same as j, but has value zero when i is not the same as j, after i and j have been replaced by numbers. This equation is seen to impose six conditions on a permissible array of direction cosines relating two sets of orthogonal axes.

1.1.2 Transformation of stress components   As both forces and areas can be treated as vectors and resolved by using direction cosines, the problem of finding stress components referred to new axes can now be taken up again. The equilibrium of a small tetrahedron OXYZ is considered, as shown in Fig. 1.4(a). The inclination of face XYZ to original axes x_1, x_2, and x_3 is determined by the normal ON which has direction cosines n_1, n_2, and n_3. If this face has area A, the areas of the other faces OYZ, OZX, and OYZ are given respectively by

$$A_1 = n_1 A \qquad A_2 = n_2 A \qquad A_3 = n_3 A$$

The direction of normal ON is identified with the direction of new axis x_1' while a second new axis x_2' is drawn in the plane of face XYZ as shown. To find the stress acting on the oblique face in the x_2'-direction, all forces acting on the other faces are resolved in this direction, which is taken to have direction cosines m_1, m_2, and m_3 relative to the original axes. Taking the particular face

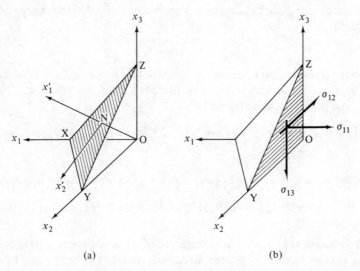

(a)   (b)

Fig. 1.4 Transformation of stress

5

OYZ, positive stresses act as indicated in Fig. 1.4(b), and contribute a force acting in the negative x_2'-direction equal to

$$A_1(m_1 \sigma_{11} + m_2 \sigma_{12} + m_3 \sigma_{13})$$

The other two faces give forces

and
$$A_2(m_1 \sigma_{21} + m_2 \sigma_{22} + m_3 \sigma_{23})$$
$$A_3(m_1 \sigma_{31} + m_2 \sigma_{32} + m_3 \sigma_{33})$$

Hence the total force F acting in the negative x_2'-direction can be expressed as the double summation

$$F = A n_k m_l \sigma_{kl}$$

The stress acting on face XYZ in the x_2'-direction is written σ_{12}', and as no other stress acting on this face contributes any force in this chosen direction, the total force acting in the positive x_2'-direction is given by $F = A\sigma_{12}'$. This force must be equal and opposite to the summation of forces acting on the other faces, giving a stress component

$$\sigma_{12}' = n_k m_l \sigma_{kl}$$

This relation has been derived for new axes x_1' and x_2' having direction cosines n_k and m_l ($k, l = 1, 2, 3$). Any two axes x_i' and x_j' belonging to some new set might have been selected, respectively related to the original axes by sets of three direction cosines which are now rewritten as

$$n_k = a_{ik} \qquad m_l = a_{jl}$$

The equation for transforming stress components now takes the concise and general form

$$\sigma_{ij}' = a_{ik} a_{jl} \sigma_{kl} \tag{1.2}$$

where the primed stress is referred to the new coordinates and a double summation is carried out with respect to subscripts k and l. For example, typical direct and shear stresses are given by

$$\sigma_{11}' = a_{11}{}^2 \sigma_{11} + a_{12}{}^2 \sigma_{22} + a_{13}{}^2 \sigma_{33} + 2a_{11} a_{12} \sigma_{12} + 2a_{12} a_{13} \sigma_{23}$$
$$+ 2a_{13} a_{11} \sigma_{31}$$

$$\sigma_{12}' = a_{11} a_{21} \sigma_{11} + a_{12} a_{22} \sigma_{22} + a_{13} a_{23} \sigma_{33} + (a_{11} a_{22} + a_{12} a_{21}) \sigma_{12}$$
$$+ (a_{12} a_{23} + a_{13} a_{22}) \sigma_{23} + (a_{11} a_{23} + a_{13} a_{21}) \sigma_{31}$$

1.1.3 Invariants of stress   A combination of stress components which remains of the same numerical value when all components are referred to any different set of axes is called an invariant. There are three independent invariants, respectively consisting of linear, quadratic, and cubic combinations, and these may be concisely written, using a new symbol s for stress

$$I_1 = s_{ii} \qquad I_2 = s_{ij} s_{ji} \qquad I_3 = s_{ij} s_{jk} s_{ki}$$

To prove that these are invariant, the method given in R. Hill's *Theory of Plasticity*, p. 343, may be followed. For example, the second invariant is transformed by means of (1.2) to obtain stress components s'_{ij} referred to some new coordinates,

$$s'_{ij} s'_{ji} = (a_{ik} a_{jl} s_{kl})(a_{jm} a_{in} s_{mn})$$

By disregarding the brackets, the order of this product may be changed,

$$s'_{ij} s'_{ji} = (a_{ik} a_{in} s_{kl})(a_{jm} a_{jl} s_{mn})$$

However, by the orthogonality relation (1.1) for direction cosines, we have

$$a_{ik} a_{in} = \delta_{kn} \qquad a_{jm} a_{jl} = \delta_{ml}$$

But, by definition of the δ-symbol,

$$\delta_{kn} s_{kl} = s_{nl} \qquad \delta_{ml} s_{mn} = s_{ln}$$

Hence, the product reduces to

$$s'_{ij} s'_{ji} = s_{nl} s_{ln} = s_{ij} s_{ji}$$

since dummy subscripts only signify a summation and can be represented by any symbol.

When expanded, the above expressions for invariants I_1, I_2, and I_3 contain 3, 9, and 27 terms, but some of these combine to give 3, 6, and 10 terms respectively. A further simplification can be made by noting that any combination of these invariants is itself an invariant. Hence the invariants can be conveniently redefined as

$$
\left.
\begin{aligned}
J_1 &= I_1 = \sigma_{xx} + \sigma_{yy} + \sigma_{zz} \\
J_2 &= \tfrac{1}{2}(I_2 - I_1^2) = -(\sigma_{xx}\sigma_{yy} + \sigma_{yy}\sigma_{zz} + \sigma_{zz}\sigma_{xx}) \\
&\quad + \sigma_{xy}^2 + \sigma_{yz}^2 + \sigma_{zx}^2 \\
J_3 &= \tfrac{1}{6}(2I_3 - 3I_2 I_1 + I_1^3) \\
&= \sigma_{xx}\sigma_{yy}\sigma_{zz} + 2\sigma_{xy}\sigma_{yz}\sigma_{zx} \\
&\quad - (\sigma_{xx}\sigma_{yz}^2 + \sigma_{yy}\sigma_{zx}^2 + \sigma_{zz}\sigma_{xy}^2)
\end{aligned}
\right\} \tag{1.3}
$$

It is necessary, but not sufficient, that an expression for an invariant should remain unchanged after permutation of the subscripts.

1.1.4 Quadric surface of stress To prove that principal axes of stress always exist, use is made of a geometrical representation of stress. Consider the surface described by the terminal point P of a vector of some variable length r drawn from a fixed point. At the fixed point, a stress acts, having components σ_{ij} referred to axes x_1, x_2, and x_3. When the vector coincides with some chosen direction x_1' which is specified in relation to the original axes by direction cosines m_i $(i = 1, 2, 3)$, the direct stress acting in this direction is given by

$$\sigma_{11}' = m_i m_j \sigma_{ij}$$

When both sides of this equation are multiplied by r^2 and it is noted that the coordinates x_i of point P are given by $x_i = rm_i$, the equation becomes

$$r^2 \sigma_{11}' = x_i x_j \sigma_{ij}$$

However, it can be shown by the methods already outlined that the product on the right is an invariant, and may be expanded and set equal to some constant C for a given stress state,

$$r^2 \sigma_{11}' = x^2 \sigma_{xx} + y^2 \sigma_{yy} + z^2 \sigma_{zz} + 2xy \sigma_{xy}$$
$$+ 2yz \sigma_{yz} + 2zx \sigma_{zx} = C \tag{1.4}$$

Geometrically, this quadric surface having stress components as coefficients in its equation must have three planes of symmetry, which are the principal planes of stress. These planes intersect along lines which are the principal axes of stress. If coordinate axes x, y, and z are chosen to coincide with these principal axes, symmetry of the quadric surface shows that the value of direct stress σ_{11}' must be unaffected by changing the sign of any of the coordinate values of any chosen point on the surface. For this to be so, (1.4) indicates that shear stress components referred to the principal axes must be zero. Having demonstrated this point, it is convenient for practical purposes to re-define principal axes as the particular set of axes on which all shear components of stress are zero. The direct stresses acting in the directions of the principal axes are the principal stresses, which may be written with single subscripts, σ_1, σ_2, and σ_3. The highest and lowest of these values σ_1 and σ_3 must be respectively higher and lower than any component of direct stress referred to other axes. A precise description of stress, together with the concept of principal axes, is usually considered to date from the work of A. L. Cauchy (1828).

To obtain a clearer visual appreciation of this quadric surface, it can be simplified without loss of generality by adding equal values to all three principal stresses so as to bring the value of the least stress σ_3 to zero. If the principal axes are marked x, y, and z, the quadric surface will now consist of a cylinder of non-circular section generated by lines parallel to the z-direction, the section in the (x,y) plane being, in fact, an ellipse. This simplified surface can be used to show that a set of right-angled axes exist along which the direct stresses are equal. Axes x' and y' may be first located in the (x,y) plane, and rotated until they make angles of $45°$ with the y-axis, as shown in Fig. 1.5(a). The set of new axes is then rotated about the original x-axis as shown in Fig. 1.5(b) until lengths OX and OZ intercepted by the surface become equal.

Evidently, rotation in the reverse direction will give an alternative position satisfying this requirement. When the lengths of the three radius vectors drawn to the quadric surface are equal, the direct stresses acting in these directions will be equal, as indicated by (1.4). Therefore, if, for any stress state, these particular reference axes are chosen, the stress appears as an isotropic stress plus three shear stress components.

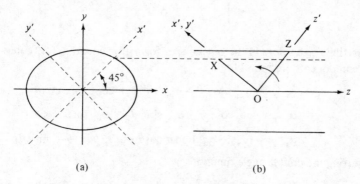

Fig. 1.5 Axes of equal direct stress

1.1.5 Values of principal stresses   The problem of finding principal stresses from any given set of six components is now considered. It might be suspected that as three invariants can be found from the three principal stresses, the reverse should be true also. Let the principal stresses be given by the roots λ of the cubic equation

$$(\lambda - \sigma_1)(\lambda - \sigma_2)(\lambda - \sigma_3) = 0$$

When expanded, this becomes

$$\lambda^3 - J_1 \lambda^2 - J_2 \lambda - J_3 = 0 \tag{1.5}$$

where $J_1 = \sigma_1 + \sigma_2 + \sigma_3$, $J_2 = -(\sigma_1\sigma_2 + \sigma_2\sigma_3 + \sigma_3\sigma_1)$, and $J_3 = \sigma_1\sigma_2\sigma_3$. These expressions have already been established in (1.3) for components related to any axes, so their numerical values can be inserted into (1.5). In solving this equation, it is of some slight assistance to eliminate the second term by reducing each of the direct stress components by an amount $\frac{1}{3}(\sigma_{xx} + \sigma_{yy} + \sigma_{zz})$, finally adding the same amount to each of the principal stresses as found by solving the equation. Mathematically, equation (1.5) must always have three real roots, as expected on physical grounds.

1.1.6 Two-dimensional transformation   Rotation of a set of axes with one of the axes fixed in direction is of special interest, as it gives equations for transforming stress components in a plane. Suppose that (x,y) axes are rotated

2

through positive anticlockwise angle θ to new positions (x',y'). The direction cosines for this transformation are as shown:

	x	y	z
x'	$\cos\theta$	$\sin\theta$	0
y'	$-\sin\theta$	$\cos\theta$	0
z'	0	0	1

By inserting these values in the general transformation equation (1.2), new stress components are found to be

$$\sigma_{x'x'} = \sigma_{xx}\cos^2\theta + \sigma_{yy}\sin^2\theta + 2\sigma_{xy}\sin\theta\cos\theta$$

$$\sigma_{y'y'} = \sigma_{yy}\cos^2\theta + \sigma_{xx}\sin^2\theta - 2\sigma_{xy}\sin\theta\cos\theta$$

$$\sigma_{x'y'} = (\sigma_{yy} - \sigma_{xx})\sin\theta\cos\theta + \sigma_{xy}(\cos^2\theta - \sin^2\theta)$$

By inserting the double angle functions

$$\cos^2\theta = \tfrac{1}{2}(1 + \cos 2\theta), \quad \sin^2\theta = \tfrac{1}{2}(1 - \cos 2\theta), \quad \sin\theta\cos\theta = \tfrac{1}{2}\sin 2\theta$$

this two-dimensional transformation can be written in the complex form

$$\left.\begin{aligned}\sigma_{x'x'} + \sigma_{y'y'} &= \sigma_{xx} + \sigma_{yy} \\ \sigma_{y'y'} - \sigma_{x'x'} + 2i\sigma_{x'y'} &= (\sigma_{yy} - \sigma_{xx} + 2i\sigma_{xy})\,e^{2i\theta}\end{aligned}\right\} \tag{1.6}$$

This suggests the representation of two-dimensional stress by a circle construction as shown in Fig. 1.6, using the invariant quantities

$$\sigma = \tfrac{1}{2}(\sigma_{xx} + \sigma_{yy}) \qquad \tau = |\sigma_{yy} - \sigma_{xx} + 2i\sigma_{xy}|$$

Drawn to a suitable scale, the circle of radius τ has its centre situated at a horizontal distance from a datum line equal to mean stress σ, this distance being measured to the right when σ is positive. It is seen from (1.6) that for a rotation of axes through angle θ, the radius vector is rotated through angle 2θ. The value

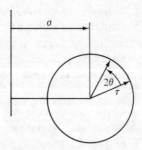

Fig. 1.6 Stress circle

τ is the maximum shear stress, and principal stresses are given by

$$\sigma_1 = \sigma + \tau \qquad \sigma_2 = \sigma - \tau$$

1.1.7 Stress gradient   Although any values may be assigned to six stress components at a point, the way in which these components may vary from one point to another is restricted by conditions of equilibrium. Figure 1.7 shows all those stresses which act in the x-direction on the various faces of an element having sides of length δx, δy, and δz. From one side of the element to the other,

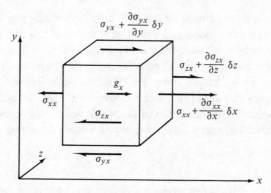

Fig. 1.7 Stress equilibrium

each stress is allowed to increase by an amount equal to its spatial rate of change multiplied by the incremental distance along the positive coordinate axis. Each stress is then multiplied by the area on which it acts, to give the total force in the x-direction,

$$\delta F_x = \left(\frac{\partial \sigma_{xx}}{\partial x} + \frac{\partial \sigma_{yx}}{\partial y} + \frac{\partial \sigma_{zx}}{\partial z} \right) \delta x \, \delta y \, \delta z$$

Account should also be taken of force acting on all the interior particles of the element. This body force might be due to the action of gravity or acceleration on the masses of the particles, but whatever the cause may be, it is assumed that the force per unit volume is a vector having component g_x acting in the x-direction. When multiplied by the volume of the element, this gives an additional force. However, the total force acting in the x-direction must be zero, for equilibrium of the element. Similarly, the sum of forces acting in the other two directions must be zero, which leads to the generalized equilibrium equation:

$$\frac{\partial \sigma_{ij}}{\partial x_i} + g_j = 0 \tag{1.7}$$

11

in which summation is carried out with respect to subscript i. Unless otherwise specified, the body force will be assumed to be zero.

When curvilinear coordinates are used, opposite faces of an element may not be parallel and may have different areas, so this form of the stress equilibrium equation is valid only for rectangular coordinates. The equation is of primary importance in the mechanics of deformable bodies, and must be satisfied by any permissible distribution of stress.

1.2 Strain

In the first place, a method must be established for specifying any kind of distortion that may be imposed on a solid body, provided this distortion is small. During distortion, all points within the body will move relative to some fixed coordinate system (x, y, z). The displacement vector for each particle will have components measured along the coordinate axes. These components may be written (u, v, w) or (u_1, u_2, u_3) as convenient. This vector field fully describes any movement or distortion of the whole body, and is not subject to any physical restriction, except that the components vary continuously through the body.

1.2.1 Strains derived from displacements   From the displacement components at a single point, it is not possible to tell whether the displacement is due to distortion of the body or to movement of the body without distortion. Distortion or strain implies a change in displacement from one point to another, and can be associated with derivatives of displacements $\partial u_i / \partial x_j$ with respect to distances x_j along the coordinate directions. This general derivative can be concisely written $u_{i,j}$ in which the comma indicates differentiation with respect to the coordinate specified by the subscript which follows it.

Derivatives of displacements are not adequate by themselves to define strain, as they may reflect to some extent a rigid-body rotation. To eliminate the effect of such a motion, the derivatives may be separated into symmetrical and anti-symmetrical sets, by writing each derivative in the form

$$u_{i,j} = \tfrac{1}{2}(u_{i,j} + u_{j,i}) + \tfrac{1}{2}(u_{i,j} - u_{j,i})$$

Of the two terms on the right, only the first provides a suitable measure of distortion, as it remains unchanged when movement of the whole body occurs without distortion. To illustrate this, Fig. 1.8 shows distortion of a rectangular element in the (x,y) plane. For small displacements u and v in the x- and y-directions, the sides of the element rotate through angles

$$\beta_1 = \frac{\partial v}{\partial x} \qquad \beta_2 = \frac{\partial u}{\partial y}$$

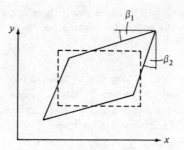

Fig. 1.8 Positive shear strain

A small rigid-body rotation will increase one of these angles and reduce the other by the same amount. The sum of these angles is therefore unaffected by pure rotation, which suggests a definition of shear strain

$$e_{xy} = \tfrac{1}{2}\left(\frac{\partial v}{\partial x} + \frac{\partial u}{\partial y}\right)$$

The two subscripts indicate the plane in which an element is distorted in outline, and the order of subscripts is of no significance. It should be noted that some engineering text-books use a slightly different definition of shear strain, in which the factor one-half is omitted.

Direct strain in the x-direction is simply given by the x-derivative of the component of displacement in the x-direction,

$$e_{xx} = \frac{\partial u}{\partial x}$$

and physically represents the fractional increase in length of an element. Hence the general component of direct or shear strain is given in terms of displacements by

$$e_{ij} = \tfrac{1}{2}(u_{i,j} + u_{j,i}) \tag{1.8}$$

1.2.2 Transformation of strains Let new coordinates x_j' be related to original coordinates x_l by a transformation of rotation. The inverse transformation and partial derivatives arising from it will be used,

$$x_l = a_{jl}x_j' \qquad \frac{\partial x_l}{\partial x_j'} = a_{jl}$$

where a_{jl} are the direction cosines of original axis x_l relative to the three new axes x_j'. At a particular point, a component of displacement u_i' measured along

13

one of the new axes, say x_i', is given in terms of three components u_k measured along original axes x_k by the vector transformation

$$u_i' = a_{ik} u_k$$

where a_{ik} are direction cosines of new axis x_i' relative to the three original axes x_k. This displacement is now differentiated along the new axis x_j'. If the component obtained by resolving a vector along a fixed axis can be differentiated as a scalar, the chain rule for differentiating gives

$$\frac{\partial u_i'}{\partial x_j'} = \frac{\partial x_l}{\partial x_j'} \frac{\partial u_i'}{\partial x_l}$$

By inserting the expressions already derived, one obtains

$$\frac{\partial u_i'}{\partial x_j'} = a_{ik} a_{jl} \frac{\partial u_k}{\partial x_l}$$

As subscript i can be interchanged with j, and k with l, the strain component referred to the new axes now follows from (1.8),

$$e_{ij}' = a_{ik} a_{jl} e_{kl} \qquad (1.9)$$

As this tranformation is precisely the same as the transformation (1.2) for stress components, similar arguments may be used for deriving principal strains.

1.2.3 Compatibility of strain   Although any set of six strain components may be imposed on a particular element, the strains in adjoining elements will then be subject to the restriction that all values are compatible with a continuous distribution of the three displacement components (u, v, w). In the (x,y) plane, we have strain components

$$e_{xx} = \frac{\partial u}{\partial x} \qquad e_{yy} = \frac{\partial v}{\partial y} \qquad e_{xy} = \tfrac{1}{2}\left(\frac{\partial u}{\partial y} + \frac{\partial v}{\partial x}\right)$$

These may be differentiated to give

$$\frac{\partial^2 e_{xx}}{\partial y^2} = \frac{\partial^3 u}{\partial x\,\partial y^2} \qquad \frac{\partial^2 e_{yy}}{\partial x^2} = \frac{\partial^3 v}{\partial x^2\,\partial y} \qquad 2\frac{\partial^2 e_{xy}}{\partial x\,\partial y} = \frac{\partial^3 u}{\partial x\,\partial y^2} + \frac{\partial^3 v}{\partial x^2\,\partial y}$$

It is seen that displacements u and v can be eliminated, leaving a differential equation

$$\frac{\partial^2 e_{xx}}{\partial y^2} + \frac{\partial^2 e_{yy}}{\partial x^2} - 2\frac{\partial^2 e_{xy}}{\partial x\,\partial y} = 0 \qquad (1.10)$$

which must be satisfied by the spatial distribution of strain components. Two more equations of this kind are obtained by permuting the three coordinate variables. There is not much practical advantage in writing these equations in a subscript notation (see I.S. Sokolnikoff, *Theory of Elasticity*). In curvilinear coordinate systems, these equations take different and rather complicated forms, but it is rarely necessary to use such equations directly.

By a similar method, it can be shown that strains must satisfy three further equations of the type

$$\frac{\partial}{\partial x}\left(\frac{\partial e_{xy}}{\partial z} - \frac{\partial e_{yz}}{\partial x} + \frac{\partial e_{zx}}{\partial y}\right) = \frac{\partial^2 e_{xx}}{\partial y \, \partial z} \tag{1.11}$$

Fortunately, equations of this second set are identically satisfied in states of plane stress and strain, in which no component varies in the z-direction and the only non-zero shear strain is e_{xy}. In other situations, it may be necessary to take account of all six compatibility equations.

1.3 Elastic Material

To make further progress in formulating the mechanics of deformable bodies, it is necessary to specify the relevant properties of some idealized material. For the present purpose, it is assumed that the material is both linear and isotropic. Linearity of the material means that if any set of stress components are all increased so that each is finally n times its original value, then each strain component in the final state will also be n times its original value. It is assumed that the body is initially free from stress before loading is applied, and that the stresses do not attain certain limiting values which will be examined later. The reservation should also be made that strains do not become so large as to change the shape of an element excessively, which might give rise to ambiguity in the coordinate directions used for describing stress and strain. It should be noted that a structure made from linear material is not always bound to give a linear response to loading, for reasons connected with the geometrical shape of the structure, as mentioned later in chapter 4.

An isotropic material is one in which a certain set of stress components will produce the same strains regardless of the orientation of the chosen coordinate axes relative to another set of axes attached to the material. In other words, it is not possible to discover the existence of any axes of the material from the response of the material to stress. The success of the simple theory of elasticity when applied to engineering problems is in some degree due to the good behaviour of metals as judged by these idealizations. If, during deformation, each molecule in an aggregate continues to interact with the same set of neighbouring molecules, it may be expected that the body will give a linear response to stress and will precisely recover its original shape on removal of stress. Alloys intended for carrying stress usually have internal structures containing dispersions of hard constituents which prevent the movement of crystal defects. Such materials are almost perfectly linear and elastic until stresses reach some fairly well-defined limiting values. Further, it is fortunate that most engineering materials consist of aggregates of small equi-axed crystals which have a random orientation relative

15

to each other, so that the directional properties of individual crystals become averaged and are not reflected in the behaviour of the aggregate.

1.3.1 Change of volume and shape   The law of linearity between stress and strain, often called Hooke's law, must be more particularly examined in the light of physical processes occurring during deformation. In spite of the complex structure of materials on a microscopic and atomic scale, it is fortunate that only two processes need be taken into account in examining small strains in a linear isotropic material. The first process is volumetric expansion or contraction and the second is shearing. This proposition was clearly stated by G. G. Stokes in 1845.

It is assumed that isotropic stress and shearing stress have no joint effect on the material other than the sum of the effects they would produce if acting separately. For solid bodies of a wide variety, this is a very appropriate idealization, provided the fractional change in volume remains small. With this assumption, changes in volume can be described without reference to shear stresses, by using only isotropic stress σ and isotropic strain e. These are given by

$$\sigma = \tfrac{1}{3}(\sigma_{xx} + \sigma_{yy} + \sigma_{zz}) \qquad e = \tfrac{1}{3}(e_{xx} + e_{yy} + e_{zz})$$

and since these quantities are invariants, they may be calculated from components referred to any axes (x, y, z). A linear relationship

$$e = \sigma/3K \tag{1.12}$$

can now be taken, where $3K$ is a constant. This retains the constant K with its customary interpretation as the bulk modulus, defined as the ratio between applied pressure p and fractional decrease in volume $\delta V/V$, so that $K = pV/\delta V$.

If a shear stress σ_{xy} is applied by itself, it will be expected to produce a proportional shear strain. When shear strain is defined according to (1.8), this linear relationship may be written $e_{xy} = \sigma_{xy}/2G$, where $2G$ is a constant. This retains the shear modulus G in its customary form, defined as the ratio of shear stress to the value of shear strain calculated using the engineering definition of shear strain. As the material is isotropic, this constant will be the same when a shear stress component is applied on any plane. It has been seen in previous discussion of the stress quadric, that for any stress state, three right-angled axes exist along which the three direct stresses are equal. By taking these axes (x_1, x_2, x_3) as reference axes, the mean stress σ may be deducted from each of these direct stresses to reduce all of them to zero. On a right-angled element having sides parallel to these axes, a set of three shear stresses τ_{kl} can now be imposed, accompanied by shear strains η_{kl}. The direct stress σ'_{rr} acting along some line of direction r can now be found by transformation, together with the corresponding direct strain,

$$\sigma'_{rr} = n_k n_l \tau_{kl} \qquad e'_{rr} = n_k n_l \eta_{kl}$$

where n_k and n_l are direction cosines of line r with respect to axes (x_1, x_2, x_3). Here, primes signify stresses and strains from which the isotropic components have been removed. According to the assumptions made so far, simple shear stresses are linearly related to shear strains by $\tau_{kl} = 2G\eta_{kl}$. By combining this with the last two expressions it is seen that $\sigma'_{rr} = 2Ge'_{rr}$. This means that reduced values of direct stress and strain are related by the same constant of proportionality as that which relates shear stresses and strains, so the general linear relationship for any stress components can be written

$$\sigma'_{ij} = 2Ge'_{ij} \tag{1.13}$$

The reduced values required for this equation are obtained by deducting isotropic stress from the direct stresses only, and by deducting isotropic strain from the direct strains only, in accordance with

$$\sigma'_{ij} = \sigma_{ij} - \delta_{ij}\sigma \qquad e'_{ij} = e_{ij} - \delta_{ij}e$$

the shear stresses and strains remaining unchanged. Equations (1.12) and (1.13) may be called the constitutive equations for the idealized material.

1.3.2 Alternative elastic constants   In practice, total stresses and strains have to be handled in calculations, and it is convenient to introduce other elastic constants which are relevant to a tensile test and can be found from such a test. When a parallel bar with its axis lying along the z-axis is subjected to a uniaxial tensile stress T, longitudinal strain e_{zz} and transverse strains e_{xx} and e_{yy} may be measured and inserted into the equations

$$e_{zz} = T/E \qquad e_{xx} = e_{yy} = -\nu e_{zz}$$

Constants E and ν are named Young's modulus and Poisson's ratio in honour of early nineteenth-century mathematicians. By noting that

$$\sigma = T/3 \qquad \sigma'_{zz} = 2T/3 \qquad \sigma'_{xx} = \sigma'_{yy} = -T/3$$

the strains can be alternatively found from (1.12) and (1.13) in terms of constants K and G. By comparing these alternative expressions for strains, two relations between the four constants may be derived,

$$G = E/2(1 - \nu) \qquad K = E/3(1 - 2\nu) \tag{1.14}$$

On inserting these values into the fundamental elasticity equations (1.12) and (1.13), six linear equations for strain components in terms of total stresses are obtained, of the type

$$Ee_{xx} = \sigma_{xx} - \nu(\sigma_{yy} + \sigma_{zz}) \qquad 2Ge_{xy} = \sigma_{xy} \tag{1.15}$$

These apply to any elastic body in any geometrical situation.

1.3.3 Strain energy   In any body, the energy absorbed by an element can be found by multiplying the average force acting by the displacement of the point of application of force. On elementary surface A_x perpendicular to the x-axis, direct stress σ_{xx} exerts a force $A_x\sigma_{xx}$. The relative movement δu_x between the two ends of the elementary block is given in terms of strain by $\delta u_x = e_{xx}\delta x$. In a linear elastic material, the work δU supplied during a deformation will be half the product of the final values of force and relative movement, i.e., $\delta U = \frac{1}{2}\sigma_{xx}e_{xx}\delta v$, where $\delta v = A_x\delta x$ is the volume of the element. Shear stresses may be considered in a somewhat similar way. Stress σ_{xy} acts on area A_x and moves through distance $\delta u_y = \delta x\, \partial u_y/\partial x$. Also, the complementary shear stress σ_{yx} acts on area A_y and moves through distance $\delta u_x = \delta y\, \partial u_x/\partial y$. The definition of shear strain (1.8) now indicates a contribution of energy $\delta U = \sigma_{xy}e_{xy}\, \delta v$. This double contribution from each shear stress component is reproduced by the general summation

$$\delta U = \tfrac{1}{2}\sigma_{ij}e_{ij}\,\delta v \tag{1.16}$$

Expressions for strain energy in terms of stresses alone or strains alone may be found by using elasticity relations of one kind or another. If the basic elastic constants are to be used, expression (1.16) is first put into the form

$$\delta U = \tfrac{1}{2}(\sigma'_{ij}e'_{ij} + 3\sigma e)\,\delta v$$

by using the fact that $\sigma'_{ii} = e'_{ii} = 0$. Relations (1.12) and (1.13) can now be used to give

$$\delta U = (\sigma'_{ij}\sigma'_{ij}/4G + \sigma^2/2K)\,\delta v \tag{1.17}$$

Alternative expressions for strain energy will be found in chapter 4.

1.4 Criterion of Elastic Breakdown

A very brief discussion is given here of some of the more elementary theories for predicting the effectiveness of combinations of stress components in producing inelastic or plastic deformation. A very large amount of experimental investigation has been described, but this is not reviewed here, and for such information and more detailed discussion of it, reference should be made to specialized works such as R. Hill's *Theory of Plasticity*. Theoretical arguments are based on the experimental observation that plastic yielding and flow are not influenced by the isotropic component of stress. This stress may influence the amount of plastic strain that precedes fracture, but such considerations are not strictly relevant here. With this starting point, it is usual to eliminate the isotropic stress by working with reduced or deviator stresses, obtained by deducting the isotropic stress from each of the direct stress components,

$$\sigma'_{ij} = \sigma_{ij} - \delta_{ij}\sigma$$

where $\sigma = \frac{1}{3}\sigma_{ii}$ and the symbol δ_{ij} has value unity if i is the same as j, but is otherwise zero. Reduced stresses are distinguished by a prime.

If the material is isotropic, it will react only to the invariants of stress, regardless of the orientation of the principal axes of stress relative to the material.

The two assumptions that the material is isotropic and is insensitive to isotropic stress lead to the conclusion that the onset of plastic yielding can be predicted from the three reduced principal stresses, written in descending order of magnitude as σ_1', σ_2', and σ_3'. However, these reduced values are defined in such a way that their sum is bound to be zero, so only the two extreme values need be considered.

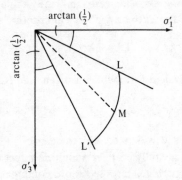

Fig. 1.9 Plastic yield locus

These extreme values must plot inside the sector in Fig. 1.9. If any two components plotted should be so changed that their ratio falls outside the sector shown, it will be found that they no longer represent the greatest and least of the three principal stresses, so the actual components must then be relabelled to fulfil this condition. Observed stresses at which inelastic yielding commences, or at which some small specified value of plastic strain is attained may be plotted in this diagram. Such points may be joined to give a yield locus marked LL'. The terminal points of this line correspond to testing conditions of uniaxial tension and compression, while point M intercepted by a radius at 45° represents pure shear such as is obtained in a torsion test. There are, of course, other ways of graphically representing a plastic yield locus. It is found experimentally that when materials free from internal stress are tested by correct methods, the plastic yield stress in compression is very nearly equal to that in tension.

1.4.1 Particular theories   Two simple theories have received substantial support. According to the maximum shear stress criterion, often called the Tresca criterion, plastic yielding begins at some limiting value of shear stress. The largest shear stress that occurs on any plane passing through a chosen point is that referred to axes which make angles of 45° with the axes of greatest and least principal stresses and which are coplanar with these axes. The maximum shear stress τ is therefore given in terms of these principal stresses by

$$\tau = \tfrac{1}{2}(\sigma_1 - \sigma_3) \tag{1.18}$$

19

The second theory postulates that yielding occurs at a limiting value of the quadratic stress invariant. Any form of this invariant may be taken, but it must be adjusted to comply with the physical requirement that its value is unchanged by superposition of isotropic stress. Further, the result may be multiplied by any convenient factor to suit the type of test to be used for obtaining an experimental value. For example, in a torsion test, a single shear stress is imposed, all other components of stress being zero, and if the value of this stress to cause plastic yielding is found to be k, the yield criterion may be written

$$k^2 = \tfrac{1}{2}\sigma'_{ij}\sigma'_{ij} \tag{1.19}$$

An alternative and equivalent form which is often used is

$$6k^2 = (\sigma_{xx} - \sigma_{yy})^2 + (\sigma_{yy} - \sigma_{zz})^2 + (\sigma_{zz} - \sigma_{xx})^2 + 6(\sigma_{xy}^2 + \sigma_{yz}^2 + \sigma_{zx}^2)$$

This function is frequently associated with the name of R. von Mises. For a wide variety of metals, it successfully correlates limiting values of stress components when tests are carried out which impose various combinations of these components. For example, it predicts a yield stress Y in uniaxial tension which is related to shear yield stress in a torsion test by $Y = \sqrt{3}\,k$. The physical considerations so far taken into account do not preclude the possibility that the third invariant of stress will have an effect, and this possibility has received attention from many experimental investigators. Such an effect might indicate a preference for one or other of the formulae (1.18) or (1.19), but it is found that predictions based on either of these criteria never diverge very widely for any imposed combination of stress components.

1.4.2 Physical interpretations

Much searching has been done in the past for physical interpretations of the second invariant criterion (1.19). Although such interpretations may appear to be rather unnecessary, they provide exercises in stress analysis. Hencky's interpretation was that yielding occurs at a critical value of shear strain energy U_0 per unit volume. This is directly obtainable from (1.17). Comparison with (1.19) indicates that $U_0 = k^2/2G$, where G is the shear modulus.

A. Nadai in *Theory of Flow and Fracture of Solids* pointed out that the quadratic invariant (1.19) has a value proportionate to the square of resultant shear stress τ_0 acting on the octahedral planes. These planes form the faces of a regular octahedron positioned so that the principal axes of stress pass through its corners. To calculate this stress, the plane intersecting the positive coordinate axes may be chosen. If axis z' is taken to be the direction of the normal, the two other axes may be taken to lie in the octahedral plane. It is convenient to position axis x' in a horizontal plane and y' in a vertical plane. The required scheme of direction cosines is as shown.

	x	y	z
x'	$\dfrac{-1}{\sqrt{2}}$	$\dfrac{1}{\sqrt{2}}$	0
y'	$\dfrac{-1}{\sqrt{6}}$	$\dfrac{-1}{\sqrt{6}}$	$\sqrt{\dfrac{2}{3}}$
z'	$\dfrac{1}{\sqrt{3}}$	$\dfrac{1}{\sqrt{3}}$	$\dfrac{1}{\sqrt{3}}$

The transformation equation (1.2) now gives shear stresses acting in the x'- and y'-directions in terms of the principal stresses. The sum of the squares of these shear stresses, denoted by τ_0^2, is found to have the same form as the invariant (1.19), and is numerically expressed by $\tau_0^2 = 2k^2/3$.

A further proposal has been made that the second invariant represents the average squared value of resultant shear stress acting on elements of the surface of a small sphere within the material. Spherical coordinates (r,θ,ϕ) may be used for calculating these stresses. A radius vector of length r makes meridian angle ϕ with the z-axis, which means that the z-axis passes through the pole of the new coordinates. This vector has a projection in the (x,y) plane which is inclined at anticlockwise angle θ to the x-axis. The required direction cosines are

	x	y	z
r	$\sin\phi\cos\theta$	$\sin\phi\sin\theta$	$\cos\phi$
θ	$-\sin\theta$	$\cos\theta$	0
ϕ	$\cos\phi\cos\theta$	$\cos\phi\sin\theta$	$-\sin\phi$

These allow shear stresses $\sigma_{r\theta}$ and $\sigma_{r\phi}$ to be calculated in terms of the principal stresses, which are taken to act in the x-, y-, and z-directions. The sum of the squares of these shear stress components is then integrated over the surface of a sphere to give the average resultant shear stress squared, denoted by τ_s^2. This expression is found to have the form of invariant (1.19), with a numerical relationship $\tau_s^2 = 2k^2/5$.

1.5 Alternative Coordinates

Hitherto, rectangular coordinates have been used for establishing basic equations connected with stress, strain, and elasticity. Alternative coordinates are often useful for describing conditions at curved boundary surfaces. Such coordinate systems are always orthogonal, that is, the three axes are at right angles to each other at every point. For such coordinate systems, any equation can be simply rewritten with the new coordinates replacing *x, y,* and *z,* provided the equation does not contain derivatives. The simple treatment of derivatives of scalar quantities given here will be adequate for most purposes, though it should be remembered that the results do not apply to derivatives of vector and tensor components, as such components are themselves variable according to the direction in which they are taken.

1.5.1 Differentiation along a curve   Suppose that scalar quantity F is a function of the position of a point in a plane. Attention is given to the particular values at points defined by distance s measured along a curve from point P as shown in Fig. 1.10. The radius of curvature of this curve at point P is some fixed

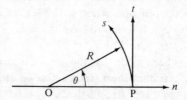

Fig. 1.10 Differentiation along a curve

radius R. For the purpose of examining first and second derivatives, this information is sufficient whatever the equation of the curve may be, but for the present purpose, the whole curve is taken to be a circular arc. At point P, alternative rectangular coordinates (n, t) are set up, normal and tangential to the curve. Geometrically, the coordinates (n, t) of a point on the curve are given by

$$n = -R\left(1 - \cos\frac{s}{R}\right) \qquad t = R\sin\frac{s}{R}$$

Derivatives with respect to s are now taken and expressed in terms of n and t,

$$\frac{\partial n}{\partial s} = -\sin\frac{s}{R} = -\frac{t}{R} \qquad \frac{\partial t}{\partial s} = \cos\frac{s}{R} = 1 + \frac{n}{R}$$

The derivative of F along the curve is now expressed by the rule for partial differentiation,

$$\frac{\partial F}{\partial s} = \frac{\partial F}{\partial n}\frac{\partial n}{\partial s} + \frac{\partial F}{\partial t}\frac{\partial t}{\partial s}$$

On inserting the partial derivatives just found, this becomes

$$\frac{\partial F}{\partial s} = \left[-\frac{t}{R}\frac{\partial}{\partial n} + \left(1 + \frac{n}{R}\right)\frac{\partial}{\partial t}\right]F$$

22

The second derivative with respect to s is found by duplicating the operator,

$$\frac{\partial^2 F}{\partial s^2} = -\frac{t}{R}\frac{\partial}{\partial n}\left[-\frac{t}{R}\frac{\partial F}{\partial n} + \left(1 + \frac{n}{R}\right)\frac{\partial F}{\partial t}\right]$$

$$+ \left(1 + \frac{n}{R}\right)\left[-\frac{1}{R}\frac{\partial F}{\partial n} - \frac{t}{R}\frac{\partial^2 F}{\partial n\,\partial t} + \left(1 + \frac{n}{R}\right)\frac{\partial^2 F}{\partial t^2}\right]$$

Although this is true at any point on the curve, the only point of interest is at $n = t = 0$, where we have

$$\frac{\partial^2 F}{\partial s^2} = \frac{1}{R}\frac{\partial F}{\partial n} + \frac{\partial^2 F}{\partial t^2}$$

The fixed radius R can now be made equal to any variable radius r. On writing arc length $s = r\theta$, the first derivatives become

$$\frac{\partial F}{\partial n} = \frac{\partial F}{\partial r} \qquad \frac{\partial F}{\partial t} = \frac{1}{r}\frac{\partial F}{\partial \theta}$$

Since differentiating with respect to r or n is carried out along the same straight line, these operations are in all respects the same. The first derivatives can therefore be differentiated to give two second derivatives at point P, and the other follows from the differentiation along the curve,

$$\frac{\partial^2 F}{\partial n^2} = \frac{\partial^2 F}{\partial r^2} \qquad \frac{\partial^2 F}{\partial n\,\partial t} = \frac{\partial}{\partial r}\left(\frac{1}{r}\frac{\partial F}{\partial \theta}\right) \qquad \frac{\partial^2 F}{\partial t^2} = \frac{1}{r}\frac{\partial F}{\partial r} + \frac{1}{r^2}\frac{\partial^2 F}{\partial \theta^2} \qquad (1.20)$$

The geometrical significance of the last of these expressions can be appreciated by considering a particular function represented by the conical surface $F = F_0 - F_1 r$, where F_0 and F_1 are constants, and r is drawn from point O in Fig. 1.10. Along the curve s, function F is constant, but along line t it has a variation following the parabolic outline of a conic section.

The Laplace operator

$$\nabla^2 = \frac{\partial^2}{\partial n^2} + \frac{\partial^2}{\partial t^2}$$

can now be rewritten in polar coordinates using equations (1.20),

$$\nabla^2 = \frac{\partial^2}{\partial r^2} + \frac{1}{r}\frac{\partial}{\partial r} + \frac{1}{r^2}\frac{\partial^2}{\partial \theta^2}$$

2. Two-Dimensional Stress

In the simple theory of beams and shafts, the strain distribution across the section is assumed to be linear. From the elastic relations between stress and strain, the stress distribution follows. The resultant force and moment of these stresses can then be compared with the externally applied force and moment, to determine the stress value. To ensure that the structure will be satisfactory, sizes of cross-sections are adjusted to give stresses that appear to be safe in relation to some test value for the material, such as tensile strength. This approach to engineering design may be adequate for structures which are made from ductile materials and carry static loads. The argument may be used at any sites of stress concentration; local plastic yielding will relieve the high stresses.

A more precise appreciation of irregularities in stress distribution is needed for successfully designing members carrying alternating loads. A fatigue crack will form in any small region where the material is subjected to an excessive range of repeated stress. Once formed, the crack may spread to cause failure, even when the general stress level in the member is quite low. Generally speaking, no machine part that develops a crack in service can be regarded as satisfactory. Stress concentrations must therefore be recognized and reduced in severity as far as possible. In large welded steel structures, stress concentrations may provide a starting point for a fast-moving crack, with catastrophic results. Elastic analysis of stresses around sharp notches and cracks has provided the guide lines along which the study of fracture mechanics has developed.

Practical methods of designing to avoid fracture are not directly discussed here, and reference may be made to specialized texts such as that of R. B. Heywood on fatigue, and that of E. R. Parker on brittle fracture. The purpose of the present chapter is to work out from first principles the general methods of stress analysis based on plane elasticity. Particular attention will be given to the versatile method of complex stress functions.

2.1 Plane Deformation

The problem of finding components of stress and displacement at points in the interior of a loaded body is simplified if it can be assumed that there is a particular direction, say, the z-direction, along which there is no change in the distribution of stress and strain over the (x,y) plane. Two idealized situations can be defined.

24

2.1.1 Plane stress Consider a sheet of uniform thickness lying in the (x,y) plane and loaded by coplanar forces acting at its edge, as illustrated in Fig. 2.1(a). Near the outer edges there may be irregularities in stress due to the way in which forces are transmitted to the sheet, but at some small distance from the edge,

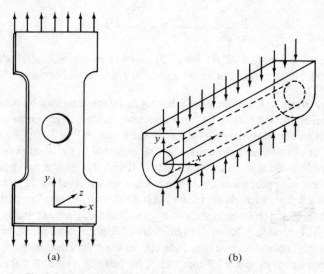

Fig. 2.1 Illustrations of (a) plane stress, (b) plane strain

comparable to the sheet thickness, it may be assumed that the sheet carries only the stress components σ_{xx}, σ_{yy}, and σ_{xy}, while the other components are negligible or zero,

$$\sigma_{zx} = \sigma_{zy} = \sigma_{zz} = 0$$

This idealized state of plane stress allows the elasticity equations (1.15) to be simplified,

$$Ee_{xx} = \sigma_{xx} - \nu\sigma_{yy} \qquad Ee_{yy} = \sigma_{yy} - \nu\sigma_{xx} \qquad Ee_{xy} = (1+\nu)\sigma_{xy}$$

These strain components can now be inserted into the equation of strain compatibility (1.10),

$$\frac{\partial^2 e_{yy}}{\partial x^2} + \frac{\partial^2 e_{xx}}{\partial y^2} - 2\frac{\partial^2 e_{xy}}{\partial x\,\partial y} = 0$$

to give a differential equation restricting the stress components,

$$\frac{\partial^2 \sigma_{yy}}{\partial x^2} + \frac{\partial^2 \sigma_{xx}}{\partial y^2} - 2\frac{\partial^2 \sigma_{xy}}{\partial x\,\partial y} - \nu\frac{\partial}{\partial x}\left(\frac{\partial\sigma_{xx}}{\partial x} + \frac{\partial\sigma_{xy}}{\partial y}\right) - \nu\frac{\partial}{\partial y}\left(\frac{\partial\sigma_{yy}}{\partial y} + \frac{\partial\sigma_{xy}}{\partial x}\right) = 0$$

However, by using the equations of static stress equilibrium (1.7), which are

$$\frac{\partial \sigma_{xx}}{\partial x} + \frac{\partial \sigma_{xy}}{\partial y} = 0 \qquad \frac{\partial \sigma_{yy}}{\partial y} + \frac{\partial \sigma_{xy}}{\partial x} = 0 \tag{2.1}$$

the second order equation reduces to

$$\frac{\partial^2 \sigma_{yy}}{\partial x^2} + \frac{\partial^2 \sigma_{xx}}{\partial y^2} - 2\frac{\partial^2 \sigma_{xy}}{\partial x \, \partial y} = 0 \tag{2.2}$$

Evidently, these equations (2.1) and (2.2) together restrict admissible distributions of stress in any situation where a state of plane stress prevails.

2.1.2 Plane strain   Consider a body having a surface generated by lines parallel to the z-direction, with a length in this direction which is large compared with dimensions measured in the (x,y) plane, as illustrated in Fig. 2.1(b). Any irregularities of stress near the ends might be expected to vanish at points taken at some small distance from the ends. Alternatively, it might be assumed that initially plane ends perpendicular to the z-direction are constrained to remain plane during deformation. It is assumed that no twisting or bending action is applied to the ends, and that if stresses σ_{xx}, σ_{yy}, and σ_{xy} are externally applied, these do not vary with z. From the symmetry of this situation it can be deduced that any initially plane section perpendicular to the z-direction must remain so after loading has been applied. From this, it follows that two of the shear strains are zero, and the elasticity equations require the corresponding stresses to be zero,

$$e_{zx} = e_{zy} = \sigma_{zx} = \sigma_{zy} = 0$$

If the body is considered to be divided by planes transverse to the z-direction into a number of parts of equal length, each part will carry the same resultant force in the z-direction and will therefore be stretched or compressed by the same amount, so the strain e_{zz} will not vary with z, and must therefore have some constant value C throughout the body. When this value is inserted in the elasticity equation

$$E e_{zz} = \sigma_{zz} - \nu(\sigma_{xx} + \sigma_{yy})$$

the longitudinal stress becomes

$$\sigma_{zz} = EC + \nu(\sigma_{xx} + \sigma_{yy})$$

This stress may be used in the remaining elasticity equations to obtain modified expressions for strains,

$$E' e_{xx} = \sigma_{xx} - \nu' \sigma_{yy} - \nu E' C \qquad E' e_{yy} = \sigma_{yy} - \nu' \sigma_{xx} - \nu E' C$$

$$E' e_{xy} = (1 + \nu') \sigma_{xy}$$

where $E' = E/(1 - \nu^2)$ and $\nu' = \nu/(1 - \nu)$. On inserting these strains into the strain compatibility equation, the constant term $\nu E' C$ and the new elastic constants all disappear, leaving equations for stress distribution identical with (2.1) and (2.2). It therefore appears that in both plane stress and plane strain, the same boundary stresses will give rise to the same internal distribution of stress. However, strains and displacements in a state of plane strain may be different from those in plane stress. It is mentioned that a state of plane strain is often understood as a situation where the longitudinal strain e_{zz} is zero.

2.1.3 Real stress function   If, in any problem, it is possible to discover expressions for stresses in terms of x and y which satisfy differential equations (2.1) and (2.2) at all interior points, as well as correctly representing the externally applied stresses, then the stress distribution would be fully determined. However, the process of solving such a boundary-value problem can be simplified by deriving all three stress components from a single scalar quantity. The problem can then be considered solved when a satisfactory expression is found for this scalar in terms of x and y. We see that the static equilibrium equations (2.1) are always satisfied if stress components are obtained by differentiating a scalar stress function F as follows,

$$\sigma_{xx} = \frac{\partial^2 F}{\partial y^2} \qquad \sigma_{yy} = \frac{\partial^2 F}{\partial x^2} \qquad \sigma_{xy} = -\frac{\partial^2 F}{\partial x \, \partial y} \qquad (2.3)$$

The second-order equation (2.2) must also be satisfied, and when the above values of stresses are inserted, an equation is obtained which restricts the function F,

$$\frac{\partial^4 F}{\partial x^4} + 2\frac{\partial^4 F}{\partial x^2 \, \partial y^2} + \frac{\partial^4 F}{\partial y^4} = 0$$

This biharmonic equation is written in various shortened forms,

$$\left[\frac{\partial^2}{\partial x^2} + \frac{\partial^2}{\partial y^2}\right]^2 F = \nabla^4 F = 0 \qquad (2.4)$$

To find this function F, often called an Airy stress function, one or more functions of x and y which satisfy the biharmonic equation may be selected and multiplied by constants. On finding stress components from (2.3), their values at the boundary may be compared with known values of externally applied stresses so as to determine the constant multipliers.

2.1.4 Complex stress functions   It may be asked whether there are alternative functions from which admissible stress distributions can be derived. If the three stress components at a point are to be identified with derivatives of some scalar A, then these derivatives must transform in the same way as components of the stress tensor σ_{ij} when the reference axes are rotated in the (x,y) plane. As there are three ways in which sets of derivatives can be written down to satisfy this condition, three scalars A, B, and F may be introduced, giving general expressions for stress components

$$\sigma_{xx} = \frac{\partial^2 A}{\partial x^2} - 2\frac{\partial^2 B}{\partial x \, \partial y} + \frac{\partial^2 F}{\partial y^2}$$

$$\sigma_{yy} = \frac{\partial^2 A}{\partial y^2} + 2\frac{\partial^2 B}{\partial x \, \partial y} + \frac{\partial^2 F}{\partial x^2}$$

$$\sigma_{xy} = \frac{\partial^2 A}{\partial x \, \partial y} - \left(\frac{\partial^2}{\partial y^2} - \frac{\partial^2}{\partial x^2}\right) B - \frac{\partial^2 F}{\partial x \, \partial y} \qquad (2.5)$$

If A and B are put equal to zero, stresses are satisfactorily given by function F, as shown previously. With F put equal to zero, the stress components may be inserted into differential equations (2.1) and (2.2) to obtain equations restricting the real scalars A and B,

$$\frac{\partial}{\partial x}\nabla^2 A - \frac{\partial}{\partial y}\nabla^2 B = 0$$

$$\frac{\partial}{\partial y}\nabla^2 A + \frac{\partial}{\partial x}\nabla^2 B = 0 \qquad (2.6)$$

It is seen that in general, both of these scalars are needed, and it might appear that this method of describing a two-dimensional stress field is less concise that that using the single function F. However, equations (2.6) happen to have the same form as the Cauchy-Riemann equations for a function of complex variable $z = x + iy$ having real and imaginary parts $\nabla^2 A$ and $\nabla^2 B$. Equations (2.6) are therefore solved by introducing two functions $\phi(z)$ and $\chi(z)$ such that

$$A + iB = \bar{z}\phi + \chi \qquad (2.7)$$

This solution is verified by differentiating,

$$\left(\frac{\partial^2}{\partial x^2} + \frac{\partial^2}{\partial y^2}\right)(A + iB) = 4\phi'(z) + 2\chi''(z)$$

giving a function of z which is analytic and which therefore satisfies (2.6). An obvious advantage of using this formulation of the equations of plane elasticity is that any analytic functions of z can be chosen, whereas admissible real functions F are restricted to solutions of the biharmonic equation.

2.1.5 Properties of analytic functions

Any function ϕ or χ as introduced in (2.7) is a function of z alone. Although the function may be expressed in x and y, these variables must occur as the combination $z = x + iy$, without any residual term in x alone or y alone. Terms in $\bar{z} = x - iy$ are also excluded. Such a function of z will have an unambiguous derivative with respect to z at any point, in the

28

same way as any function of a single variable. Such differentiable functions of z are called regular or analytic functions, and the functions ϕ and χ are invariably assumed to be of this kind.

A prime indicates differentiation with respect to the variable in brackets,

$$\phi'(z) = \frac{d\phi}{dz}$$

but if the bracketed variable is omitted, differentiation is understood to be with respect to z. Differentiation of ϕ with respect to the conjugate variable $\bar{z}$ is impossible, and although the conjugate function $\bar{\phi}$ may be differentiated with respect to $\bar{z}$, this is, in fact, never done. The complex conjugate $\bar{\phi}$ is obtained by changing the sign of i wherever it occurs in the function ϕ, thus changing the sign of the imaginary part of the function. The notation

$$\bar{\phi}' = \overline{\left(\frac{d\phi}{dz}\right)}$$

conveys that differentiation with respect to z is carried out first, and the conjugate is then taken.

The above restrictions on differentiating with respect to z do not apply to partial differentiation with respect to x or y, which may be carried out on any function, whether it is analytic or not. However, as the basic function ϕ is assumed to be analytic, rules for differentiating it can be formulated as follows. Since the variable x always occurs as part of the combination $z = x + iy$, the derivative with respect to x is the same as the derivative with respect to z,

$$\frac{\partial \phi}{\partial x} = \frac{\partial z}{\partial x}\frac{d\phi}{dz} = \frac{d\phi}{dz} = \phi'$$

Differentiation with respect to y introduces a factor i,

$$\frac{\partial \phi}{\partial y} = \frac{\partial z}{\partial y}\frac{d\phi}{dz} = i\frac{d\phi}{dz} = i\phi'$$

For differentiating a conjugate, the conjugate bar may be ignored while differentiation is carried out, and afterwards replaced,

$$\frac{\partial \bar{\phi}}{\partial x} = \overline{\left(\frac{\partial \phi}{\partial x}\right)} = \overline{\left(\frac{d\phi}{dz}\right)} = \bar{\phi}'$$

Differentiation of the conjugate with respect to y introduces a further minus sign,

$$\frac{\partial \bar{\phi}}{\partial y} = \overline{\left(\frac{\partial z}{\partial y}\frac{d\phi}{dz}\right)} = \overline{\left(\frac{\partial z}{\partial y}\right)}\overline{\left(\frac{d\phi}{dz}\right)} = -i\bar{\phi}'$$

2.1.6 Expressions for stresses   The stress components follow from (2.5) by replacing the real scalars A and B by complex functions (2.7). For example,

$$\sigma_{xx} = \frac{\partial^2 A}{\partial x^2} - 2\frac{\partial^2 B}{\partial x\,\partial y}$$

$$= \frac{\partial^2}{\partial x^2}\mathrm{Re}\,(\bar{z}\phi + \chi) - 2\frac{\partial^2}{\partial x\,\partial y}\mathrm{Im}\,(\bar{z}\phi + \chi)$$

where Re signifies 'real part' and Im signifies 'imaginary part'. Using the rules already derived for differentiating, and noting that Im $i\phi$ = Re ϕ, the three stress components become

$$\sigma_{xx} = 2\,\mathrm{Re}\,\phi' - \mathrm{Re}\,(\bar{z}\phi'' + \psi')$$
$$\sigma_{yy} = 2\,\mathrm{Re}\,\phi' + \mathrm{Re}\,(\bar{z}\phi'' + \psi')$$
$$\sigma_{xy} = \mathrm{Im}\,(\bar{z}\phi'' + \psi') \tag{2.8}$$

where a new function $\psi(z) = \chi'(z)$ has been introduced The following combinations of these components can now be found,

$$\sigma_{xx} + \sigma_{yy} = 4\,\mathrm{Re}\,\phi' \qquad \sigma_{yy} - \sigma_{xx} + 2i\sigma_{xy} = 2(\bar{z}\phi'' + \psi') \tag{2.9}$$

2.1.7 Displacements   Of the various ways of discussing displacements, perhaps the simplest procedure is to write down the correct expression in terms of the stress functions and then verify it by differentiating. Displacement components u_x and u_y in the x- and y-directions are given by

$$2G(u_x + iu_y) = \kappa\phi - z\bar{\phi}' - \bar{\psi} \tag{2.10}$$

where G is the shear modulus and the constant κ is given, for a state of plane stress, by $\kappa = (3 - \nu)/(1 + \nu)$.

Differentiation now gives

$$2G\frac{\partial}{\partial x}(u_x + iu_y) = \kappa\phi' - \bar{\phi}' - z\bar{\phi}'' - \bar{\psi}'$$

$$2G\frac{\partial}{\partial y}(u_x + iu_y) = i(\kappa\phi' - \bar{\phi}' + z\bar{\phi}'' + \bar{\psi}')$$

Strains are now obtainable from

$$e_{xx} = \frac{\partial u_x}{\partial x} \qquad e_{yy} = \frac{\partial u_y}{\partial y} \qquad 2e_{xy} = \frac{\partial u_y}{\partial x} + \frac{\partial u_x}{\partial y}$$

by separating real and imaginary parts of the last two expressions, using such relations as

$$\mathrm{Re}\,(i\phi') = -\mathrm{Im}\,\phi' \qquad \mathrm{Im}\,(z\bar{\phi}'') = -\mathrm{Im}\,(\bar{z}\phi'')$$

30

One thus obtains the strain combinations

$$e_{xx} + e_{yy} = 4\frac{1-\nu}{E}\operatorname{Re}\phi'$$

$$e_{yy} - e_{xx} + 2ie_{xy} = 2\frac{1+\nu}{E}(\bar{z}\phi'' + \bar{\psi}')$$

by making use of the relation between the elastic constants, $G = E/2(1 + \nu)$. On replacing these combinations of stress functions by combinations of stresses from (2.9), the elasticity equations for plane stress are reproduced. Hence, the assumed expression for displacements is shown to be correct.

For plane strain, the elastic constants may be modified by writing $E/(1 - \nu^2)$ in place of E, and $\nu/(1 - \nu)$ in place of ν. The expression for displacements (2.10) therefore remains unchanged, except that the value $\kappa = 3 - 4\nu$ should be used.

2.1.8 Rigid body motion and uniform stress   Consider the functions

$$\phi = L + imz \qquad \psi = N$$

where L and N are complex constants and m is a real constant. From (2.8) it is seen that all stress components are zero, and the displacement vector is found from (2.10),

$$2G(u_x + iu_y) = \kappa L - \bar{N} + i(\kappa + 1)mz$$

This indicates small movements of the whole body in the x- and y-directions, plus rotation about the origin through a small angle ω given by $\omega = (\kappa + 1)m/2G$. Therefore, if these stress functions are added to any other functions ϕ and ψ, the stresses and strains remain unchanged, the only effect being to superimpose a rigid-body motion.

When a sheet is subjected to uniform stress, with components S_x and S_y in the x- and y-directions and a shear stress $\sigma_{xy} = Q$, this stress is described by the stress functions

$$\phi = \tfrac{1}{4}(S_x + S_y)z \qquad \psi = \tfrac{1}{2}(S_y - S_x + 2iQ)z \qquad (2.11)$$

as may be verified by applying equations (2.8).

2.1.9 Rotation of reference axes   Consider a state of biaxial stress, for which the stress functions are

$$\phi(z) = \tfrac{1}{2}Sz \qquad \psi(z) = 0$$

Suppose now that new coordinate axes x_1 and y_1 are chosen, inclined at a positive angle η to the (x,y) axes. The new complex variable $z_1 = x_1 + iy_1$ can be expressed as $z_1 = ze^{-i\eta}$. If the last stress function is now written in terms of this new complex variable, we have $\phi(z_1) = \frac{1}{2}Sz_1e^{i\eta}$, which gives on differentiation with respect to z_1 a stress sum

$$\sigma_x + \sigma_y = 4\,\mathrm{Re}\,\phi'(z_1) = 2S\cos\eta$$

This is obviously incorrect, as the stress sum must be invariant with respect to rotation of axes. It seems that complex stress functions must be understood to be referred to (x,y) axes that remain fixed relative to the elastic body. For any other axes, new functions must be calculated before their derivatives will correctly give the stresses. However, this is unnecessary, for if the functions referred to the original (x,y) axes are retained, and stresses referred to these axes are determined by differentiating with respect to z, the stress components may be subsequently transformed in order to find the components referred to any other axes.

2.2 Circular Boundaries

For bodies bounded by circular arcs, it is convenient to refer stress components to polar coordinates (r,θ), the angle θ being measured from the x-axis in an anticlockwise direction. The new stress components are obtained by transformation, using (1.6),

$$\sigma_{rr} + \sigma_{\theta\theta} = \sigma_{xx} + \sigma_{yy}$$

$$\sigma_{\theta\theta} - \sigma_{rr} + 2i\sigma_{r\theta} = (\sigma_{yy} - \sigma_{xx} + 2i\sigma_{xy})e^{2i\theta}$$

As stresses referred to (x,y) axes are obtainable from the stress functions by means of (2.9), the above expressions become

$$\sigma_{rr} + \sigma_{\theta\theta} = 4\,\mathrm{Re}\,\phi'(z)$$

$$\sigma_{\theta\theta} - \sigma_{rr} + 2i\sigma_{r\theta} = 2[\bar{z}\phi''(z) + \psi'(z)]e^{2i\theta} \tag{2.12}$$

By separating real and imaginary parts, the individual components are obtained,

$$\sigma_{rr} = 2\,\mathrm{Re}\,\phi' - \mathrm{Re}\,[e^{2i\theta}(\bar{z}\phi'' + \psi')]$$

$$\sigma_{\theta\theta} = 2\,\mathrm{Re}\,\phi' + \mathrm{Re}\,[e^{2i\theta}(\bar{z}\phi'' + \psi')] \tag{2.13}$$

$$\sigma_{r\theta} = \mathrm{Im}\,[e^{2i\theta}(\bar{z}\phi'' + \psi')]$$

2.2.1 General series solution   Each of the two stress functions is now taken to consist of one or more terms of a series containing positive and negative powers of z,

$$\phi = (a_m + ib_m)z^m \qquad \psi = (p_m + iq_m)z^m$$

When stress components are calculated by means of (2.13) it is found that radial stress is proportionate to $\cos n\theta$ or $\sin n\theta$ where n has some integral value which is not equal to m. If a stress distribution corresponding to some integer n is required, it is necessary to select terms for the stress function having appropriate

32

m-values. Table 2.1 shows how this selection may be made. Terms in brackets in this table produce zero stress.

Table 2.1   Stress given by each term of a power series stress function

ϕ	ψ	Radial stress
a_{n+1}	p_{n-1}	$\cos n\theta$ $(n \geqslant 2)$
a_2	(p_0)	$\cos \theta$
a_1	p_{-1}	Axisymmetrical
(a_0)	p_{-2}	$\cos \theta$
a_{-n+1}	p_{-n-1}	$\cos n\theta$ $(n \geqslant 2)$
b_{n+1}	q_{n-1}	$\sin n\theta$ $(n \geqslant 2)$
b_2	(q_0)	$\sin \theta$
(b_1)	q_{-1}	Axisymmetrical
(b_0)	q_{-2}	$\sin \theta$
b_{-n+1}	q_{-n-1}	$\sin n\theta$ $(n \geqslant 2)$

The solution for axisymmetrical stress may be considered first. When a uniform shear stress is applied to the inner edge of an annulus, the shear stress that is applied to the outer edge cannot be arbitrarily chosen, as a condition of equilibrium of moments must be maintained. There is no such restriction on the radial stresses that may be applied to outer and inner boundaries. Table 2.1 shows that three terms are available, giving general stress functions

$$\phi = a_1 z \qquad \psi = (p_{-1} + iq_{-1})z^{-1}$$

By differentiating with respect to z as required, the stresses may be found from (2.13). The can then be written in terms of radius r by making the substitutions $z = re^{i\theta}$ and $\bar{z} = re^{-i\theta}$ to obtain stresses

$$\sigma_{rr} = 2a_1 + p_{-1} r^{-2} \qquad \sigma_{\theta\theta} = 2a_1 - p_{-1} r^{-2} \qquad \sigma_{r\theta} = -q_{-1} r^{-2}$$

The first two of these equations are the well-known equations of G. Lamé (1852), as used for thick-walled cylinders sustaining internal or external pressure. The constants for any given loading conditions are determined by inserting the known values of radial stress at the radius of each boundary.

Passing over the stresses varying as $\cos \theta$ for the moment, attention is now given to stresses varying as $\cos n\theta$ $(n = 2, 3, 4, \ldots)$. With any amplitude of radial stress applied to the outer boundary, any different amplitude may be applied to the inner boundary. Further, any shear stress varying as $\sin n\theta$ may be applied to the outer boundary with a further shear stress of some different value applied to the inner boundary. As each of these four stresses individually produces no resultant force or moment, any amplitudes may be selected. Table 2.1 shows that four constants are available for the stress functions, $a_{n+1}, a_{-n+1}, p_{n-1}$, and p_{-n-1}. These may be determined from any prescribed amplitudes of radial and shear stress applied at the boundaries. Two simplified situations can be distinguished. When the disc has no central hole, negative powers of z are excluded from the stress functions. Further, when stress is applied to the surface of a hole in an infinite sheet, with stress vanishing at infinity, terms of positive power in z are excluded. In principle, any distribution of radial load applied to the curved surface of a disc can be resolved into a Fourier series of harmonic components of stress. After finding the coefficients of the stress function for each harmonic component of applied stress, the stress at any interior point produced by each of these components can be found. Superposition of all such stresses then gives the total stress at the chosen point due to the total applied load.

2.2.2 Circular hole in an infinite plate

This problem may be solved by selecting terms for stress functions from Table 2.1 which give axisymmetrical stresses and stresses varying as $\cos 2\theta$. At large distances from the hole, the applied stress is taken to be uniaxial,

$$\sigma_y = T \qquad \sigma_x = \sigma_{xy} = 0$$

At the surface of the hole, at radius $r = a$, the required stresses are

$$\sigma_r = \sigma_{r\theta} = 0$$

for all values of θ. The stress functions which satisfy these conditions are

$$\phi = \tfrac{1}{4}T(z - 2a^2 z^{-1}) \qquad \psi = \tfrac{1}{2}T(z - a^2 z^{-1} - a^4 z^{-3})$$

If it is required to find the maximum shear stress acting in the plane of the plate at any point, this is given, in general, by

$$\tau_m = \left[\left(\frac{\sigma_y - \sigma_x}{2}\right)^2 + \sigma_{xy}^2\right]^{\frac{1}{2}} = |\bar{z}\phi''(z) + \psi'(z)|$$

or alternatively by

$$\tau_m = \left[\left(\frac{\sigma_\theta - \sigma_r}{2}\right)^2 + \sigma_{r\theta}^2\right]^{\frac{1}{2}} = |e^{2i\theta}(\bar{z}\phi'' + \psi')|$$

In the present problem, this expression is found to be

$$\tau_m = \tfrac{1}{2}T|e^{2i\theta} + \rho^{-2} - (2\rho^{-2} - 3\rho^{-4})e^{-2i\theta}|$$

where $\rho = r/a$. The calculated distribution of τ_m over a quadrant of the plate is shown as a contour diagram in Fig. 2.2. These contours coincide with the isochromatic lines in a photoelastic fringe pattern. (See, for instance Frocht, 1941.)

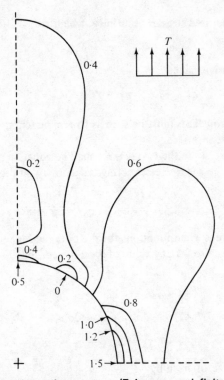

Fig. 2.2 Contours of maximum shear stress τ_m/T drawn on an infinite plate containing a circular hole, for uniaxial applied stress T

2.2.3 Elastic inclusion   As the stress functions ϕ and ψ give both stresses and displacements, they can be conveniently used for solving mixed boundary value problems. For example, consider a cylindrical inclusion which is welded into an infinite plate in such a way that the composite body is initially stress-free. The problem is to find the stress at any point when a uniaxial stress is applied to the outer edges of the plate. The inner part, of unit radius, is permitted to have different elastic constants from the outer part.

Stress functions are chosen to include all the terms giving axisymmetrical stress and stress varying as $\cos 2\theta$, as indicated by Table 2.1,

$$\phi_1 = a_{-1}\,z^{-1} + a_1\,z + a_3\,z^3 \qquad \psi_1 = p_{-3}\,z^{-3} + p_{-1}\,z^{-1} + p_1\,z$$
$$\phi_2 = a'_{-1}\,z^{-1} + a'_1\,z + a'_3 z^3 \qquad \psi_2 = p'_{-3}\,z^{-3} + p'_{-1}\,z^{-1} + p'_1\,z$$

The subscript (1) refers to the inner region while subscript (2) and primed constants refer to the outer region. As the functions for the inner region cannot contain terms of negative power, we have

$$a_{-1} = p_{-3} = p_{-1} = 0$$

35

Also, to give the required stresses at infinity, which are

$$\sigma_y = T \qquad \sigma_x = \sigma_{xy} = 0$$

it is necessary to assign values

$$a_1' = \tfrac{1}{4}T \qquad p_1' = \tfrac{1}{2}T \qquad a_3' = 0$$

The six remaining constants must be such as to give matching of stresses and displacements at the interface.

So that radial stresses in the two regions shall be equal on all points on the interface, and also the shear stresses acting tangentially on the interface, the combination

$$\sigma_{rr} - i\sigma_{r\theta} = 2\,\mathrm{Re}\,\phi' - e^{2i\theta}(\bar{z}\phi'' + \psi')$$

must have equal values when functions for the outer and inner regions are inserted. This must be so for any value of θ on the contour $z = e^{i\theta}$, which gives three equations

$$\tfrac{1}{2}T = 2a_1 - p_{-1}'$$

$$\tfrac{1}{2}T = -4a_{-1}' + p_1 + 3p_{-3}'$$

$$\tfrac{1}{2}T = 6a_3 + 2a_{-1}' + p_1 - 3p_{-3}'$$

The displacement vector given by

$$2G(u_x + iu_y) = \kappa\phi - z\overline{\phi'} - \bar{\psi}$$

must also have a common value for both regions at the radius of the interface. Elastic constants for the two regions are distinguished by subscripts in the additional three equations

$$G_2 \kappa_1 a_3 = G_1(a_{-1}' - p_{-3}')$$

$$G_2(\kappa_1 - 1)a_1 + G_1 p_{-1}' = \tfrac{1}{4}G_1(\kappa_2 - 1)T$$

$$G_1 \kappa_2 a_{-1}' + G_2(3a_3 + p_1) = \tfrac{1}{2}G_1 T$$

The above six equations are satisfied by the values

$$a_1 = \tfrac{1}{4}T\,\frac{(\kappa_2 + 1)G_1}{2G_1 + (\kappa_1 - 1)G_2} \qquad a_{-1}' = p_{-3}' = \tfrac{1}{2}T\,\frac{G_1 - G_2}{G_2 + \kappa_2 G_1}$$

$$a_3 = 0 \qquad p_1 = a_{-1}' + \tfrac{1}{2}T \qquad p_{-1}' = 2a_1 - \tfrac{1}{2}T$$

This solution may be used for finding stresses in an infinite medium at points where it joins with a long cylindrical inclusion, and for such a state of plane strain, the constant κ is expressed in terms of Poisson's ratio by $\kappa = 3 - 4\nu$. A value $\nu = 0\cdot3$ was chosen for plotting Fig. 2.3, which shows stresses at two particular points for various ratios of shear moduli. When the inner region becomes

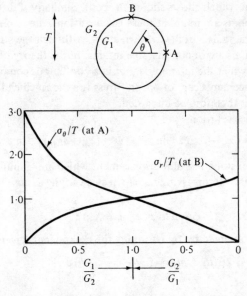

Fig. 2.3 Stress concentration factor in a matrix of modulus G_2 at junction with a cylinder of modulus G_1 in a field of uniaxial stress

an empty hole, modulus $G_1 = 0$, the hoop stress at the edge of the hole is given by

$$\sigma_\theta = T(1 + 2\cos 2\theta)$$

for a state of either plane strain or plane stress, and reaches the well-known maximum value $3T$ at points $\theta = 0$ and $\theta = \pi$.

2.2.4 Multi-valued displacements   Continuing the discussion of stresses in an annulus, some special points arise when radial stress varies as $\cos\theta$, or shear stress varies as $\sin\theta$. Such stresses each exert a resultant force on the annulus. This means that of the four amplitudes of radial and shear stress at the inner and outer boundaries, only three can be assigned arbitrarily, as the fourth must take the value which maintains overall equilibrium. It would appear from Table 2.1 that only two terms yield non-zero stresses. To obtain a third, a stress function of a different kind must be examined.

Suppose that in some particular problem, an expression for radial displacement has been derived, and it is found that the displacement at position $\theta = 2\pi$ is not the same as that at $\theta = 0$. If the annulus is incomplete, and covers a range of angles $0 < \theta < \theta_1$, such a multi-valued displacement would give no cause for concern, as the values for angles greater than θ_1 would be fictitious, corresponding to points outside the body. However, if the annulus covers an angle 2π, any difference in radial displacements at positions $\theta = 2\pi$ and $\theta = 0$ implies that a radial cut is present, which allows sliding to occur. Similarly if different circumferential displacements are found at these two positions, the implication is that the two edges of the radial cut have drawn apart. Although the stresses associated with such dislocations are of considerable interest in the theory of materials, the conclusion drawn at the moment is that in a continuous complete annulus, multi-valued displacements cannot occur. Stress functions which lead to such displacements must therefore be discarded.

The general series solution

$$\phi = (\alpha_m + i\beta_m)z^m \ln z \qquad \psi = (\lambda_m + i\mu_m)z^m \ln z$$

provides distributions of stress and displacement which are all multi-valued, with a few exceptions. Two terms giving radial stresses varying as $\cos\theta$ are obtained by writing $m = 0$,

$$\phi = \alpha \ln z \qquad \psi = \lambda \ln z$$

Displacements are obtained by transforming the components given by (2.10),

$$2G(u_r + iu_\theta) = (\kappa\phi - z\overline{\phi'} - \overline{\psi})e^{-i\theta}$$

These components are

$$2Gu_r = [\alpha(\kappa \ln r - 1) - \lambda]\cos\theta + [\alpha\kappa + \lambda]\theta\sin\theta$$

$$2Gu_\theta = [-\alpha(\kappa \ln r + 1) + \lambda]\sin\theta + [\alpha\kappa + \lambda]\theta\cos\theta$$

It can be seen that the parts of these expressions giving multi-valued displacements may be eliminated by stipulating that $\lambda = -\alpha\kappa$. With these constants, the new stress function is apparently satisfactory for use in the problem of a continuous annulus. Taking the available terms from Table 2.1, the general stress functions may be written

$$\phi = a_2 z^2 + \alpha \ln z \qquad \psi = p_{-2} z^{-2} - \alpha\kappa \ln z \qquad (2.14)$$

Three constants are now available for representing the three amplitudes of applied stress that can be arbitrarily assigned.

2.2.5 Concentrated force in a plane The solution just found can be utilized for solving the problem of a concentrated coplanar force P applied at point $z = 0$ on an infinite sheet, and directed along the x-axis. Consider a circular contour of some radius r surrounding the origin. Positive stresses acting on this contour are shown in Fig. 2.4. Equilibrium of the portion of the sheet enclosed by this contour is expressed by

$$P + \int_0^{2\pi} (\sigma_{rr}\cos\theta - \sigma_{r\theta}\sin\theta)r\,d\theta = 0$$

As this is to be true for any value of r, only those stresses varying as $1/r$ can be permitted, and these are furnished by the logarithmic terms in (2.14). When stress components are derived and inserted into the equilibrium equation, the

38

single constant is determined as $\alpha = -P/2\pi(1 + \kappa)$. For a state of plane stress, the elastic constant is written $\kappa = (3 - \nu)/(1 + \nu)$. This leads to a radial stress

$$\sigma_{rr} = -\frac{3 + \nu}{4\pi r} P \cos\theta$$

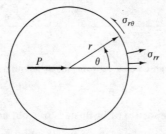

Fig. 2.4 Stresses due to a concentrated force

the other components being numerically smaller. If the force P is applied at the surface of a hole of finite size, or if the outer boundary of the sheet has a finite radius, the solution is not quite so simple, as all three constants in the stress function (2.14) must be determined from the given distributions of stress on the boundaries. Physically, a load cannot be applied at a point, and must be transmitted through a finite area of contact.

2.3 Direct Calculation of Stress Functions

In dealing with an annulus, it has been possible to select certain terms of a stress function in series form which could then be fitted to known stresses and displacements at the boundaries. However, when the stress function does not consist of terms of any regular series, a method is needed for calculating it from given boundary stresses. These stresses consist of the normal and shear stresses at each point on the surface, as nothing is initially known about the direct stress acting in a direction tangential to the surface.

2.3.1 Traction on an element of a contour   Suppose that the normal to a line segment BB′ lies in the n-direction, which is inclined at angle η to the x-direction as shown in Fig. 2.5(a). The tangential direction t makes a positive angle of 90° with direction n. For the moment, it is immaterial whether this line segment lies within the body or whether it forms part of a boundary. Tractions N_n and N_t acting on the contour are identical with positive stresses σ_{nn} and σ_{nt}, and are therefore given in terms of the stress functions by (2.12),

$$N_n - iN_t = \phi' + \overline{\phi}' - e^{2i\eta}(\overline{z}\phi'' + \psi')$$

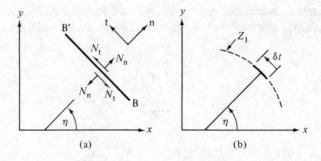

Fig. 2.5 (a) Traction on a contour, (b) differentiation along a contour

By taking complex conjugates of both sides one obtains

$$N_n + iN_t = \overline{\phi}' + \phi' - e^{-2i\eta}(z\overline{\phi}'' + \overline{\psi}') \tag{2.15}$$

2.3.2 Differentiation along a contour   A slight diversion is made at this point to consider how a function of x and y can be differentiated along a specified path in the (x,y) plane. If a function of z happens to be analytic, it will have a unique derivative at any chosen point regardless of the direction in which the point is approached. However, if the function is not analytic, it can still be differentiated along a specified path. If this path is one on which z has values z_1 the derivative of function f may be written

$$\frac{df}{dz_1} = \frac{\partial x}{\partial z_1}\frac{\partial f}{\partial x} + \frac{\partial y}{\partial z_1}\frac{\partial f}{\partial y}$$

If a segment δt of a contour has a normal which makes angle η with the x-axis, as shown in Fig. 2.5(b), this incremental length corresponds with an increment δz_1 in the complex variable, given by

$$\delta z_1 = \delta t\, e^{i(\eta + \pi/2)}$$

The segment also has components in the x- and y-directions given by

$$\delta x = -\,\delta t \sin\eta \qquad \delta y = \delta t\, \cos\eta$$

With the aid of these relations, the derivative can be expressed

$$\frac{df}{dz_1} = ie^{-i\eta}\left[\sin\eta\,\frac{\partial f}{\partial x} - \cos\eta\,\frac{\partial f}{\partial y}\right] \tag{2.16}$$

A particular combination of the stress functions ϕ and ψ is now examined with view to comparing the derivative of this combination with boundary values of

40

stress. The combination chosen is

$$f = \phi(z) + z\overline{\phi'}(z) + \overline{\psi}(z)$$

The derivative, as given by (2.16), is that taken along line z_1 while traversing the contour in an anticlockwise direction, and by using the rules already discussed for differentiating with respect to x and y, one obtains

$$\frac{df}{dz_1} = ie^{-i\eta}\sin\eta(\phi' + \overline{\phi'} + z\overline{\phi''} + \overline{\psi'}) + e^{-i\eta}\cos\eta(\phi' + \overline{\phi'} - z\overline{\phi''} - \overline{\psi'})$$

This immediately simplifies to

$$\frac{df}{dz_1} = \phi' + \overline{\phi'} - e^{-2i\eta}(z\overline{\phi''} + \overline{\psi'})$$

By referring back to (2.15) it is seen that this combination of stress functions expresses the tractions N_n and N_t acting at a point on the contour. When these tractions are inserted in place of the right-hand side of the last equation, integration of both sides gives the important relationship

$$f = \int (N_n + iN_t)\,dz_1 = [\phi + z\overline{\phi'} + \overline{\psi}]_B \qquad (2.17)$$

This equation shows that by integrating the known stresses along boundary B, one obtains the values of this combination of stress functions at various points along B. Any constant of integration is unimportant, since the addition of a constant to ϕ or ψ will not affect the stresses subsequently derived from these functions. It is seen that (2.17) yields only the particular values of the functions on the boundary, and further arguments must be used to obtain general expressions at interior points. Special methods will be considered later for reconstituting functions in general form from their boundary values. However, in many problems, it is possible to deduce the form of the function without the aid of rigorous methods, by observing that the general form of the function must be analytic, containing the coordinate variables in the form of complex variables only.

2.3.3 Sheet containing a circular hole

Some particular types of loading are selected for the purpose of illustrating the above methods for deriving stress functions, although the series solutions already discussed could be used here. Consider first a circular hole of unit radius in an infinite sheet, with stresses

$$N_r = S\cos n\theta \qquad N_\theta = Q\sin n\theta \qquad (n \geqslant 2)$$

applied to the surface of the hole. By writing $dz_1 = ie^{i\theta}\,d\theta$, and expressing the sine and cosine as exponentials, these stresses may be inserted into (2.17) and integrated to give

$$[\phi + z\overline{\phi'} + \overline{\psi}]_B = \frac{S+Q}{2(n+1)}e^{i(n+1)\theta} - \frac{S-Q}{2(n-1)}e^{-i(n-1)\theta} \qquad (2.18)$$

In this problem, it is known that the functions ϕ and ψ will take the form z^m where m is some negative integer, as the stresses must vanish at infinity. Hence the term in the last equation having a negative exponent may be associated with function ϕ, while the term of positive exponent may be attributed to the conjugates $\bar{\phi}'$ and $\bar{\psi}$. Also, as ϕ must be analytic, its boundary value as shown by (2.18) is satisfied by taking a general expression

$$\phi = -\frac{S-Q}{2(n-1)} z^{-(n-1)} \tag{2.19}$$

It is now a simple matter to replace this value, together with its derivative, in the original equation (2.18) to arrive at the general expression for the other stress function,

$$\psi = \frac{(n+2)Q - nS}{2(n_{+1})} z^{-(n+1)} \tag{2.20}$$

A more systematic way of determining ψ will be mentioned later.

The problem is now considered of an infinite sheet containing a stress-free hole, with a particular kind of stress

$$\sigma_x = -S' \qquad \sigma_y = S' \qquad \sigma_{xy} = 0$$

applied to the outer boundaries of the sheet, given by stress functions

$$\phi = 0 \qquad \psi = S'z \tag{2.21}$$

at a large distance from the hole. This problem may be solved by superimposition of the uniform distribution and another distribution which renders the hole stress-free. In the absence of a hole, the uniform distribution would give stresses on a circular contour

$$\sigma_r = -S'\cos 2\theta \qquad \sigma_{r\theta} = S'\sin 2\theta$$

These stresses, with their signs changed, are used as boundary values in the problem already solved, by inserting into (2.19) and (2.20) the values $n = 2$, $S = S'$, and $Q = -S'$. On adding these functions to those in (2.21), the required solution is obtained,

$$\phi = -S'z^{-1} \qquad \psi = S'(z - z^{-3})$$

As a final step in this exercise, the hoop stress at the edge of the hole may be found from these stress functions, using the standard equations (2.13), and this is given by $\sigma_\theta = 4S' \cos 2\theta$. Hence for this skew-symmetrical direct stress S' applied to the outer boundary of the sheet, a circular hole gives a stress concentration factor of 4. By rotating the reference axes through $45°$ it is seen that the stress applied at the outer boundary appears as a shear stress $\sigma_{xy} = S'$ acting by itself. The stress distribution already found therefore applies for this type of loading.

42

2.4 Solution for an Elliptical Hole

Although it is quite possible to calculate stresses around notches having various contours other than elliptical contours, the general solution for an elliptical cavity leads to a series of extremely useful results, and therefore a fairly full treatment is warranted. For conveniently describing stresses at the surface of an elliptical hole in a loaded sheet, a suitable coordinate system is required. Coordinates α and β are used, these being the real and imaginary parts of a complex variable $\xi = \alpha + i\beta$. The required relationship between this variable and the variable $z = x + iy$ in rectangular coordinates is given by

$$z = a \cosh \xi \qquad (2.22)$$

This may be verified by separating real and imaginary parts of each side,

$$x = a \cosh \alpha \cos \beta \quad y = a \sinh \alpha \sin \beta$$

Eliminating β, we get

$$\frac{x^2}{\cosh^2 \alpha} + \frac{y^2}{\sinh^2 \alpha} = a^2$$

If some constant value of α is chosen, this relation between x and y represents an ellipse drawn in the (x,y) plane. Similarly, lines of constant β are found to be hyperbolic curves intersecting the ellipses at right angles at all points in the plane, as shown in Fig. 2.6. As α tends to zero, the corresponding ellipse shrinks to a straight line of length $2a$, but as α becomes very large, the ellipse approaches a circular shape.

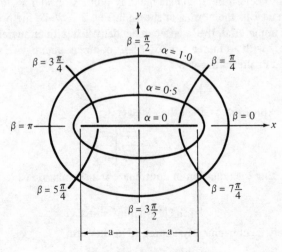

Fig. 2.6 Elliptical coordinates

2.4.1 Uniform stress   Suppose that a sheet without any hole has an all-round stress S applied to its edges. Superimposed on this is a stress S' which is tensile in the y-direction and compressive in the x-direction. In addition a positive shear stress Q is applied to the edges of the sheet. It can be seen that any general state of uniform stress can be resolved into these three parts, so that at any point, the stress components referred to (x,y) coordinates are

$$\sigma_x = S - S' \qquad \sigma_y = S + S' \qquad \sigma_{xy} = Q$$

The α-direction is taken to be normal to a line of constant α, that is, normal to an elliptical contour at some chosen point on the contour. This normal makes a

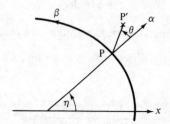

Fig. 2.7 Transformation of stress components

positive angle η with the x-direction, as shown in Fig. 2.7. Stresses referred to the new coordinates are now obtained by transforming the above (x,y) components,

$$\sigma_\alpha + \sigma_\beta = 2S$$
$$\sigma_\beta - \sigma_\alpha + 2i\sigma_{\alpha\beta} = 2(S' + iQ)\, e^{2i\eta} \tag{2.23}$$

It remains to determine the angle η at any point. Consider a short line PP' inclined at angle θ to the normal as shown in Fig. 2.7. While the precise length of this line is immaterial, the length can be identified with an increment $|\,\delta z\,|$ or alternatively with an increment $|\,\delta \xi\,|$. The complex quantity $(P' - P)$ can now be represented in alternative ways,

$$\delta z = |\delta z|\, e^{i(\eta + \theta)} \qquad \delta \xi = |\delta \xi|\, e^{i\theta}$$

Dividing one expression by the other, we get

$$e^{i\eta} = \frac{dz}{d\xi} \bigg/ \left| \frac{dz}{d\xi} \right|$$

By differentiating the equation of coordinate transformation (2.22) and using the fact that

$$|\sinh \xi|^2 = \sinh \xi \, \sinh \overline{\xi}$$

the angle η may be conveniently expressed in the form

$$e^{2i\eta} = \sinh \xi / \sinh \overline{\xi} \tag{2.24}$$

44

If they should be needed, the stress functions in terms of z for a state of uniform stress as given in (2.11) can be rewritten in terms of ξ by using (2.22),

$$\phi = \tfrac{1}{2} Sa \cosh \xi \qquad \psi = (S' + iQ) a \cosh \xi \qquad (2.25)$$

2.4.2 Infinite sheet with a stress-free hole   A problem of some practical interest is to find the stresses near an elliptical hole when some uniform state of stress exists at points remote from the hole. The approach used here is to write down the stresses on an elliptical contour drawn on a sheet which has no hole. With the signs of these stresses reversed, they are used as boundary values for solving a separate problem in which the surface of the hole is loaded while the stress is zero at the outer boundary of the sheet. When this solution is super-imposed on that for uniform stress the solution to the original problem is obtained.

We proceed now with the solution for a loaded hole, with zero stress at infinity. The tractions applied to the surface of the hole are obtained by reversing the signs of stresses given by (2.23),

$$N_\alpha + iN_\beta = - [S - (S' - iQ) e^{-2i\eta}]$$

From (2.21) it is seen that $dz = a \sinh \xi \, d\xi$, and along an elliptical contour, $d\xi = i \, d\beta$. With the expression for angle η from (2.24), the boundary traction integral as given by (2.17) can now be written

$$f = -ia \int_B [S \sinh \xi - (S' - iQ) \sinh \overline{\xi}] \, d\beta$$

As this integration is to be carried out on the contour on which α has constant value α_1 it is convenient to write $e^{\alpha_1} = \lambda$, so that

$$2 \sinh \xi = \lambda e^{i\beta} - \lambda^{-1} e^{-i\beta} \qquad 2 \sinh \overline{\xi} = \lambda e^{-i\beta} - \lambda^{-1} e^{i\beta}$$

Integration can now be carried out in accordance with (2.17),

$$[\phi + z\overline{\phi}' + \overline{\psi}]_B =$$
$$- \tfrac{1}{2} a [S(\lambda \, e^{i\beta} + \lambda^{-1} \, e^{-i\beta}) + (S' - iQ) \, (\lambda e^{-i\beta} + \lambda^{-1} e^{i\beta})]$$

As the stresses vanish at infinity, terms on the right having positive β exponents must be associated with the conjugate functions. After setting aside these terms, there remains the expression

$$[\phi]_B = - \tfrac{1}{2} a [S\lambda^{-1} \, e^{-i\beta} + (S' - iQ) \lambda e^{-i\beta}]$$

At any general point within the body, ϕ must be a function of the complex variable ξ. It is seen that the function which reduces to the correct value at the boundary is

$$\phi = - \tfrac{1}{2} a [Se^{-\xi} + (S' - iQ)\lambda^2 e^{-\xi}]$$

45

This function can now be added to that in (2.25), to give the function for a sheet loaded at its outer boundary and containing a stress-free hole,

$$\phi = \tfrac{1}{2} Sa \sinh \xi - \tfrac{1}{2} a (S' - iQ) \lambda^2 e^{-\xi} \qquad (2.26)$$

The solution is not yet complete, as function ψ must be determined. When the boundary is stress-free, the boundary traction equation takes the simple form

$$[\psi]_B = - [\bar{\phi} + \bar{z} \frac{d\phi}{dz}]_B$$

For differentiating with respect to z, the transformation equation (2.22) indicates that $dz/d\xi = a \sinh \xi$, so we have

$$\frac{d\phi}{dz} = \frac{d\xi}{dz} \frac{d\phi}{d\xi} = \frac{1}{a \sinh \xi} \frac{d\phi}{d\xi} \qquad (2.27)$$

Hence ψ is found to be

$$\psi = \tfrac{1}{2} a \left[(S' + iQ) e^{\xi} - \frac{S(\lambda^2 + \lambda^{-2}) + (S' - iQ)(\lambda^4 e^{-2\xi} + 1)}{2 \sinh \xi} \right] \qquad (2.28)$$

Stress functions (2.26) and (2.28) now allow stresses at any point to be calculated, although this in itself is quite a lengthy process. It must be remembered that differentiation with respect to z must be in accordance with (2.27).

2.4.3 Stress concentration at a notch root

The stress functions take a slightly simpler form when only a single component of stress T is applied to the outer edges of the sheet in a direction perpendicular to the major axis of the hole, i.e., the y-direction. In place of the stresses previously used, we write $S = S' = T/2$, so that S and S' combine in the y-direction and cancel in the x-direction to give a uniaxial stress T. The stress functions from (2.26) and (2.28) now become

$$\phi = \tfrac{1}{4} Ta (\sinh \xi - \lambda^2 e^{-\xi})$$

$$\psi = \frac{Ta}{8} \frac{e^{2\xi} - \lambda^4 e^{-2\xi} - (\lambda + \lambda^{-1})^2}{\sinh \xi} \qquad (2.29)$$

It is of particular interest to determine the stress acting in a tangential direction at the surface of the hole, since the highest tensile stress is expected to occur there. To calculate this, it is noted that as the normal stress is zero, the tangential stress is equal to the stress sum at any point on this contour,

$$\sigma_\beta = 4 \mathrm{Re} \frac{d\phi}{dz} = T \frac{\sinh 2\alpha_1 + \lambda^2 \cos 2\beta - 1}{\cosh 2\alpha_1 - \cos 2\beta}$$

The greatest tensile value occurs at each end of the major axis of the ellipse, and is obtained by putting $\beta = 0$,

$$\sigma_\beta = T(1 + 2 \coth \alpha_1) \qquad (2.30)$$

46

In practical situations, the solution for an elliptical hole is often applied to an open-ended notch. As shown in Fig. 2.8, the semi-major axis of the ellipse is identical with the total depth d of the notch below the vertical edge of the plate, and is therefore given by $d = a \cosh \alpha_1$. In fact, when a cut is made in the plate along the y-axis, normal stresses are relieved along this edge, but this is

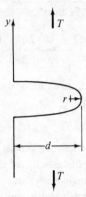

Fig. 2.8 Stress concentration at an edge notch

assumed to have only a negligible effect on stress at the notch root. The root radius r is given by

$$r = a \sinh^2 \alpha_1 / \cosh \alpha_1$$

Hence the maximum stress, which acts in the y-direction, may be found from (2.30) to be

$$\sigma_y = T \left(1 + \sqrt{\frac{d}{r}}\right)$$

This remarkably simple formula, due to C. E. Inglis (1913), is often used in situations where the sides of the notch do not follow an elliptical contour.

2.4.4 Crack under uniaxial stress
It is now assumed that the elliptical cavity has its height in the y-direction reduced to zero so that it becomes a slit of length $2a$ in the x-direction. The stress functions (2.29) are rewritten with $\alpha_1 = 0$ and $\lambda = 1$,

$$\phi = Ta(e^\xi - 3e^{-\xi})/8 \qquad \psi = Ta(\sinh 2\xi - 2)/4 \sinh \xi \qquad (2.31)$$

These lead to stresses

$$\sigma_\alpha + \sigma_\beta = \frac{2T \sinh 2\alpha}{\cosh 2\alpha - \cos 2\beta} - T$$

$$\sigma_\beta - \sigma_\alpha + 2i\sigma_{\alpha\beta}$$
$$= 2T \left[\frac{\cosh 2\alpha \cos 2\beta - 1 + i \sinh 2\alpha \sin 2\beta}{2(\cosh 2\alpha - \cos 2\beta)} \right.$$
$$\left. + \frac{\sinh 2\alpha (1 - \cos 2\beta) + i(\cosh 2\alpha - 1) \sin 2\beta}{(\cosh 2\alpha - \cos 2\beta)^2} \right]$$

The second expression can be conveniently used for computing the maximum shear stress at various points, given by

$$\tau_m = \tfrac{1}{2} |(\sigma_\beta - \sigma_\alpha) + 2i\sigma_{\alpha\beta}|$$

and contours of maximum shear stress are shown in Fig. 2.9. These lines correspond to isochromatic lines in photoelasticity. If radius r is measured from the tip of a crack of length $2a$, with angle θ measured from the x-axis, one obtains, for small values of r, a simple expression for maximum shear stress,

$$\tau_m = \frac{T}{2} \left(\frac{a}{2r} \right)^{1/2} \sin \theta$$

As this value is infinitely large when r tends to zero, any tensile stress T would be expected to cause plastic yielding in the two opposite regions of high stress at the crack tip, one of which is shown in Fig. 2.9. The region of plastic yielding at the end of a crack in a thick plate is, in fact, observed to be two-lobed. However, as yielding disturbs the stress distribution in the elastic part of the plate, one

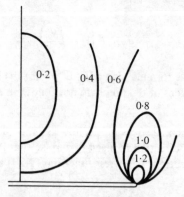

Fig. 2.9 Maximum shear stress in the region of a crack, expressed as a multiple of the unidirectional stress applied to the plate

should not expect the outline of the yielded region to coincide exactly with an isochromatic line determined for a purely elastic state.

When the plate is thin in relation to the crack length, any appreciable tensile force applied to the plate will cause yielding by shearing on planes at 45° to the plane of the plate. A single narrow region of yielding is then observed to extend along the direction of the crack.

It is finally noted that for an internal crack of length $2a$, the stress functions (2.31) can be written in rectangular coordinates,

$$\phi = \tfrac{1}{4}T[2(z^2 - a^2)^{\frac{1}{2}} - z] \qquad \psi = \tfrac{1}{2}T[z - a^2/(z^2 - a^2)^{\frac{1}{2}}] \qquad (2.32)$$

At points on the x-axis, for distances $x > a$, the stresses are found to be

$$\sigma_x = \frac{Tx}{(x^2 - a^2)^{\frac{1}{2}}} - T \qquad \sigma_y = \frac{Tx}{(x^2 - a^2)^{\frac{1}{2}}}$$

It is seen that as the end of the crack is closely approached, the state of stress becomes virtually an all-round stress.

At points on the x-axis at distances $x < a$, the stress functions (2.32) indicate stresses $\sigma_x = -T$ and $\sigma_y = 0$. This compressive stress in the x-direction may cause the edges of a slit to buckle out of the plane of the sheet. An alternative method for analysing stresses around a straight crack having stress-free surfaces is provided by Westergaard stress functions, to be considered later.

2.5 Cauchy Integrals

So far, continuous distributions of stress have been applied to boundary surfaces. The method used was to write down boundary values of a stress function in terms of distance measured along the boundary. As the stress function at interior points must be an analytic function of z, this function was deduced by inspection of its boundary value. However, this simple method is not adequate when boundary stresses are discontinuously distributed, and a more general theory is then required for relating an analytic function to its boundary values. This is provided by Cauchy's integral theorem, which will now be briefly examined.

2.5.1 General properties　　Consider first an external boundary consisting of a closed loop in the (x,y) plane, as indicated in Fig. 2.10(a). On this boundary B, the variable z takes a series of values z_1. It is supposed that some function $f(z_1)$ is known at all points on B. The theorem then states that the function $f(z)$ which has this boundary value is given in its general form by

$$f(z) = \frac{1}{2\pi i} \int \frac{f(z_1)\,dz_1}{z_1 - z} \qquad (2.33)$$

This integral is taken around the complete contour in an anticlockwise direction. The proof of the theorem, as discussed in texts on complex variables, depends

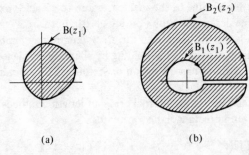

Fig. 2.10 Integrals around (a) external boundary, (b) internal boundary

on the absence of any singular points within the contour, at which $f(z)$ attains an infinite value.

To investigate the working of this theorem, consider a function having a boundary value of polynomial form

$$f(z_1) = \sum_1^\infty A_m z_1^{-m} + A_0 + \sum_1^\infty A_n z_1^n$$

First, a general term of positive index is selected. When this is inserted into (2.33), the integrand may be expanded into a series

$$\frac{A_n z_1^n}{z_1 - z} = A_n \left[z_1^{n-1} + z_1^{n-2} z + \ldots + z^{n-1} + \frac{z^n}{z_1 - z} \right] \tag{2.34}$$

When these terms are integrated around the contour, all give a zero integral except the last,

$$A_n z^n \int_B \frac{dz_1}{z_1 - z} = 2\pi i A_n z^n$$

If a general term of negative index is now selected, it may be expanded into the series

$$\frac{A_m z^{-m}}{z_1 - z} = A_m \left[-\frac{1}{z} \frac{1}{z_1^m} - \frac{1}{z^2} \frac{1}{z_1^{m-1}} - \ldots - \frac{1}{z^m} \frac{1}{z_1} + \frac{1}{z^m} \frac{1}{z_1 - z} \right] \tag{2.35}$$

When integrated around the contour, all terms give a zero integral except the last two, which mutually cancel,

$$-\int_B \frac{dz_1}{z_1} + \int_B \frac{dz_1}{z_1 - z} = 0$$

50

It appears that if we have a function $f(z_1)$ in the form of a series, another series with identical coefficients is obtained for the function $f(z)$, but with the terms of negative power omitted, i.e.,

$$f(z) = A_0 + A_1 z + A_2 z^2 + \ldots$$

Therefore, when the boundary value $f(z_1)$ is a continuous function, z_1 may be replaced by z in this function to obtain $f(z)$, provided this function does not reach an infinite value at any point within the contour.

Attention is now given to the problem of determining the analytic function $f(z)$ in an infinite sheet containing a hole of boundary B_1 when function $f(z_1)$ is given on this boundary. Another outer contour B_2 is drawn to enclose some point z as shown in Fig. 2.10(b). These two contours are now considered to be connected so as to form a single contour which makes an anticlockwise circuit of point z. The integral theorem as given in (2.33) can be applied to this composite contour to give

$$f(z) = \frac{1}{2\pi i} \left\{ \int_{B_2} \frac{f(z_2) \, dz_2}{z_2 - z} - \int_{B_1} \frac{f(z_1) \, dz_1}{z_1 - z} \right\}$$

Each integral is assumed to be taken in an anticlockwise direction, but as the composite contour requires B_1 to be followed in the reverse direction, this integral is given a negative sign. The outer contour B_2 is now allowed to expand indefinitely so that all parts of it are eventually at a very large distance from the origin. If it is specified that the function $f(z)$ vanishes at infinity, the value of $f(z_2)$ will vanish on contour B_2 so that this integral is of zero value, leaving

$$f(z) = -\frac{1}{2\pi i} \int_{B_1} \frac{f(z_1) \, dz_1}{z_1 - z} \tag{2.36}$$

This equation is the same as (2.33) except for the minus sign. To investigate the significance of this sign, function $f(z_1)$ may be supposed to consist of a series of terms of non-zero negative powers. A typical term may be expanded as in (2.35) and integrated around the contour. However, it is now noted that

$$\int_{B_1} \frac{dz_1}{z_1 - z} = 0$$

because the point z now lies outside the contour. The single remaining term having a non-zero integral gives

$$-\frac{A_m}{z^m} \int_{B_1} \frac{dz_1}{z_1} = -2\pi i \frac{A_m}{z^m}$$

This minus sign cancels the minus sign in (2.36), so the coefficient in the series $f(z)$ is the same as that in the series $f(z_1)$. By a similar argument, terms in $f(z_1)$

of zero or positive power give a zero value of the contour integral. Hence a function $f(z_1)$ given on the boundary of a hole leads to a function

$$f(z) = A_{-1} z^{-1} + A_{-2} z^{-2} + \ldots$$

at any point in the infinite z-plane outside the hole contour.

This is the form that must be taken by a stress function when stress is continuously distributed at the surface of the hole and vanishes at a large distance from the hole. However, the formal application of the integral theorem (2.36) remains valid when the distribution of boundary stress is not expressible as a continuous function of z_1.

2.5.2 Calculation of stress functions

The Cauchy integral theorem is now used for finding stress functions from the known stresses applied at a boundary. While this method can be used for an external or an internal boundary, the following discussion relates primarily to an infinite sheet containing an internal cavity. The function f is identified with the boundary stress integral (2.17) obtained previously,

$$f = \int (N_n + iN_t) dz_1$$

This integral was found to correspond with the value of a combination of the stress functions at the hole surface B.

$$f(z_1) = [\phi(z) + z\bar{\phi}'(z) + \bar{\psi}(z)]_B \tag{2.37}$$

As a further restriction, it is supposed that the hole is circular, so that z_1 becomes equal to a variable $\epsilon = e^{i\theta}$ and can be replaced by this new variable. As stress is assumed to vanish at infinity, the functions ϕ and ψ must each consist of series of negative powers of z or of functions that can be so expressed. Terms of negative power is the known function $f(\epsilon)$ are therefore associated with $\phi(z)$ alone, while the remaining terms of positive power in $f(\epsilon)$ are associated with the complex conjugates on the right of (2.37). Hence $\phi(z)$ may be determined by applying a Cauchy integral to $f(\epsilon)$, as this process will reject the unwanted terms of positive power,

$$\phi(z) = -\frac{1}{2\pi i} \int_C \frac{f(\epsilon) d\epsilon}{\epsilon - z} \tag{2.38}$$

the integral being taken around the whole circular boundary C.

To determine the function ψ it is appropriate to take complex conjugates of both sides of (2.37). Writing $\phi'(z) = \Phi(z)$ and noting that $\bar{z}_1 = \bar{\epsilon} = \epsilon^{-1}$, we get

$$[\psi(z)]_B = \bar{f}(\epsilon) - \bar{\phi}(\epsilon) - \frac{\Phi(\epsilon)}{\epsilon}$$

When Cauchy integrals are taken of each of the terms on the right, it is seen that as $\bar{\phi}(\epsilon)$ is now equivalent to a series of positive powers of ϵ, the integral is zero.

The last term on the right gives an integral equal to the same expression with z written for ϵ. Hence the required function is given by

$$\psi(z) = -\frac{1}{2\pi i}\int_C \frac{\bar{f}(\epsilon)\,d\epsilon}{\epsilon - z} - \frac{\phi'(z)}{z} \tag{2.39}$$

The systematic way in which the two complex stress functions can be calculated makes this method of analysis superior to the method using a real stress function, as no analogous method is available for finding the real stress function.

2.5.3 Partly loaded circular hole

As an illustration of the use of Cauchy integrals, the stress is calculated in an infinite sheet containing a hole of unit radius. Consider a symmetrical distribution of tensile stress applied to the surface, having uniform value S over two sectors as shown in Fig. 2.11, the value on the intervening sectors being zero. The boundary stress integral

$$f = \int N_r\,d\epsilon$$

is first evaluated, where $\epsilon = e^{i\theta}$, and N_r is put equal to S or zero as appropriate. Writing $e^{i\theta_1} = \epsilon_1$ where θ_1 is the semi-angle of the loaded sector, symmetry gives the values of ϵ at the four points shown in Fig. 2.11,

$$\epsilon_A = \epsilon_1^{-1} \qquad \epsilon_B = \epsilon_1 \qquad \epsilon_C = -\epsilon_1^{-1} \qquad \epsilon_D = -\epsilon_1$$

The integral is now found to have values in the four sectors

$$f(AB) = S\epsilon \qquad\qquad f(BC) = S\epsilon_1$$

$$f(CD) = S(\epsilon_1 + \epsilon_1^{-1} + \epsilon) \quad f(DA) = S\epsilon_1^{-1}$$

The addition of a constant to all of these values leaves the subsequent calculation unchanged, so the starting point of the above integration is unimportant.

To find ϕ, a Cauchy integral is taken around the whole circumference of the

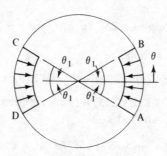

Fig. 2.11 Partly loaded circular hole

unit circle, as in (2.38). With a slight rearrangement where necessary, this integral becomes

$$\phi = -\frac{S}{2\pi i}\left\{\int_A^B\left(1+\frac{z}{\epsilon-z}\right)d\epsilon + \int_B^C\frac{\epsilon_1\,d\epsilon}{\epsilon-z}\right.$$

$$\left. + \int_C^D\left(1+\frac{z+\epsilon_1+\epsilon_1^{-1}}{\epsilon-z}\right)d\epsilon + \int_D^A\frac{\epsilon_1^{-1}\,d\epsilon}{\epsilon-z}\right\}$$

This integral may be evaluated and simplified to

$$\phi = -\frac{S}{2\pi i}[(z+\epsilon_1)\ln(z+\epsilon_1)+(z-\epsilon_1)\ln(z-\epsilon_1)$$

$$-(z-\epsilon_1^{-1})\ln(z-\epsilon_1^{-1})-(z+\epsilon_1^{-1})\ln(z+\epsilon_1^{-1})]$$

Differentiation now gives

$$\phi'(z)=-\frac{S}{2\pi i}\ln\frac{z^2-\epsilon_1^2}{z^2-\epsilon_1^{-2}}\tag{2.40}$$

The other stress function follows from (2.39),

$$\psi = -\frac{S}{2\pi i}\left[-4i\theta_1\,z^{-1}+\epsilon_1\ln\frac{z-\epsilon_1^{-1}}{z+\epsilon_1^{-1}}-\epsilon_1^{-1}\ln\frac{z-\epsilon_1}{z+\epsilon_1}\right]$$

Attention is now restricted to stresses at the surface of the hole. It follows from (2.40) that

$$2\,\mathrm{Re}\,\phi'(z)=-\frac{S}{\pi}\mathrm{Im}[2i\theta_1+\ln\sin(\theta-\theta_1)-\ln\sin(\theta+\theta_1)]$$

where Re and Im signify real and imaginary parts. This may be written

$$2\,\mathrm{Re}\,\phi' = S-2S\theta_1/\pi\quad(0<\theta<\theta_1)$$

$$2\,\mathrm{Re}\,\phi' = -2S\theta_1/\pi\quad(\theta_1<\theta<\pi/2)$$

Also, on the surface of the hole,

$$[\bar{z}\phi''(z)+\psi'(z)]e^{2i\theta}=-2S\theta_1/\pi$$

From the standard expressions (2.13) for deriving stresses, the stresses at the surface of the hole are found to be

$$\sigma_r = S\quad \sigma_\theta = S-4S\theta_1/\pi\quad(0<\theta<\theta_1)$$

$$\sigma_r = 0\quad \sigma_\theta = -4S\theta_1/\pi\quad(\theta_1<\theta<\pi/2)$$

At point $\theta = \pi/2$, which is on the stress-free part of the hole, hoop stress approaches a value $-2S$ as the angle θ_1 subtended by the load distribution approaches $\pi/2$. However, at points on the loaded part of the circumference, the hoop stress approaches $-S$ as θ_1 approaches $\pi/2$. This means that the effect of leaving a small gap in an otherwise uniform distribution of radial stress is to double the maximum hoop stress.

2.6 Transformation of Body Contour

When the boundary of a body is non-circular, it is convenient for the purpose of evaluating Cauchy integrals to transform the boundary contour into a circle. An equation of conformal transformation is required, $z = \Omega(\zeta)$, which expresses $z = x + iy$ as some analytic function Ω of another complex variable $\zeta = \rho\, e^{i\beta}$. This transformation will ensure that for any chosen point z in the z-plane, there is a corresponding point ζ in the ζ-plane. Dealing particularly with a hole in the z-plane, it is required that the transformation should give a series of points on the circle of unit radius $\rho = 1$ in the ζ-plane as a series of points $z = z_1$ are followed around the contour of the hole in the z-plane, as shown in Fig. 2.12.

By means of this transformation equation, it is possible to rewrite any function, say $\phi(z)$, in terms of the new variable ζ. To clarify the meaning of the notation used, it may be noted that if $\phi(z)$ has a certain real part at some chosen z-point, then at the corresponding ζ-point, the real part of function $\phi(\zeta)$ will have precisely the same value. Similarly, the imaginary parts are the same. This means that ϕ has the same numerical identity at a chosen point irrespective of the variable used to describe the position of the point. The equality $\phi(z) = \phi(\zeta)$ at any point does not convey that in the algebraic expressions for ϕ, the variables z and ζ are interchangeable.

In particular, the value of ϕ at any selected point on contour $z = z_1$ is equal to the value at the corresponding point $\zeta = e^{i\beta}$ on the unit circle, defined by angle β. For convenience, we write $e^{i\beta} = \epsilon$, so that at any chosen point on the contour, $\phi(z_1) = \phi(\epsilon)$. To obtain the second form, it is merely necessary to rewrite $\phi(z_1)$ replacing z_1 by $\Omega(\epsilon)$ in accordance with the transformation equation. From this function $\phi(\epsilon)$ on the hole contour, it is required to find the general expression

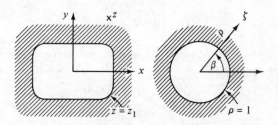

Fig. 2.12 Transformation of hole contour

55

$\phi(\zeta)$ which is valid at points outside the hole contour. This may be found by taking a Cauchy integral around the unit circle.

When this is done, it would be quite permissible to revert from the form $\phi(\zeta)$ to the form $\phi(z)$ by using the inverse of the transformation equation, as the essential purpose of the transformation has been fulfilled. However, this step may not be necessary, and it is usually satisfactory to leave the function as an expression in ζ. It must be remembered that when stress functions are differentiated to give stress components, differentiation must be with respect to variable z. This does not present any difficulty, as we have

$$\phi'(z) = \frac{\partial \phi}{\partial \zeta} \frac{\partial \zeta}{\partial z} = \frac{\phi'(\zeta)}{\Omega'(\zeta)}$$

2.6.1 Elliptical contour An important transformation is that for mapping an elliptical contour on to a unit circle. Let an elliptical contour in the z-plane be described by

$$z_1 = \tfrac{1}{2}a[e^{\alpha_1 + i\beta} + e^{-(\alpha_1 + i\beta)}]$$

in accordance with the previously used equation (2.22). Writing $e^{\alpha_1} = \lambda$, the conformal transformation

$$z = \tfrac{1}{2}a(\lambda\zeta + \lambda^{-1}\zeta^{-1}) \tag{2.41}$$

shows that if $\zeta = \rho e^{i\beta}$, the circle $\rho = 1$ in the ζ-plane maps on to the ellipse $z = z_1$ in the z-plane. It so happens that angle β will serve as a variable in both planes. If we now write $e^{i\beta} = \epsilon$, corresponding points on the ellipse and unit circle are related by

$$z_1 = \tfrac{1}{2}a(\lambda\epsilon + \lambda^{-1}\epsilon^{-1}) \tag{2.42}$$

For calculating stress functions, the boundary stress integral (2.17),

$$f = \int (N_\alpha + iN_\beta)\,dz_1$$

is first calculated using the known tractions on the elliptical boundary $z = z_1$. When z_1 is expressed in terms of ϵ by means of (2.42), function f represents the boundary value of a combination of stress functions as previously found in (2.17),

$$f(\epsilon) = \left[\phi + z\left(\overline{\frac{d\phi}{dz}}\right) + \bar{\psi}\right] \tag{2.43}$$

If these functions are now taken to be expressed in terms of ζ, the first takes a form already established in (2.38),

$$\phi(\zeta) = -\frac{1}{2\pi i}\int_C \frac{f(\epsilon)\,d\epsilon}{\epsilon - \zeta} \tag{2.44}$$

the integral being taken around the unit circle C.

56

For calculating the stress function ψ we take complex conjugates of both sides of (2.43),

$$[\psi]_C = \bar{f}(\epsilon) - \bar{\phi}(\epsilon) - \left[\bar{z}\frac{d\phi}{dz}\right]_C \qquad (2.45)$$

The derivative may be expressed

$$\frac{d\phi}{dz} = \frac{d\phi}{d\zeta}\frac{d\zeta}{dz} = \phi'(\zeta)\bigg/\left(\frac{dz}{d\zeta}\right) = \Phi(\zeta)\bigg/\left(\frac{dz}{d\zeta}\right)$$

The transformation equations (2.41) and (2.42) now provide the required values of $\bar{z}$ and $dz/d\zeta$ on circle C, giving

$$\left[\bar{z}\frac{d\phi}{dz}\right]_C = \frac{\lambda\epsilon^{-1} + \lambda^{-1}\,\epsilon}{\lambda - \lambda^{-1}\,\epsilon^{-2}}\Phi(\epsilon)$$

Cauchy integrals can now be taken of the terms on the right of (2.45). The last term gives an expression unchanged in form, but with ζ written in place of ϵ, while the term $\bar{\phi}(\epsilon)$ gives a zero integral, the result being

$$\psi(\zeta) = -\frac{1}{2\pi i}\int_C \frac{\bar{f}(\epsilon)\,d\epsilon}{\epsilon - \zeta} - \frac{\lambda\zeta^{-1} + \lambda^{-1}\,\zeta}{\lambda - \lambda^{-1}\,\zeta^{-2}}\phi'(\zeta) \qquad (2.46)$$

It has been assumed that the stresses acting on the surface of the hole have no force resultant and that stresses vanish at points remote from the hole. More general methods have been described by N. I. Muskhelishvili (1953).

2.6.2 Partly loaded elliptical hole
Uniform tensile stress S is applied to two parts of an elliptical boundary so that the loading is symmetrical with respect to x and y axes, as shown in Fig. 2.13. The stress at infinity is taken to be zero.

For obtaining the integral of tractions on boundary z_1 a value of dz_1 may be taken from (2.42),

$$f = \tfrac{1}{2}a\int N_\alpha(\lambda - \lambda^{-1}\,\epsilon^{-2})\,d\epsilon$$

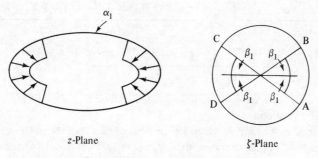

z-Plane

ζ-Plane

Fig. 2.13 Partly loaded elliptical hole

From the symmetrical disposition of points A, B, C, and D in Fig. 2.10, the limits of integration are given by

$$\epsilon_A = \epsilon_1^{-1} \qquad \epsilon_B = \epsilon_1 \qquad \epsilon_C = -\epsilon_1^{-1} \qquad \epsilon_D = -\epsilon_1$$

where $\epsilon_1 = e^{i\beta_1}$ and β_1 is the half-angle of the loaded part of the surface. By putting traction N_α equal to S or zero along each segment, values of f on the unit circle are given by

$$f(AB) = \tfrac{1}{2}Sa[\lambda\epsilon + \lambda^{-1}\epsilon^{-1}]$$

$$f(BC) = \tfrac{1}{2}Sa[\lambda\epsilon_1 + \lambda^{-1}\epsilon_1^{-1}]$$

$$f(CD) = \tfrac{1}{2}Sa[\lambda(\epsilon_1 + \epsilon_1^{-1} + \epsilon) + \lambda^{-1}(\epsilon_1 + \epsilon_1^{-1} + \epsilon^{-1})]$$

$$f(DA) = \tfrac{1}{2}Sa[\lambda\epsilon_1^{-1} + \lambda^{-1}\epsilon_1]$$

A Cauchy integral is now taken around the whole circumference of the unit circle in accordance with (2.44) to obtain a rather lengthy expression for $\phi(\zeta)$. When differentiated with respect to ζ this expression gives

$$\phi'(\zeta) = -\frac{Sa}{4\pi_i}\left[\frac{4i\beta_1}{\lambda\zeta^2} + \left(\lambda - \frac{\lambda^{-1}}{\zeta^2}\right)\ln\frac{\zeta^2 - \epsilon_1^2}{\zeta^2 - \epsilon_1^{-2}}\right] \tag{2.47}$$

The other function ψ may be found from (2.46), and when differentiated gives

$$\psi'(\zeta) = -\frac{Sa}{2\pi i}\left[\frac{2i\beta_1(\lambda^2 + \lambda^{-2})(\lambda + \lambda^{-1}\zeta^{-2})}{(\lambda\zeta - \lambda^{-1}\zeta^{-1})^2} - \frac{(\epsilon_1^2 - \epsilon_1^{-2})(\lambda + \lambda^{-1}\zeta^2)}{(\zeta^2 - \epsilon_1^2)(\zeta^2 - \epsilon_1^{-2})}\right]$$

For finding stresses, angle η between the normal to the boundary and the x-axis, as already found in (2.24), may be expressed as

$$e^{2i\eta} = (\lambda\zeta - \lambda^{-1}\zeta^{-1})/(\lambda\bar{\zeta} - \lambda^{-1}\bar{\zeta}^{-1})$$

Using this expression, and the method previously discussed for differentiating with respect to z, the following combination is obtained,

$$[\bar{z}\phi''(z) + \psi'(z)]e^{2i\eta} =$$

$$-\frac{S}{i\pi(\lambda\bar{\zeta} - \lambda^{-1}\bar{\zeta}^{-1})}\left[2i\beta_1\frac{(\lambda^2 + \lambda^{-2})(\lambda\zeta + \lambda^{-1}\zeta^{-1}) - 2(\lambda\bar{\zeta} + \lambda^{-1}\bar{\zeta}^{-1})}{(\lambda\zeta - \lambda^{-1}\zeta^{-1})^2}\right.$$

$$\left. + \frac{\zeta^2(\epsilon_1^2 - \epsilon_1^{-2})(\lambda\bar{\zeta} + \lambda^{-1}\bar{\zeta}^{-1} - \lambda\zeta^{-1} - \lambda^{-1}\zeta)}{(\zeta^2 - \epsilon_1^2)(\zeta^2 - \epsilon_1^{-2})}\right] \tag{2.48}$$

This expression, together with (2.47) allows the stresses applied to the surface of the hole to be verified.

Particular attention is now given to stresses acting at points on the x-axis when the elliptical hole has degenerated into a slit ($\lambda = 1$). Fortunately expression (2.48) becomes equal to zero under these conditions which means that the stress at points on this line is an all-round stress. Writing $\zeta = \rho$ in (2.47) this stress is found to be

$$\sigma_y = 2 \, \mathrm{Re} \, \phi'(z) = -\frac{S}{\pi} \left[\frac{4\beta_1}{\rho^2 - 1} + \mathrm{Im} \ln \frac{\rho^2 - \epsilon_1^{\,2}}{\rho^2 - \epsilon_1^{\,-2}} \right]$$

This expression readily reduces to

$$\sigma_y = -\frac{S}{\pi} \left[\frac{4\beta_1}{\rho^2 - 1} - 2 \arctan \frac{\sin 2\beta_1}{\rho^2 - \cos 2\beta_1} \right] \tag{2.49}$$

At points on the x-axis, the transformation (2.41) reduces to

$$\begin{align} x &= \tfrac{1}{2}a(\rho + \rho^{-1}) \qquad (x > a) \\ x &= a \cos \beta \qquad\qquad (x < a) \end{align} \tag{2.50}$$

Hence, the part of the stress which increases to infinity at the crack tip is found to be

$$\sigma_y = -\frac{2S\beta_1}{\pi} \frac{x}{(x^2 - a^2)^{\frac{1}{2}}}$$

where β_1 defines the part of the crack surface which carries tensile applied stress S.

2.6.3 Crack opened by a wedge   By differentiating (2.49) with respect to β_1 and then putting $\beta_1 = \pi/2$, we get the stress due to pressure p acting on the crack surface over small intervals $\delta\beta_1$ on each side of the y-axis,

$$\sigma_y = \frac{\partial \sigma_y}{\partial \beta_1} \delta\beta_1 = \frac{4p}{\pi} \left(\frac{1}{\rho^2 - 1} + \frac{1}{\rho^2 + 1} \right) \delta\beta_1$$

These pressures are equivalent to resultant forces P applied to the crack surfaces, and acting along the positive and negative y-axes, the magnitude being $P = 2p\delta x = 2ap\delta\beta_1$. When variable ρ is expressed in terms of x by means of (2.50), the stress becomes

$$\sigma_y = \frac{Pa}{\pi x (x^2 - a^2)^{\frac{1}{2}}}$$

This stress, such as might be produced by driving a wedge into a crack at its mid-point, was calculated by Westergaard (1939) by an alternative method which will be considered next.

2.7 Westergaard's Stress Functions

These are essentially real stress functions of the biharmonic or Airy type, referred to coordinates (x,y). They are particularly convenient for analysing stresses in bodies containing discontinuities such as cracks. Since there are no singular points in the coordinate system itself, physical discontinuities in stress are represented by discontinuous changes in the stress function at particular points or surfaces. These functions may be generated from analytic functions of complex variable z in the following way.

If Z is any analytic function of z, its real and imaginary parts satisfy the relations

$$\nabla^2 \operatorname{Re} Z = 0 \qquad \nabla^2 \operatorname{Im} Z = 0$$

Further, if some function $f(x,y)$ satisfies the equation $\nabla^2 f = 0$, then solutions F of the biharmonic equation $\nabla^4 F = 0$ are provided by the alternatives

$$F = f \qquad F = xf \qquad F = yf$$

This can be verified by differentiating. Hence admissible biharmonic functions F are provided by either $\operatorname{Re} Z$ or $\operatorname{Im} Z$ or products of either of these with x or y. Suppose that a limited number of these alternatives is chosen, giving a stress function

$$F = \operatorname{Re} Z + y \operatorname{Im} Z' \tag{2.51}$$

It is noted that function Z' which is differentiated with respect to z is still an analytic function, and the two terms are dimensionally homogeneous. Stresses are now obtained from

$$\sigma_{xx} = \frac{\partial^2 F}{\partial y^2} \qquad \sigma_{yy} = \frac{\partial^2 F}{\partial x^2} \qquad \sigma_{xy} = -\frac{\partial^2 F}{\partial x\, \partial y}$$

In order to differentiate real and imaginary parts with respect to x and y, the following rules may be used,

$$\frac{\partial}{\partial x} \operatorname{Re} Z = \frac{\partial}{\partial y} \operatorname{Im} Z = \operatorname{Re} Z'$$

$$\frac{\partial}{\partial x} \operatorname{Im} Z = -\frac{\partial}{\partial y} \operatorname{Re} Z = \operatorname{Im} Z' \tag{2.52}$$

These rules may be verified by using the chain rule for differentiating, e.g.,

$$\frac{\partial}{\partial y} \operatorname{Im} Z = \operatorname{Im} \left(\frac{\partial z}{\partial y} \frac{\partial Z}{\partial z} \right) = \operatorname{Im} iZ' = \operatorname{Re} Z'$$

By differentiating according to (2.52), the chosen stress function (2.51) yields the following stress components,

$$\sigma_{xx} = \operatorname{Re} Z'' - y \operatorname{Im} Z'''$$

$$\sigma_{yy} = \operatorname{Re} Z'' + y \operatorname{Im} Z''' \tag{2.53}$$

$$\sigma_{xy} = -y \operatorname{Re} Z'''$$

This implies that $\sigma_{xx} = \sigma_{yy}$ and that $\sigma_{xy} = 0$ at all points on the line $y = 0$, so the stress function (2.51) is adequate only for problems in which this happens to be true. Whether this restricted form of stress function is adequate or not will come to light when it is called upon to satisfy all the boundary values of the problem. If an alternative form of the function (2.51) is selected, different stress equations (2.53) will be needed, as shown later.

2.7.1 Crack with biaxial tension   As an example of the very direct solution obtainable by this method, consider an infinite plate containing a straight crack of length $2a$ situated on the x-axis such that $-a < x < a$, with a remote stress field consisting of all-round tensile stress S. For conditions of either plane stress or plane strain, the appropriate stress function is

$$Z' = S(z^2 - a^2)^{\frac{1}{2}} \qquad (2.54)$$

Differentiation gives

$$Z'' = Sz/(z^2 - a^2)^{\frac{1}{2}} \qquad Z''' = -Sa^2/(z^2 - a^2)^{\frac{3}{2}}$$

Stresses can now be found from (2.53), and the known values on the surface of the crack and at very large values of z can be verified. For expressing the singularity value of σ_y it is useful to introduce the small distance r measured along the x-axis from the tip of the crack, when it may be found that

$$\sigma_y = S(a/2r)^{\frac{1}{2}} \qquad (2.55)$$

The stress function used for biaxial stress cannot be modified to account for uniaxial stress, because this condition violates the assumption that $\sigma_x = \sigma_y$ on $y = 0$. However, it is easy to find this stress distribution by superimposing two distributions. First, by making biaxial stress S numerically equal to uniaxial applied stress T, the distribution is as already found. Then, an additional applied stress $\sigma_x = -T$ simply gives this stress acting at all points just as if the crack was not present. When these distributions are added, we get the required stresses at remote points, $\sigma_y = T$ and $\sigma_x = 0$. If attention is restricted to the singularity value of σ_y as given by (2.55), the magnitude of any applied stress σ_x does nothing to change this value, which therefore depends only on the value of the component σ_y applied at the outer edge of the plate.

The further problem of a crack subjected to internal pressure p with the outer edges of the plate stress-free can be solved either by modifying the stress function or by means of a superimposition. If, on the distribution for biaxial tension $S = p$ we superimpose uniform stress at all points $\sigma_x = \sigma_y = -p$, the required stresses are given both at the outer edges of the plate and on the crack surfaces. The singularity value as given by (2.55) remains tensile.

2.7.2 Contact stress between cylinders   This problem is often associated with the name of H. Hertz (1881). Although there are alternative methods of analysis, perhaps none is so direct as that using Westergaard's stress functions. It is first necessary to establish expressions for displacements u and v in the x- and

y-directions. For conditions of plane strain, these can be written in terms of shear modulus G and Poisson's ratio ν,

$$2Gu = (1 - 2\nu)\,\text{Re}\,Z' - y\,\text{Im}\,Z''$$
$$2Gv = 2(1 - \nu)\,\text{Im}\,Z' - y\,\text{Re}\,Z'' \tag{2.56}$$

These may be proved correct by differentiating according to (2.52) to obtain strains. These strains can then be compared with stresses from (2.53) by means of the elasticity equations for plane strain,

$$2Ge_{xx} = (1 - \nu)\,\sigma_{xx} - \nu\sigma_{yy} \qquad 2Ge_{yy} = (1 - \nu)\,\sigma_{yy} - \nu\sigma_{xx} \qquad 2Ge_{xy} = \sigma_{xy}$$

Suppose that the initially flat surface of a semi-infinite body is displaced to form a cylindrical depression of radius R, as shown in Fig. 2.14(a). The stress function

$$Z'' = -A[(a^2 - z^2)^{\frac{1}{2}} + iz]$$

is chosen, where A is a real constant. Equation (2.56) may be differentiated to give the curvature of the surface over the region $-a < x < a$,

$$2G\frac{\partial^2 v}{\partial x^2} = 2(1 - \nu)\,\text{Im}\,Z'''$$

Hence the radius of curvature is found,

$$\frac{1}{R} = -\frac{\partial^2 v}{\partial x^2} = \frac{A(1 - \nu)}{G}$$

The stress function also gives vertical stress, from (2.53),

$$\sigma_{yy} = -A(a^2 - x^2)^{\frac{1}{2}}$$

acting over the area of contact. Taking unit length along the axis of the cylindrical impression, the upward force P is found by integrating stress with respect

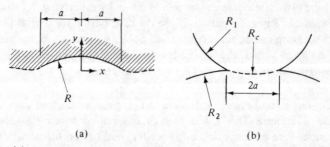

(a) (b)

Fig. 2.14 (a) Impression in a plane surface, (b) two cylinders pressed into contact

to x, giving $P = \frac{1}{2}\pi A a^2$. By eliminating constant A, a relation is now available between force and width of contact band,

$$\frac{1}{R} = \frac{4(1 - \nu^2)P}{\pi E a^2} \tag{2.57}$$

Although this result has been obtained for a change in curvature of an initially flat surface, it may be applied to situations in which two surfaces of curvatures $1/R_1$ and $1/R_2$ are pressed together. The common surface is assumed to take up some curvature $1/R_c$ as shown in Fig. 2.14(b). Changes in curvature may be expressed by adapting (2.57), allowing for the possibility that the bodies may have different elastic constants,

$$\frac{1}{R_1} - \frac{1}{R_c} = \frac{4P}{\pi a^2} \frac{1 - \nu_1^2}{E_1}$$

$$\frac{1}{R_2} + \frac{1}{R_c} = \frac{4P}{\pi a^2} \frac{1 - \nu_2^2}{E_2}$$

By adding these expressions to eliminate R_c, a relation is obtained which defines the half-width a of the band of contact in terms of the load P pressing the two cylinders together,

$$\frac{1}{R_1} + \frac{1}{R_2} = \frac{4P}{\pi a^2} \left(\frac{1 - \nu_1^2}{E_1} + \frac{1 - \nu_2^2}{E_2} \right)$$

Knowing the width of the contact region, the maximum compressive stress can be found from $\sigma_y = -2P/\pi a$.

2.7.3 Crack under shear   Consider a sheet containing a crack of total length $2a$ on the x-axis, with shear stress $\sigma_{xy} = Q$ applied to the infinitely remote boundary of the sheet. For solving this problem, equation (2.51) for the stress function is replaced by an alternative,

$$F = -y \operatorname{Re} Z'$$

New expressions for stresses then follow,

$$\sigma_{xx} = 2\operatorname{Im} Z'' + y \operatorname{Re} Z''' \qquad \sigma_{yy} = -y \operatorname{Re} Z''' \quad , \qquad \sigma_{xy} = \operatorname{Re} Z'' - y \operatorname{Im} Z'''$$

If the stress function is taken to be

$$Z' = Q(z^2 - a^2)^{\frac{1}{2}}$$

it may be verified that when z is large, the direct stresses vanish, leaving the required shear stress, and that the surface of the crack is stress-free. The shear stress near the tip of the crack is of special interest. If r is a small distance measured from the tip of the crack along the x-axis, this stress becomes

$$\sigma_{xy} = Q(a/2r)^{\frac{1}{2}} \tag{2.58}$$

2.8 Terminology of Fracture Mechanics

Although the theory of linear elasticity predicts an infinite stress at the end of a crack in a loaded member, tests of cracked members show that some finite load is sustained before failure occurs through extension of the crack. For correlating test results it would appear that some finite quantity is required which takes into account the spatial extent of the region of high stress. Such a quantity may be obtained by selecting a point at some small distance r from the tip of the crack, and multiplying a chosen stress component at this point by the square root of the distance. The selected point is taken on the x-axis, that is on the line on which the straight crack is situated. The quantity so obtained, multiplied by a conventional constant, is called the stress intensity factor. Various factors are distinguished according to the stress component considered. When this component is the direct stress perpendicular to the line of the crack, we have a stress intensity factor of the first kind, written K_I and defined by

$$\sigma_{yy} = K_I/(2\pi r)^{\frac{1}{2}}$$

In principle, this factor can be calculated for any body of a specified geometrical shape carrying a specified distribution of forces and stresses on its outer boundary. A simple example is that of an infinite sheet containing a single crack of total length $2a$, with tensile stress T acting in the y-direction on the outer edges. At points on the x-axis near the crack tip, the required stress is given by (2.55),

$$\sigma_y = T(a/2r)^{\frac{1}{2}}$$

On comparing this with the above expression defining K_I it is seen that in this particular problem,

$$K_I = T(\pi a)^{\frac{1}{2}}$$

Suppose that this sheet is now split along the y-axis. This will result in the release of stress σ_x at all points along the cut. We now have a sheet with stress $\sigma_y = T$ at remote points as before, but the crack is of length a and is at right angles to a stress-free edge. The stress intensity factor for this geometry would be expected to be almost the same as for a crack of length $2a$ in the interior of the sheet. However, it is not exactly the same, the value for the edge crack being about 12% higher than that for the internal crack, according to Eshelby (1968).

A second type of stress intensity factor K_{II} is defined in a similar way, except that it refers to the shear stress component in the (x,y) plane

$$\sigma_{xy} = K_{II}/(2\pi r)^{\frac{1}{2}}$$

For example, consider an infinite sheet containing a crack of total length $2a$, with shear stress $\sigma_{xy} = Q$ acting on the outer boundary. Shear stress at points on the x-axis at distance r from the crack tip is given in (2.58),

$$\sigma_{xy} = Q(a/2r)^{\frac{1}{2}}$$

Hence the stress intensity factor for this particular situation is

$$K_{II} = Q(\pi a)^{\frac{1}{2}}$$

2.8.1 Anti-plane strain   A third stress intensity factor K_{III} refers to shear stress σ_{yz}. This stress is zero under the ideal stress conditions of plane stress and plane strain, but becomes prominent in the further idealized stress state of anti-plane strain, which will now be briefly discussed. As shown in Fig. 2.15, a crack

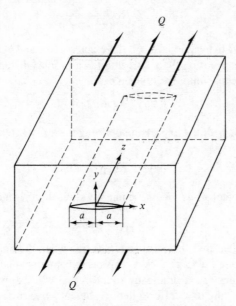

Fig. 2.15 Crack under conditions of anti-plane strain

of uniform width $2a$ is assumed to extend through an indefinitely large body which has uniform shear stress $\sigma_{yz} = Q$ applied to its outer surface. Displacements u and v in the x- and y-directions are taken to be zero while all displacements and stresses are constant with respect to distance along the z-axis. The idealized deformation is therefore defined by the distribution $w(x,y)$ of displacement in the z-direction. From these assumptions it follows that the only non-zero stresses are

$$\sigma_{zx} = G\frac{\partial w}{\partial x} \qquad \sigma_{zy} = G\frac{\partial w}{\partial y}$$

where G is the shear modulus of elasticity. The stress equilibrium equations (1.7) reduce to

$$\frac{\partial \sigma_{zx}}{\partial x} + \frac{\partial \sigma_{zy}}{\partial y} = 0$$

65

so it is seen that the distribution $w(x,y)$ must satisfy the equation $\nabla^2 w = 0$. This is satisfied if w is the real or imaginary part of an analytic complex function.

This particular problem is solved by the function

$$w = (Q/G)\operatorname{Re}(z^2 - a^2)^{\frac{1}{2}} \tag{2.59}$$

Here, $z = x + iy$ is the complex variable, and not the third coordinate. By using rules (2.52) for differentiating, it is found that

$$\sigma_{zy} = Q\operatorname{Re}\frac{z}{(z^2 - a^2)^{\frac{1}{2}}}$$

It can be seen that the values both on the crack surface and at infinity are as required. At some point on the x-axis situated at a small distance r from the crack tip, this stress component is given by

$$\sigma_{zy} = Q(a/2r)^{\frac{1}{2}}$$

By comparing this with the expression defining the stress intensity factor, which is

$$\sigma_{zy} = K_{III}/(2\pi r)^{\frac{1}{2}}$$

it is seen that the factor for the particular situation that has been specified is

$$K_{III} = Q(\pi a)^{\frac{1}{2}}$$

It may be verified that this result remains true when the body is split through the plane containing the (y,z) axes, as this plane happens to be stress-free. Therefore the stress intensity factor already found is that which applies to an open-ended crack extending to a uniform depth a below a free surface. For example, these conditions are fulfilled by a longitudinal or transverse crack in a bar subjected to torsion, provided the crack depth is small relative to the measurements of the cross-section.

2.8.2 Crack opening displacement   In the experimental tests sometimes carried out for assessing the resistance of a material to brittle fracture, that is, its fracture toughness, a measurement is made of the amount by which a crack opens under load, before the test-piece finally fractures. In a ductile material, plastic deformation will occur around the tip of the crack, and the crack will remain open after load is removed from the test-piece. However, for small loads, the extent of this plastic yielding is limited, and on removing the load, the crack opening is almost completely reversed. It is of interest to know the amount of purely elastic opening that occurs, and this is very easily obtained by using Westergaard's function (2.54) for plane strain, together with equation (2.56) for vertical displacement of the crack surface along line $y = 0$,

$$Gv = (1 - \nu)T(a^2 - x^2)^{\frac{1}{2}}$$

Hence the maximum value in the middle of the crack, $x = 0$, is given by $v = 2Ta/E'$, where $E' = E/(1 - v^2)$. For conditions of plane stress, that is, for a crack in a very thin sheet, E' is replaced by E in the expression for displacement. It should be noted that the crack opening is twice the displacement of the crack surface above the x-axis. Although this result has been derived for biaxial tensile stress T at the outer boundary of the plate, the same result holds for uniaxial stress T in the y-direction, as externally applied stress in the x-direction produces no additional opening.

3. Flexure of Flat Plates

If $w(x,y)$ is the amount by which an elastic plate deflects away from the (x,y) plane, curvatures can be found in terms of w, and hence strains and stresses at any point. Equilibrium then gives a differential equation restricting the function $w(x,y)$. General solutions of this equation may be fitted to the known conditions at the edges of the plate, which may be specified as deflections, gradients, bending moments, or shear forces.

In the present chapter, this procedure is worked out to a stage where some practical problems connected with round and rectangular plates can be solved. Often, similar results can be obtained by using strain energy methods, to be discussed in the next chapter. Not much will be said about the flexure of beams, although this topic provides an obvious introduction to the analysis of plates, as the equations for plates are essentially two-dimensional forms of the one-dimensional equations for beams.

3.1 Thick and Thin Plates

Bending of a beam or plate of uniform thickness usually brings to mind a linear distribution of stress and strain through the thickness, with values on opposite sides that are numerically equal but opposite in sign. Before elaborating this idea into a method of analysis, its relevance to practical situations should be briefly examined, and attention should be drawn to its limitations.

3.1.1 Membrane stress   The total stress in a plate may be thought of as a superposition of bending stress and membrane stress. Membrane stress is the average stress through the thickness, as might be set up by the stretching effect of force vectors lying in the plane of the plate.

As an example, consider the square-section duct of internal dimension d and wall thickness t, as shown in Fig. 3.1(a), carrying internal pressure p. By making any cut across the section parallel to one of the sides, equilibrium of the parts indicates a membrane stress $pd/2t$. Considering one side as a beam with built-in ends, the bending moment at point A near one corner can be evaluated as $pd^2/12$. This total stress in the inner surface therefore consists of the following two terms, membrane stress being written first,

$$\sigma = \tfrac{1}{2}p\left(\frac{d}{t} + \frac{d^2}{t^2}\right)$$

As a further example, consider the duct shown in Fig. 3.1(b), of internal diameter d. Membrane stress is clearly equal to $pd/2t$. When the cylinder expands,

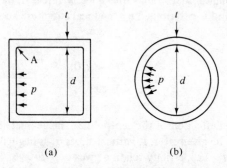

(a)   (b)

Fig. 3.1 Membrane and bending stress

curvature of the wall is reduced, leading to a bending stress given by $\tfrac{1}{2}p$, provided the ratio t/d is small. On the inner surface, the total stress is therefore

$$\sigma = \tfrac{1}{2}p\left(\frac{d}{t} + 1\right)$$

When stresses in these two ducts are compared, quite opposite tendencies are seen as wall thickness ratio t/d is reduced. In the round duct, the stress becomes almost entirely a membrane stress. On the other hand, the square-section duct shows itself to be weak in resisting internal pressure due to high bending stress in the flat sides. No general assertion can therefore be made that membrane stress is either more or less important in very thin plates, as this depends on whether the plate is flat or curved.

It might appear that if a flat plate is loaded with forces acting at right angles to its flat surfaces, so that the forces have no components acting on the plane of the plate, there would be no membrane stress set up. To see that this need not be true, consider a circular plate loaded in such a way that its edge assumes a waved shape. As the circumference is then longer than it was originally, there will be a tensile hoop stress set up in the rim. This stress varies as the square of deflection at any chosen point, but is independent of plate thickness. Meanwhile, bending stress will vary linearly with both deflection and plate thickness. Hence, for a certain deflection, membrane stress becomes relatively more important as the plate is made very thin. Calculation of this membrane stress will be considered in the next chapter. For the present, it is noted that predictions based

69

on bending stresses alone may become inaccurate if deflections exceed some value comparable with the plate thickness. This limitation does not apply to deflections of beams.

3.1.2 Flexure of beams Consider a beam in the (x,z) plane with its centre-line originally coinciding with the x-axis. The basic assumption of the simple theory of flexure is that longitudinal strain varies linearly through the depth, this is, in the z-direction, and is determined by the local curvature according to the relation

$$e_x = -z \frac{d^2 w(x,0)}{dx^2} \tag{3.1}$$

where $w(x,0)$ is the deflection of the centre-line. The geometrical justification of this formula will be given later. If vertical stress σ_z is negligible in relation to longitudinal stress σ_x the elasticity relation gives $\sigma_x = Ee_x$, so the longitudinal stress is also linearly distributed through the depth. When the beam has unit thickness in the direction perpendicular to the (x,z) plane, this longitudinal stress will have a resultant moment

$$M = -\int_{-h}^{h} z\sigma_x \, dz$$

where h is the half-depth of the beam. The sign of the bending moment is arranged to be positive when the moment tends to produce positive curvature, giving the basic differential equation

$$M = EI \frac{d^2 w}{dx^2} \tag{3.2}$$

where $I = 2h^3/3$.

3.1.3 Transverse shear stress To investigate how far this formulation of the flexure problem is adequate, some simple examples can be examined. In the elementary situation where a beam is bent by couples M at its ends, and the longitudinal stress is externally applied as a linear distribution over the ends, no other component of stress is present, and the relations (3.1) and (3.2) are precisely correct.

Now consider a beam carrying shear force S_0 at end $x = 0$, as shown in Fig. 3.2. An exact solution is available for this plane stress problem, as may be

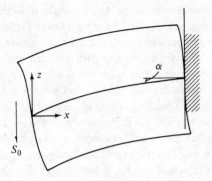

Fig. 3.2 Beam carrying shear force

verified by starting from the following components of displacement u and w in the x- and z-directions,

$$u = \frac{S_0}{2EI}[x^2 z - \tfrac{1}{3}(2 + v)z^3 + 2(1 + v)h^2 z + A + Cz]$$

$$w = -\frac{S_0}{2EI}[\tfrac{1}{3}x^3 + vxz^2 + B + Cx]$$

(3.3)

These displacements are shown to an exaggerated extend in Fig. 3.2, but they will actually be small in comparison with the dimensions of the body. Constants A, B, and C define a rigid-body motion which will not affect the strains, which are given by

$$e_x = \frac{\partial u}{\partial x} \qquad e_z = \frac{\partial w}{\partial z} \qquad e_{xy} = \tfrac{1}{2}\left(\frac{\partial u}{\partial z} + \frac{\partial w}{\partial x}\right)$$

The elastic relations for plane stress ($\sigma_y = 0$) are

$$\sigma_x = E'(e_x + ve_z) \qquad \sigma_z = E'(e_z + ve_x) \qquad \sigma_{xz} = (1 - v)E'e_{xz}$$

where $E' = E/(1 - v^2)$. The stresses thus found are

$$\sigma_x = S_0 xz/I \qquad \sigma_z = 0 \qquad \sigma_{xz} = 3S_0(1 - z^2/h^2)/4h$$

These stresses are seen to satisfy the requirement that upper and lower surfaces of the beam are stress-free. As the bending moment at any distance x is given by $M = -S_0 x$, it is seen that the differential equation (3.2) is exactly correct for this type of loading, but one notes that the shear force is applied by shear stresses acting in a prescribed way on the end $x = 0$, that is, with a parabolic distribution.

Inaccuracies in the relation between bending moment and deflection may arise either through the differential equation (3.2) becoming inexact, or through difficulties in integrating this equation. For example, if, in the last problem, the shear stresses applied at end $x = 0$ have the required resultant force S_0, but are not distributed in the prescribed way, the displacements and the other stress components near this end will become irregular, so that over a short distance from $x = 0$ comparable to beam depth $2h$, the relation (3.2) between moment and curvature of the centre-line will not be exactly correct. However, at larger distances from the point of loading, one might argue that according to Saint Venant's principle, the stresses will depend only on the resultant applied force and moment sustained at the point considered, and not on details of the way in which loads are transmitted to the beam. Obviously, an argument of this kind is admissible only if the beam has a fairly large length in relation to the size of the domains of irregular stress distribution around the points of loading.

A further difficulty arises in deciding upon the components of rigid-body displacement in (3.3). When a shear force is sustained, the end of the beam does not remain flat. Also, the centre-line is not at right angles to the end surface, but is rotated from this position through angle $\alpha = 3(1 - \nu)S_0/2Eh$, as shown in Fig. 3.2. If it is specified that $\partial u/\partial z = 0$ on the centre-line at a fixed end, as indicated by a vertical tangent line in Fig. 3.2, then it will be necessary to allow for slope α when integrating the flexure equation (3.2). However, it is more convenient to solve the flexure problem by considering the centre-line to have zero gradient at a fixed end. This is equivalent to assuming that sections which are originally plane and perpendicular to the centre-line remain so after flexure of the beam. A correction for shearing can then be applied.

When the beam carries distributed load, the transverse stress σ_z is not zero, and the differential equation (3.2) is not exactly satisfied at points near the end of the beam, but the error becomes relatively small at larger distances. Again, it has to be assumed that the beam has a length equal to several times its depth, so that such irregularities become negligible.

3.1.4 Shear deflection in beams
When a beam sustains a shear force, an originally plane cross-section distorts in such a way that deflection under load becomes somewhat greater than that calculated from the simple theory of flexure. A strain energy method can be used for assessing whether this additional deflection is significant. The argument used is that deflection w under applied load P is such that the work done is equal to strain energy U_B due to bending stress, plus strain energy U_S due to shear stress, i.e. $\frac{1}{2}Pw = U_B + U_S$.

Consider a beam of rectangular section, of depth $2h$ in the z-direction and width b in the y-direction, with the axis of the beam lying along the x-axis. The only component of direct stress present is the longitudinal stress,

$$\sigma_x = 3Mz/2h^3$$

where M is the bending moment at distance x from the end of the beam. The only component of shear stress present is the transverse shear stress, which has a parabolic distribution over the depth,

$$\sigma_{zx} = \frac{3S}{4bh}\left(1 - \frac{z^2}{h^2}\right)$$

where S is the shear force, which may also vary with x. As discussed more fully in the next chapter, the two strain energies are given by

$$U_B = \frac{1}{2E}\iiint \sigma_x^2\, dx\,dy\,dz \qquad U_S = \frac{1}{2G}\iiint \sigma_{zx}^2\, dx\,dy\,dz$$

On inserting the above stresses and integrating over the cross-section, the ratio between the energies is found to be

$$\frac{U_S}{U_B} = \frac{4(1+\nu)}{5}h^2\left(\int S^2\, dx \Big/ \int M^2\, dx\right)$$

If deflection w_B due to bending alone is first calculated, the total deflection w_T which includes the contribution of shear strain can be expressed as

$$w_T = w_B(1 + U_S/U_B)$$

For example, this multiplying factor may be found for a cantilever of length L. With load P acting at the end from which distance x is measured, we have shear force $S = P$ and bending moment $M = Px$. On carrying out the integrations and using a value $\nu = 0\cdot3$, the multiplying factor for end deflection is found to be

$$w_T = w_B\left(1 + 0\cdot78t^2/L^2\right)$$

where $t = 2h$ is the depth of the beam. This indicates an additional deflection due to shear stresses of about 5% when $L = 4t$. However, a simply supported beam loaded at mid-span may be considered equivalent to two cantilevers with their fixed ends placed together. In this case, an increment of deflection of 5% is found when $L = 8t$, where L is the total length. This example gives a numerical indication of what is meant by a slender beam, in which shear deflection can be ignored, though it should be noted that the numerical values depend on the shape of the cross-section.

3.1.5 Equilibrium of a beam

The shear force S at any point at distance x measured along the beam is taken as the sum of all external forces acting downwards on the left of this point. Upward forces give a negative contribution. Alternatively, the shear force can be taken as the sum of all forces acting upwards on the right of the point considered, any downward forces being counted as negative. Because of vertical equilibrium, the alternative methods must give the same answer. Bending moment M at any distance x along the beam is taken to be the sum of all externally applied moments and moments of applied forces acting on the part of the beam to the left of the point considered, such moments

6

being counted positive when they act in a clockwise direction. Alternatively, moments on the right may be summed, when anticlockwise moments are taken as positive as this will produce the same result for a beam in equilibrium under its applied loads. It is seen that a positive shear force thus defined is the resultant of positive shear stresses acting on any transverse section, while a positive bending moment tends to produce positive curvature, that is, a convexity towards the x-axis. Although alternative sign conventions have been used, it is very necessary to adhere strictly to one sign convention or another in the analysis of beams and plates.

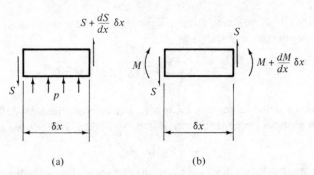

Fig. 3.3 Equilibrium of (a) shear force, (b) bending moment

The loading applied to a beam can be idealized as distributions of shear force and bending moment, and diagrams may be drawn showing how these quantities vary along the length of the beam. However, these two distributions are not independent, and are restricted by considerations of equilibrium. Consider an element carrying pressure p, which acts in the positive z-direction, as shown in Fig. 3.3(a). With positive shear forces S acting on the ends of the element, it is seen that vertical equilibrium of forces gives

$$\frac{dS}{dx} = -p$$

Now consider the rate of change of positive bending moment, as shown in Fig. 3.3(b). By taking moments about any point on the element, we get

$$\frac{dM}{dx} = -S$$

These two equilibrium equations can now be combined with the differential equation (3.2) to give two further equations

$$S = -EI\frac{d^3w}{dx^3} \qquad EI\frac{d^4w}{dx^4} = p \qquad (3.4)$$

74

These equations for a beam are given chiefly as an introduction to the derivation of similar equations for a plate, which will now be undertaken.

3.2 Deflection Derivatives and Strains

First, some relevant features of the distorted shape of a plate are examined. For the moment, the plate may be idealized as a surface lying initially in the (x,y) plane, so that the position of any point is specified by two coordinates x and y. After distortion, the height of any point above the datum plane is denoted w, so the distorted shape may be specified by a scalar function $w(x,y)$.

3.2.1 Geometrical ideas
The gradient of a surface at a point is the gradient of the tangent to the surface drawn in the required direction. It is defined as the increase in height w for unit increase in the projected length of the tangent. Therefore, the gradient in the x-direction is simply given by the first derivative $\partial w/\partial x$.

The significance of the second derivatives in now examined. Suppose that at any point P on the surface it is possible to draw a circle having a radius R which is the same as the radius of curvature of the surface in the x-direction, as shown in Fig. 3.4. At distance x in front of point P, the height of a point on the circle is w, while the bottom of the circle is at height w_0. Any arc of this circle is described in coordinates (x,w) by the equation

$$(x_0 + x)^2 + (R + w_0 - w)^2 = R^2$$

where x_0 is the horizontal distance of the centre of the circle behind point P.

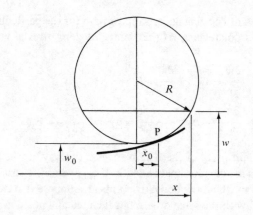

Fig. 3.4 Radius of curvature

By differentiating this expression twice and eliminating x_0 and w_0 we get a relation

$$\frac{d^2 w}{dx^2} = \frac{1}{R}\left[1 + \left(\frac{dw}{dx}\right)^2\right]^{3/2}$$

which is valid at the point $x = 0$. If it can be assumed that the gradient dw/dx is small at all points, it follows that radius R, taken in any selected x-direction, is expressed by

$$\frac{1}{R} = \frac{\partial^2 w}{\partial x^2} \tag{3.5}$$

This assumption is equivalent to positioning the centre of the circle vertically over point P, which simplifies the derivation of the last equation. The second derivative of w with respect to x may be called the curvature of the surface in the x-direction, and is equal to the reciprocal of the radius of curvature.

Turning now to the derivative $\partial^2 w/\partial x\,\partial y$, a non-zero value of this derivative means that gradient $\partial w/\partial y$ changes with increasing distance x. Alternatively, gradient $\partial w/\partial x$ changes with distance y. It is seen that the rate of change of $\partial w/\partial y$ with distance x must be the same as the rate of change of $\partial w/\partial x$ with distance y, at any point on a continuous surface. The quantity $\partial^2 w/\partial x\,\partial y$ may be called the twist of the surface with respect to coordinate axes x and y.

3.2.2 Transformation of derivatives   If the origin of coordinates (x,y) is located at some point P on a surface, the function $w(x,y)$ specifying the shape of the surface in the vicinity of this point may be expressed as a series

$$w = a_0 + a_1 x + \tfrac{1}{2}a_2 x^2 + \cdots + b_1 y + \tfrac{1}{2}b_2 y^2 + \cdots + c_2 xy + \cdots \tag{3.6}$$

in which the constant coefficients are appropriate for the particular point and for the chosen coordinate axes. At this point, the derivatives of w are given by

$$\frac{\partial w}{\partial x} = a_1 \qquad \frac{\partial w}{\partial y} = b_1$$

$$\frac{\partial^2 w}{\partial x^2} = a_2 \qquad \frac{\partial^2 w}{\partial y^2} = b_2 \qquad \frac{\partial^2 w}{\partial x\,\partial y} = c_2 \tag{3.7}$$

A fundamental property of these derivatives is the way in which they transform for rotations of the coordinate axes, as this governs the structure of any physical equation in which the derivatives appear. Suppose that new coordinates (x',y') are chosen, with the same origin, but inclined at a positive anticlockwise angle θ to the original coordinates (x,y) as shown in Fig. 3.5. Distances of a

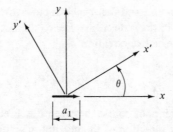

Fig. 3.5 Alternative coordinates

point from the origin are alternatively expressed in terms of the two coordinate systems by the relations

$$x = x' \cos \theta - y' \sin \theta \qquad y = y' \cos \theta + x' \sin \theta \qquad (3.8)$$

By inserting these values of x and y into (3.6), the upward deflection of the surface is obtained in terms of the new coordinates. Differentiating this new expression with respect to x' and y' gives the gradients at the point considered,

$$\frac{\partial w}{\partial x'} = a_1 \cos \theta + b_1 \sin \theta \qquad \frac{\partial w}{\partial y'} = b_1 \cos \theta - a_1 \sin \theta$$

Suppose now that the original x-axis is chosen to lie along the direction of greatest slope of the surface, as shown in Fig. 3.5. This slope will be represented by a vector of length a_1, the constant b_1 being zero. The gradient of the surface along any other axis x' is then equal to the component of the vector in the x'-direction, i.e.,

$$\frac{\partial w}{\partial x'} = a_1 \cos \theta \qquad (3.9)$$

Turning now to the second derivatives, these may also be obtained by inserting expressions (3.8) into (3.6) and differentiating with respect to x' and y' to obtain equations of the kind

$$\frac{\partial^2 w}{\partial x'^2} = a_2 \cos^2 \theta + b_2 \sin^2 \theta + 2c_2 \sin \theta \cos \theta$$

This equation shows that curvatures transform in the same way as two-dimensional stress and strain components, as discussed in chapter 1. Hence the curvature of a surface may be classed as a tensor quantity of rank two. At any point, it may be assumed that principal axes of curvature exist. If the original x-axis is chosen to coincide with the axis of the positively larger principal curvature at some particular point, the twist of the surface, given by $\partial^2 w / \partial x \, \partial y$,

77

will be zero at this point, and the principal curvatures become equal to co-efficents a_2 and b_2 in the series (3.6). Hence the last equation gives the curvature in any alternative direction x',

$$\frac{\partial^2 w}{\partial x'^2} = \tfrac{1}{2}(a_2 + b_2) + \tfrac{1}{2}(a_2 - b_2)\cos 2\theta \tag{3.10}$$

There is no relationship, of course, between the directions of the principal axes of curvature and the direction of the maximum gradient at any point.

3.2.3 Direct strain in terms of curvature
An element of the plate viewed in the y-direction is shown in Fig. 3.6. After bending, the originally straight centre-

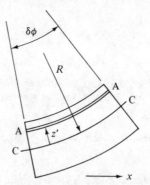

Fig. 3.6 Direct strain due to bending

line CC takes the shape of an arc of radius R. This cental surface is assumed to remain unstretched and of constant length $R\,\delta\phi$ where $\delta\phi$ is the angle finally subtended by the element. Some fibre AA can now be considered, at height z' above the central surface. Using the assumption made earlier that planes perpendicular to the centre-line remain plane and perpendicular after bending, the final length of fibre AA will be $(R - z')\,\delta\phi$. However, before bending, this fibre had the same length as the centre-line CC. Therefore, its fractional change in length, which is equal to the direct strain, is given by $e_{xx} = -z'/R$. The radius of curvature of the central surface has been expressed in (3.5) in terms of deflection w, so the strain in the x-direction becomes

$$e_{xx} = -z'\frac{\partial^2 w}{\partial x^2} \tag{3.11}$$

Strain in the y-direction may be obtained by inserting y in this expression in place of x.

3.2.4 Shear strain in terms of twist
Shear strain e_{xy} is associated with a change in shape of an elementary rectangle lying parallel to the (x,y) plane.

78

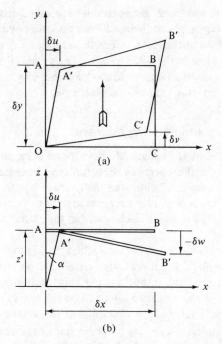

Fig. 3.7 Shear strain due to flexure (a) plan view of elementary layer, (b) view in direction of arrow

Suppose that lamina OABC in Fig. 3.7(a) distorts to shape OA'B'C'. Looking in the direction of the y-axis, it is seen that due to negative deflection $-\delta w$ at B, the edge AB tilts through angle $\alpha = -\delta w/\delta x$, as shown in Fig. 3.7(b). As point A is at some height z' above the central surface, it moves a distance $\delta u = \alpha z'$ in the x-direction. The angle of shear AOA' in Fig. 3.7(a) is now given by $\delta u/\delta y$, which may be expressed as the derivative

$$\frac{\partial u}{\partial y} = z'\frac{\partial \alpha}{\partial y} = -z'\frac{\partial^2 w}{\partial x\,\partial y}$$

By a similar argument, it may be shown that $\partial v/\partial x$ has the same value. Hence the shear strain in the (x,y) plane is given by

$$e_{xy} = \tfrac{1}{2}\left(\frac{\partial u}{\partial y} + \frac{\partial v}{\partial x}\right) = -z'\frac{\partial^2 w}{\partial x\,\partial y} \tag{3.12}$$

It is seen that this in-plane shear strain changes sign from one side of the plate to the other, in the same way as direct strain, and is zero at the centre of the thickness. The corresponding stress σ_{xy} is not to be confused with transverse shear stresses σ_{zx} and σ_{zy}, which are assumed to be of a lower order of magnitude, and which have maximum values at the mid-point of the thickness.

79

Equations (3.11) and (3.12) giving strains in terms of derivatives are not independent, and one might be deduced from the other by rotating the reference axes in the plane of the plate, since the equations for transforming second derivatives are the same as those for transforming strains. It might be expected that the principal axes of strain and stress due to bending will coincide with the principal axes of curvature at any point on the plate.

3.3 Shear Forces and Bending Moments

Strains in a distorted plate have already been found using purely geometrical arguments. The essential property of shear forces and bending moments is that they satisfy equilibrium conditions, and the relationships that will be derived all follow from this property. It is not necessary for the moment to specify how these forces are related to stresses acting within the plate.

3.3.1 Shear force   This acts in the z-direction, i.e., in the direction perpendicular to the plane of the plate. The force acting on an edge perpendicular to the x-direction is written S_x and is the force per unit length of edge. A positive shear force S_x acts in the positive z-direction on the side of an element which is situated at the greater value of the x-coordinate, i.e., on the outer side relative to the origin of coordinate axes. On the inner side of the element, this positive shear force acts in the negative z-direction. If a pressure p acts on the element on a face parallel to the (x,y) plane, this pressure is considered positive when it acts in the positive z-direction. Positive pressure and shear forces are shown in Fig. 3.8.

The pressure p acts on area $\delta x\,\delta y$, and the increments in shear force are respectively multiplied by the lengths of edge on which they act. The total force acting vertically is then equated to zero to give the equilibrium equation

$$\frac{\partial S_x}{\partial x} + \frac{\partial S_y}{\partial y} + p = 0 \tag{3.13}$$

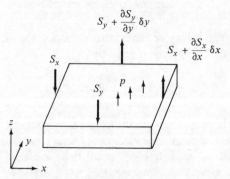

Fig. 3.8 Vertical equilibrium

80

The value of pressure p is that acting at some particular point, and it will in general vary from point to point. The incremental pressure from one side of the element to the other need not be considered, as it gives a force of a secondary degree of smallness.

It is not difficult to prove that shear force is a vector quantity, and transforms for alternative coordinate axes in the same way as the gradient of a surface. (See equation (3.9).)

3.3.2 Bending and twisting moments

External moments and torques acting on a body may be represented as vectors in a space defined by right-handed rectangular axes, and then an equilibrium condition connotes a zero resultant vector. A little thought will show that bending moment within a beam or plate cannot be regarded as a vector quantity, as the resultant moments on either side of a cut must be of opposite sign. Care is therefore taken to describe internal moments as bending or twisting moments, and not merely as moments or torques. A positive bending moment is taken to be one which tends to produce positive curvature with reference to the x or y axes, with z measured upwards. Hence a positive bending moment tends to produce a sagging deflection in the (x,z) or (y,z) planes. A positive twisting moment is taken as one which tends to produce positive twist, that is, an elevation of the corner of an element which is outermost with respect to the (x,y) axes. At a boundary or at any hypothetical contour drawn on the plate, the externally applied moment or torque is numerically the same as the internal bending moment or twisting moment. However, for the purpose of specifying an equilibrium condition, the external moment is treated as a vector, and is considered positive when acting in an anticlockwise direction. Hence, a bending moment which is positive within an element requires a positive external moment to be applied on the right-hand side of the element, and a negative external moment to be applied on the left.

Bending and twisting moments acting within a plate are now defined more particularly. Bending moment M_{xx} is the moment acting about an axis perpendicular to the x-direction, and is applied to an edge which is also perpendicular to the x-direction. This quantity is the bending moment per unit length of edge. Twisting moment M_{xy} is similarly defined as the couple per unit length of edge acting about an axis perpendicular to the x-direction and acting on an edge perpendicular to the y-direction. The subscripts indicate perpendiculars to the axis and the edge respectively.

3.3.3 Equilibrium of moments

All the moments per unit length that act about axes perpendicular to the x-axis are shown in Fig. 3.9, and they act on an element of sides δx and δy. Relevant shear forces per unit length are also shown, and all bending and twisting moments and shear forces are shown as positive values. Normal pressure acting on the plate and variations of shear force from one side of the element to the other are found to give moments of a second order of smallness, so these are neglected. The shear forces give an anticlockwise moment

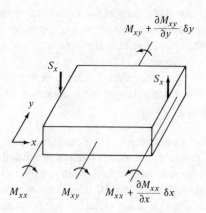

Fig. 3.9 Equilibrium of moments

$S_x \, \delta x \, \delta y$. All the applied moments are now summed and the total is equated to zero to give the equilibrium equation

$$\frac{\partial M_{xx}}{\partial x} + \frac{\partial M_{xy}}{\partial y} + S_x = 0 \tag{3.14}$$

A further summation of moments about axes perpendicular to the y-direction provides an equation similar to the last, but with x and y interchanged. Equation (3.14) is now differentiated with respect to x, and its counterpart is differentiated with respect to y, giving

$$\frac{\partial^2 M_{xx}}{\partial x^2} + \frac{\partial^2 M_{xy}}{\partial x \, \partial y} + \frac{\partial S_x}{\partial x} = 0$$

$$\frac{\partial^2 M_{yy}}{\partial y^2} + \frac{\partial^2 M_{yx}}{\partial y \, \partial x} + \frac{\partial S_y}{\partial y} = 0$$

As shown later, twisting moments M_{xy} and M_{yx} are always equal. When these equations are added together, and (3.13) is used for eliminating shear forces, a differential equation for bending moments is obtained,

$$\frac{\partial^2 M_{xx}}{\partial x^2} + 2\frac{\partial^2 M_{xy}}{\partial x \, \partial y} + \frac{\partial^2 M_{yy}}{\partial y^2} = p \tag{3.15}$$

3.3.4 Transformation of moments
This item does not form an essential step in the development of the differential equations, but serves to illustrate the physical meaning of bending moments in a plate. The procedure is similar to that used in chapter 1 for transforming stress components, where stresses acting on specified surfaces were treated as force vectors. In Fig. 3.10, alternative axes

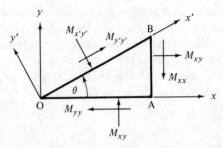

Fig. 3.10 Transformation of bending and twisting moments

x' and y' are shown inclined to original (x,y) axes at positive anticlockwise angle θ. The external moments corresponding to positive bending and twisting moments are represented as vectors. To do this, a right-hand screw rule is used, so that the direction of action is clockwise when one looks along the vector in its positive direction. All the moment vectors acting on this wedge-shaped portion of plate are now resolved in the x'-direction to give an equilibrium condition

$$\text{OB}\,M_{y'y'} + \text{AB}\,M_{xy}\cos\theta - \text{AB}\,M_{xx}\sin\theta + \text{OA}\,M_{xy}\sin\theta - \text{OA}\,M_{yy}\cos\theta = 0$$

With OA = OB cos θ and AB = OB sin θ, one obtains

$$M_{y'y'} = M_{yy}\cos^2\theta + M_{xx}\sin^2\theta - 2\,M_{xy}\sin\theta\cos\theta \qquad (3.16)$$

Without going further, it is seen that transformation equations of this kind are strictly analogous to those of (1.6) for stress and strain components, which means that bending moments at a point on a plate must be treated as tensor components.

3.3.5 Moments related to stresses   Moments and torques applied to the edge of any element of plate must give rise to direct and shear stresses on the edge of the element. Figure 3.11 shows positive bending and twisting moments M_{xx} and M_{xy}. Positive stresses σ_{xx} and σ_{xy} are also shown, these being the stresses

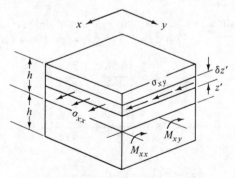

Fig. 3.11 Positive moments and stresses

that tend to produce positive strains. It is assumed that in the interior of a plate, shear stresses act horizontally, the vertical shear stresses being negligibly small. The force exerted by the stress on a strip of thickness $\delta z'$ is multiplied by the height z' of the strip above the central plane of the plate. Over unit length of edge, these moments of stresses are integrated through the thickness of the plate, which is taken as $2h$, to give equations

$$M_{xx} = -\int\limits_{-h}^{h} z'\sigma_{xx}\,dz' \qquad M_{xy} = -\int\limits_{-h}^{h} z'\sigma_{xy}\,dz' \qquad (3.17)$$

Two further equations may be obtained by interchanging x and y.

It is necessary that at any height z' above the central plane, the shear stresses σ_{xy} and σ_{yx} must be equal, as discussed in chapter 1. Hence the last equation shows that twisting moments M_{xy} and M_{yx} must also be equal, which justifies an assumption made earlier.

3.4 Equations for Elastic Plate

The equations so far derived for strains and bending moments do not particularly rely on any properties of the material. If an elastic plate is considered, strains are related to stresses by the general equations

$$Ee_{xx} = \sigma_{xx} - \nu(\sigma_{yy} + \sigma_{zz}) \qquad Ee_{yy} = \sigma_{yy} - \nu(\sigma_{zz} + \sigma_{xx})$$
$$Ee_{xy} = (1+\nu)\,\sigma_{xy}$$

where E is Young's modulus and ν is Poisson's ratio. For a thin plate, direct stress in the z-direction is assumed to be negligibly small compared with bending stresses in the x- and y-directions, so with $\sigma_{zz} = 0$, the last equations can be transposed to give

$$\sigma_{xx} = E'(e_{xx} + \nu e_{yy}) \qquad \sigma_{yy} = E'(e_{yy} + \nu e_{xx})$$
$$\sigma_{xy} = (1-\nu)E'\,e_{xy}$$

where $E' = E/(1 - \nu^2)$.

Using these equations, values of strains in terms of deflection derivatives can now be taken from (3.11) and (3.12) to give expressions for stresses

$$\sigma_{xx} = -z'E'\left[\frac{\partial^2 w}{\partial x^2} + \nu\frac{\partial^2 w}{\partial y^2}\right]$$

$$\sigma_{yy} = -z'E'\left[\frac{\partial^2 w}{\partial y^2} + \nu\frac{\partial^2 w}{\partial x^2}\right]$$

$$\sigma_{xy} = -(1-\nu)z'E'\frac{\partial^2 w}{\partial x\,\partial y}$$

It is seen that all these components of stress vary linearly with height z' above the central surface, and have opposite signs on opposite sides of the plate.

3.4.1 Differential equation of flexure

When these expressions for stresses are inserted into integrals (3.17), integration with respect to z' may be carried out to obtain bending moments

$$M_{xx} = D\left(\frac{\partial^2 w}{\partial x^2} + \nu\frac{\partial^2 w}{\partial y^2}\right) \qquad M_{yy} = D\left(\frac{\partial^2 w}{\partial y^2} + \nu\frac{\partial^2 w}{\partial x^2}\right)$$

$$M_{xy} = (1-\nu)\,D\,\frac{\partial^2 w}{\partial x\,\partial y} \tag{3.18}$$

Here, the constant $D = 2h^3 E'/3$ represents the intrinsic flexural stiffness of the plate.

It will now be recalled that equation (3.15) specified the equilibrium condition satisfied by bending moments acting in a plate. When expressions (3.18) for the moments are inserted into this equilibrium equation, we obtain a differential equation that must be satisfied by deflection w,

$$\frac{\partial^4 w}{\partial y^4} + 2\frac{\partial^4 w}{\partial x^2\partial y^2} + \frac{\partial^4 w}{\partial y^4} = \frac{p}{D}$$

This biharmonic equation may be alternatively written

$$\left(\frac{\partial^2}{\partial x^2} + \frac{\partial^2}{\partial y^2}\right)^2 w = \nabla^4 w = \frac{p}{D} \tag{3.19}$$

The original derivation of this equation is attributed to Lagrange (1811). To solve any particular problem of flexure, functions of w that are known to satisfy this equation may be selected. Values of arbitrary constants can then be assigned in such a way as to bring the solution into agreement with any deflections, gradients, bending moments, or shear forces that are known to occur on the boundaries of the plate. Moments imposed at the boundary may be expressed in terms of w by means of (3.18). Shear forces follow from the equilibrium equation (3.14), with moments taken from (3.18),

$$S_x = -D\frac{\partial}{\partial x}\nabla^2 w \qquad S_y = -D\frac{\partial}{\partial y}\nabla^2 w \tag{3.20}$$

These are vector components, being first derivatives of the scalar $\nabla^2 w$. The shear force on an edge perpendicular to some other direction, say the n-direction, is obtained by simply taking the derivative in this direction. These equations give shear force at internal points under any circumstances, but when this force is to be equated to applied force at the boundary of the plate, there may be some reservations, as mentioned later.

Generally speaking, flexure problems in rectangular coordinates often require complicated series solutions. On the other hand, polar coordinates provide a means for solving many simple and useful problems.

3.4.2 Equations in polar coordinates   Coordinates (r,θ) are suitable for dealing with circular plates, as the boundary conditions can be conveniently specified. First and second derivatives are simply replaced by the alternative forms given in equations (1.20) in chapter 1.

The biharmonic equation which must be satisfied by any expression for deflection now takes the form

$$\nabla^4 w = \left(\frac{\partial^2}{\partial r^2} + \frac{1}{r}\frac{\partial}{\partial r} + \frac{1}{r^2}\frac{\partial^2}{\partial \theta^2}\right)^2 w = \frac{p}{D} \tag{3.21}$$

As before, subscripts for bending moments indicate perpendiculars to the axis of the moment and the edge on which it acts. These bending moments follow form (3.18),

$$M_{rr} = D\left[\frac{\partial^2 w}{\partial r^2} + v\left(\frac{1}{r}\frac{\partial w}{\partial r} + \frac{1}{r^2}\frac{\partial^2 w}{\partial \theta^2}\right)\right]$$

$$M_{\theta\theta} = D\left[\frac{1}{r}\frac{\partial w}{\partial r} + \frac{1}{r^2}\frac{\partial^2 w}{\partial \theta^2} + v\frac{\partial^2 w}{\partial r^2}\right] \tag{3.22}$$

$$M_{r\theta} = (1 - v)\,D\,\frac{\partial^2}{\partial r\,\partial \theta}\left(\frac{w}{r}\right)$$

If bending stresses are needed, these follow in the usual way from bending moments, values on the upper surface being

$$\sigma_{rr} = -3M_{rr}/2h^2 \qquad \sigma_{\theta\theta} = -3M_{\theta\theta}/2h^2 \qquad \sigma_{r\theta} = -3M_{r\theta}/2h^2 \tag{3.23}$$

where h is the half-thickness of the plate.

Shear forces per unit length of edge, acting on edges perpendicular to radial and tangential lines are given by

$$S_r = -D\frac{\partial}{\partial r}\nabla^2 w \qquad S_\theta = -\frac{D}{r}\frac{\partial}{\partial \theta}\nabla^2 w \tag{3.24}$$

3.4.3 Discontinuous shear force   Suppose that a cut is made around a circle of radius r_1, so that a section through a plate is as shown in Fig. 3.12. On the inner part, the exposed edge carries shear force S_r and on the outer part, the edge carries shear force S_r'. In the absence of any external load at this place, these shear forces will be numerically equal, and are shown acting in the directions for a positive shear force. However, if a line distribution of shear force acts at this radius, equal to F, and considered positive when acting upward,

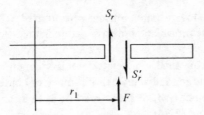

Fig. 3.12 Discontinuous shear force

the externally applied shear forces S_r and S_r' must have a resultant F per unit length of edge. This must be so at any position around the circle, giving

$$S_r(r_1, \theta) - S_r'(r_1, \theta) = F(\theta) \qquad (3.25)$$

When this is so, the deflections of the inner and outer portions of the plate must be expressed by different functions $w_1(r,\theta)$ and $w_2(r,\theta)$. From each expression the appropriate shear force at radius r_1 may be found from (3.24). On inserting these values into (3.25), an equation is obtained restricting the expressions for deflections. This procedure is valid for loading applied on any contour which does not coincide with the outside edge of the plate. When transverse loading is externally applied at the outer boundary, or on an inner boundary if this is present, some restrictions will be mentioned which apply when the boundary has a variable twist, but these restrictions do not apply for an axisymmetrical deflection which necessarily gives zero twist. Therefore, for axisymmetrical bending, shear force equation (3.24) can be used directly.

3.5 Axisymmetrical Loading of Discs

When a body of revolution carries boundary stresses that are symmetrical with respect to its axis, general methods of stress analysis are available (see, for instance, Timoshenko and Goodier, *Theory of Elasticity*, p. 343). However, if the body has a small uniform thickness in the axial direction, it is usually easier to treat the problem by means of the theory of plates, as will be done here.

Deflections will not vary with angular position, so the differential equation (3.21) simplifies to

$$\left(\frac{1}{r}\frac{\partial}{\partial r}r\frac{\partial}{\partial r}\right)\left(\frac{1}{r}\frac{\partial}{\partial r}r\frac{\partial}{\partial r}\right)w = \frac{p}{D}$$

This can be integrated at once to give a general deflection

$$w = A_0 + B_0 \ln \rho + C_0 \rho^2 + D_0 \rho^2 \ln \rho + pR^4 \rho^4/64\,D \qquad (3.26)$$

where capital letters with subscripts are arbitrary constants and ρ is the ratio of radius r to some fixed radius R, which is usually the outside radius of the disc. For a solid disc having no central hole or clamping collars and no concentrated load at the centre, the logarithmic terms may be omitted. If the solution is used for part of a plate on which no distributed pressure p acts, the final term drops out.

For an annulus having axial symmetry, boundary conditions may consist of specified deflections and radial gradients at inner and outer edges, or alternatively, specified moments M_r or shear forces S_r at these edges, which have been given in (3.22) and (3.24),

$$M_r = D\left(\frac{d^2 w}{dr^2} + \frac{v}{r}\frac{dw}{dr}\right) \qquad S_r = -D\frac{d}{dr}\left(\frac{d^2 w}{dr^2} + \frac{1}{r}\frac{dw}{dr}\right) \qquad (3.27)$$

where D is the flexural stiffness constant $2h^3E'/3$, and h is the half-thickness of the plate. When bending moments have been found, stresses follow from equations (3.23).

With specified conditions at the outer and inner edges of the annulus, the flexure problem can be solved for a line distribution of force on a circle of general radius. By superposition of an indefinite number of such solutions, the deflections and stresses can be found for any given axisymmetrical distribution of loading applied to the surface of the plate.

3.5.1 Diaphragm with built-in edge
Consider a solid plate of outside radius R carrying an upward force F per unit length on a circle of radius αR. In the regions inside and outside of the radius of loading, the deflections denoted w_1 and w_2 may be generally expressed in a form following (3.26),

$$\left.\begin{array}{ll} w_1 = D_2(A_1 + C_1\rho^2) & (\rho < \alpha) \\ w_2 = D_2(A_2 + B_2\ln\rho + C_2\rho^2 + \rho^2\ln\rho) & (\rho > \alpha) \end{array}\right\} \qquad (3.28)$$

where ρ is radius ratio r/R, and capital letters represent constants. Five conditions at the boundaries of these two regions are now used. At outside radius $\rho = 1$, both deflection w_2 and radial gradient dw_2/dr must vanish. Also, at the radius of loading, $\rho = \alpha$, there must be agreement between the two parts as regards deflection, gradient, and bending moment M_r. It is seen from (3.27) that this is equivalent to the requirements that deflection and its first and second derivatives should match at this radius. These conditions determine five of the constants,

$$\begin{array}{ll} A_1 = \frac{1}{2}(1 - \alpha^2 + 2\alpha^2\ln\alpha) & C_1 = \frac{1}{2}(1 - \alpha^2 + 2\ln\alpha) \\ A_2 = \frac{1}{2}(1 + \alpha^2) \qquad B_2 = \alpha^2 & C_2 = -A_2 \end{array}$$

Finally, the shear force S_r in the outer portion at the radius of loading is equal to $(-F)$, so it follows from (3.24) that $D_2 = \alpha FR^3/4D$, where D is the stiffness constant for the uniform plate.

The problem of a line distribution of force can now be considered as solved, and it is now of interest to see whether the solution can be used for finding the effect of a distributed pressure $p(r)$ applied to some part of the surface. For this purpose, the line distribution is considered to be the resultant of pressure acting on an elemental annulus of radial width $R\,\delta\alpha$. This means that $F = pR\,\delta\alpha$, and the constant D_2 is now given by

$$D_2 = pR^4\,\alpha\delta\alpha/4D \qquad (3.29)$$

As a particular example, consider a uniform pressure p distributed over the whole area of a circle of radius βR. This is considered as the sum of a series of line distributions at various radial positions, that is, at various α-values. Due to

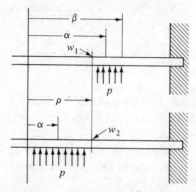

Fig. 3.13 Ranges of integration for finding deflection of a disc

the limits of validity of expressions (3.28), it is necessary to split the loading into two parts as shown in Fig. 3.13, one part outside of the radius ρR at which deflection is to be found, the other part inside of this radius. For the outer part of the loading, deflection is considered as that of the inner part, i.e., w_1. For the inner part of the loading, deflection at the required radius is considered as that of the outer part of the disc, i.e., w_2. Integrals between the appropriate limits can now be formed for these two parts of the total deflection, using (3.28) with (3.29),

$$w_1 = \frac{pR^4}{4D} \int_\rho^\beta (A_1 + C_1\,\rho^2)\,\alpha\,d\alpha$$

$$w_2 = \frac{pR^4}{4D} \int_0^\rho (A_2 + B_2\ln\rho + C_2\rho^2 + \rho^2\ln\rho)\,\alpha\,d\alpha$$

These expressions are valid for any radius ρ within the loaded portion, the total deflection being the sum of the two parts. For finding deflection at any larger radius $(\rho > \beta)$, the expression for w_1 is ignored, while the integral for w_2 takes β for its upper limit. The capitals with subscripts have already been found as

7

functions of α, and when these are inserted, integration can be carried out to give the following deflections at radius ρ,

$$w = \frac{pR^4}{64\,D}[4\beta^2 - 3\beta^4 + \rho^4 - 2\beta^4\rho^2 + (4\beta^4 + 8\beta^2\rho^2)\ln\beta] \qquad (\rho < \beta)$$

$$w = \frac{pR^4}{64\,D}[4\beta^2 + 2\beta^4 - 4\beta^2\rho^2 - 2\beta^4\rho^2 + (4\beta^4 + 8\beta^2\rho^2)\ln\rho] \quad (\rho > \beta)$$

Deflection at the centre ($\rho = 0$) is of particular interest, and may be written in terms of total applied load $P = \pi\beta^2 R^2 p$,

$$w(0) = \frac{PR^2}{64\,\pi D}(4 - 3\beta^2 + 4\beta^2\ln\beta)$$

This shows that deflection due to a concentrated load P acting at the centre ($\beta = 0$) is four times the deflection due to the same load uniformly distributed over the whole plate ($\beta = 1$).

Bending moment and radial stress may be found by differentiating displacements (3.28) with respect to ρ, as in (3.27). Integration can then be carried out in the two stages illustrated in Fig. 3.13,

$$\sigma_r(\theta) = -\frac{3pR^2}{8h^2}\left\{\int_\rho^\beta 2(1+\nu)\,C_1\,\alpha\,d\alpha \right.$$

$$\left. + \int_0^\rho \left[-(1-\nu)\frac{B_2}{\rho^2} + 2(1+\nu)\,C_2 + 2(1+\nu)\ln\rho + 3 + \nu\right]\alpha\,d\alpha\right\}$$

These integrals are correct for $\rho < \beta$. For obtaining stresses at points $\rho > \beta$, the first integral is ignored, and the second is given an upper limit β,

$$\sigma_r = \frac{3P}{32\pi h^2}\left[(1+\nu)(\beta^2 - 4\ln\beta) - (3+\nu)\frac{\rho^2}{\beta^2}\right] \qquad (\rho < \beta)$$

$$\sigma_r = \frac{3P}{32\pi h^2}\left[(1+\nu)(\beta^2 - 4\ln\rho) - 4 + (1-\nu)\frac{\beta^2}{\rho^2}\right] \qquad (\rho > \beta)$$

It is seen that stress is infinite under a concentrated central load P. However, when load is uniformly distributed over the whole disc ($\beta = 1$), the radial stress has its numerically highest value $\sigma_r = -3P/16\pi h^2$ at the outer edge (see also Prescott, *Applied Elasticity*, p. 403). When pressure is spread over a circle such that $\beta = 2/3$, radial stresses at the centre and outer edge are almost of the same numerical value, as shown by the curves in Fig. 3.14, which are drawn for a value $\nu = 0\cdot3$.

90

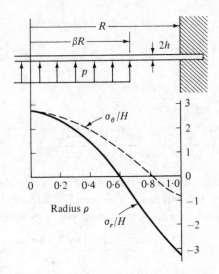

Fig. 3.14 Bending stresses in a diaphragm for a pressure p acting over a circle of radius $\beta = 2/3$ $(H = 3\beta^2 R^2 p/32\ h^2)$

Alternative methods might be envisaged for solving problems with uniform pressure acting within a circle, but the above method lends itself to numerical integration when pressure varies with radius in some specified way. In this case, pressure p must be brought inside the integral sign when integrating with respect to α.

3.5.2 Disc clamped at its centre   As a further example of axisymmetrical deformation, consider a thin uniform plate of outside radius R which is rigidly clamped between thick collars of radius αR. Alternatively, the plate may be supposed to have an integral boss of such a thickness that the boss is virtually rigid. With the centre fixed, a uniform distribution of force F per unit length is applied to the rim in an upward direction, that is, at right angles to the plane of the plate. At any radius ratio $\rho = r/R$, the upward deflection may be written in a form following (3.26),

$$w = D_0(A_0 + B_0 \ln \rho + C_0\rho^2 + \rho^2 \ln \rho)$$

Three boundary conditions are first used. There is zero moment M_r applied at the rim, gradient dw/dr is zero at the radius of the clamping collars, and at this radius the deflection is also zero. Three equations are thus obtained,

$$-(1 - v)B_0 + 2(1 + v)C_0 + 3 + v = 0$$

$$B_0/\alpha + 2\alpha C_0 + \alpha + 2\alpha \ln \alpha = 0$$

$$A_0 + B_0 \ln \alpha + C_0 \alpha^2 + \alpha^2 \ln \alpha = 0$$

91

These are satisfied when the constants have the values

$$A_0 = \tfrac{1}{2}\alpha^2 \frac{3 + \nu + (1 - \nu)\alpha^2 - 2[3 + \nu - 2(1 + \nu)\ln\alpha]\ln\alpha}{1 + \nu + (1 - \nu)\alpha^2}$$

$$B_0 = 2\alpha^2 \frac{1 - (1 + \nu)\ln\alpha}{1 + \nu + (1 - \nu)\alpha^2}$$

$$C_0 = -\tfrac{1}{2} \frac{3 + \nu + (1 - \nu)\alpha^2(1 + 2\ln\alpha)}{1 + \nu + (1 - \nu)\alpha^2}$$

Finally we note that positive shear force $S_r = F$ acts at the rim, so the constant D_0 is determined by the shear force equation (3.27), $D_0 = -FR^3/4D$. The rim deflection $w(1) = D_0(A_0 + C_0)$ can now be obtained in the form

$$w(1) = \frac{FR^3}{8D} \frac{3 + \nu - 2(1 + \nu)\alpha^2 - (1 - \nu)\alpha^4 + 4\alpha^2[2 - (1 + \nu)\ln\alpha]\ln\alpha}{1 + \nu + (1 - \nu)\alpha^2}$$

$$(3.30)$$

It is noted that as α tends to zero, the deflection is that of a solid disc simply supported at its rim with a central point load $P = 2\pi RF$, given by

$$w = \frac{1}{16\pi} \frac{3 + \nu}{1 + \nu} \frac{PR^2}{D}$$

Bending moments and stresses follow from (3.22) and (3.23). The largest value of stress is that of radial stress at the radius of the clamping collars, which is found to be

$$\sigma_r = -\frac{3}{4} \frac{FR}{h^2} \frac{(1 - \nu)(1 - \alpha^2) - 2(1 + \nu)\ln\alpha}{1 + \nu + (1 - \nu)\alpha^2}$$

3.6 Conditions at a Twisted Edge

Flexure problems so far discussed have been of a special kind, in that principal axes of curvature have been normal and tangential to the boundary contour at all points. In general, the plate may be twisted. Taking axes (x, y) parallel and perpendicular to an edge, it has been seen that flexure sets up a shear stress σ_{xy} that is proportional to twist $\partial^2 w/\partial x\, \partial y$. It might appear that if one of the edges of a plate is stress-free, the plate cannot simultaneously have a twisted shape. This is not so, for the reason that a rapid change in shear stress is found as a free edge is approached. Attention is now fixed on the narrow margin of a plate in which the shear stress varies in an irregular way.

3.6.1 Torsion of a strip   Consider a long strip of rectangular section, subjected to torsion about its longitudinal axis, which is parallel to the y-direction.

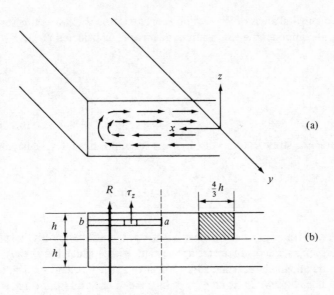

Fig. 3.15 Section of strip having edges parallel to the y-direction, showing (a) shear stresses acting on the section, (b) resultant of vertical components

Attention is fixed on a transverse section which is outward-facing with respect to the positive y-direction. Shear stresses acting on a part of this face are shown in Fig. 3.15(a). As the shear stress components acting on the (x,z) plane can be regarded as vector components, they may be represented as first derivatives of a stress function ϕ,

$$\tau_x = \frac{\partial \phi}{\partial z} \qquad \tau_z = -\frac{\partial \phi}{\partial x}$$

At any point on the contour of the section, the shear stress vector must be directed along the contour, as the lateral surfaces of the strip are stress-free. This requirement is satisfied by giving function ϕ a zero value on the contour of the section. Equilibrium within an elastic body requires that the function should be plane harmonic, which will be mentioned again in the next chapter. It follows that although the surface representing the stress function may have an irregular shape near the ends of the rectangular section, it will have a parabolic profile at some distance from the end, say at line (a) in Fig. 3.15(b). Here, the stress function may be arranged to give a shear stress of numerical value q at the upper and lower edges of the section,

$$\phi(a, z) = q(h^2 - z^2)/2h$$

On elemental areas of the section near $x = 0$, the shear stress vector will have a vertical component, giving a resultant force R, considered positive when acting upwards,

$$R = \int_{-h}^{h} \int_{a}^{b} \tau_z \, dx \, dz$$

By expressing stress in terms of the stress function, this integral may be written

$$R = -\int_{-h}^{h} [\phi(b, z) - \phi(a, z)] \, dz$$

However, at the ends of any elementary strip, as indicated in Fig. 3.15(b), the function $\phi(b,z)$ is zero. By inserting the value $\phi(a,z)$ which has already been written down, and integrating with respect to z, the resultant vertical force is found to be $R = 2qh^2/3$. It may be of interest to note that this force is equal to the resultant horizontal force acting on any portion of the half-section of width $4h/3$, as indicated in Fig. 3.15(b), provided this portion is taken well back from the end of the section.

As the maximum shear stress q is given in terms of twisting moment per unit length M_{xy} by $q = 3M_{xy}/2h^2$, the resultant vertical force can be expressed as $R = M_{xy}$. Further, as this twisting moment can be expressed in terms of deflection w, as in (3.18), the resultant force can be similarly expressed,

$$R = (1 - \nu) D \frac{\partial^2 w}{\partial x \, \partial y} \tag{3.31}$$

where D is the stiffness constant, equal to $2h^3 E'/3$.

3.6.2 Equilibrium of plate boundary   The possible presence of a shear stress resultant R was not envisaged in establishing the relationship (3.25) between an applied line distribution of force and the shear forces in the interior of a plate. It appears that a special relationship must be used when an element of the plate is selected adjacent to a boundary which is free from in-plane shear stress. Consider the element shown in Fig. 3.16, with rectangular axes (n,t) normal and tangential to the boundary at the chosen point. The width of the element in the n-direction is assumed to be sufficient to cover irregularities in shear stress due to the presence of an edge that is free from horizontal shear stress, but it is supposed that the width is not so large as to be influenced by the large-scale variations of stress in the body of the plate. Applied force V per unit length is shown acting at the extreme edge. Shear force S_n per unit length is shown acting

94

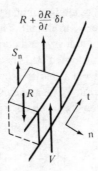

Fig. 3.16 Equilibrium of boundary

on the inside of the element, where the stress distribution is regular, and where this shear force can be calculated from the deflected shape by the usual relation

$$S_n = -D \frac{\partial}{\partial n} \nabla^2 w \qquad (3.32)$$

As with shear forces, a positive resultant force R acts upwards on the end of the element which is outermost with respect to the t-axis.

When the twist of the edge varies from one point to another, resultant force R will also vary. Vertical forces are obtained by multiplying V and S_n by length of element δt, but it is noted that resultant R is itself a force. Vertical equilibrium then requires that

$$V = S_n - \frac{dR}{dt}$$

By inserting expressions (3.31) and (3.32), a relation is obtained in terms of deflection w,

$$V = -D \frac{\partial}{\partial n} \nabla^2 w - (1-v) \, D \frac{d}{dt} \left(\frac{\partial^2 w}{\partial n \, \partial t} \right) \qquad (3.33)$$

This formulation of the boundary condition for plates is attributed to Kirchhoff (1850), but the method used here for deriving the last relationship is similar to that given by J. Prescott, *Applied Elasticity*, p. 396.

To adapt this formula for polar coordinates, the second derivative is expressed, in accordance with (1.20),

$$\frac{\partial^2 w}{\partial n \, \partial t} = \left[\frac{\partial^2}{\partial r \, \partial \theta} \frac{w}{r} \right]_B$$

This function has a series of values along boundary B of a complete or incomplete annulus, but its values elsewhere are irrelevant. This function is simply differentiated in the direction of the boundary, and curvature of the boundary will not affect this first derivative. It is noted that the whole term for force R does not transform as a third derivative. Particular expressions can now be written down for applied force distribution V_r on an edge perpendicular to a radial line and applied force distribution V_θ on an edge perpendicular to the tangential direction,

$$V_r = -D\left(\left[\frac{\partial}{\partial r}\nabla^2 w\right]_B + (1-\nu)\frac{1}{r}\frac{d}{d\theta}\left[\frac{\partial^2}{\partial r\,\partial\theta}\frac{w}{r}\right]_B\right)$$

$$V_\theta = -D\left(\left[\frac{1}{r}\frac{\partial}{\partial\theta}\nabla^2 w\right]_B + (1-\nu)\frac{d}{dr}\left[\frac{\partial^2}{\partial r\,\partial\theta}\frac{w}{r}\right]_B\right) \qquad (3.34)$$

Similar expressions were used by Rayleigh. If a boundary B consists of arcs and radial lines, special considerations apply at the right-angled corners, to be mentioned later.

3.7 Unsymmetrical Loading Problems

The foregoing equations in polar coordinates are now applied to plates bounded by circular contours which become twisted during flexure. If the circular boundaries are continuous, this occurs in all modes of deflection except those having axial symmetry.

3.7.1 **Disc with edge loading varying as** $\cos\theta$ Consider a disc carrying a line distribution of force $F = F_1\cos\theta$ at its outer edge, of radius R. As this type of loading will tend to tilt the plate about a diameter, it is supposed that the centre is rigidly clamped between two thick collars of radius αR. The general expression for deflection at any radius $r = \rho R$ is such as will satisfy the differential equation (3.21),

$$w = D_1(A_1\rho + B_1\rho^{-1} + C_1\rho^3 + \rho\ln\rho)\cos\theta \qquad (3.35)$$

For relating deflection to applied load F, it is quite possible to use expression (3.34), but when loads vary as $\cos\theta$, an alternative method can be used. The moments of all forces acting on a complete annulus of inside radius ρR are taken about diameter AA, as shown in Fig. 3.17. The positive moments and shear force per unit length that act on the inside edge are shown. An equilibrium equation is thus obtained,

$$\int_0^{2\pi} [(R^2 F - \rho^2 R^2 S_r - \rho R M_r)\cos\theta + \rho R M_{r\theta}\sin\theta]\,d\theta = 0$$

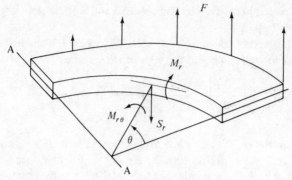

Fig. 3.17 Equilibrium of moments

By using the general expression (3.35) for deflection, and taking derivatives as necessary, equations (3.22) and (3.24) give the two moments and the shear force in terms of deflection. With these inserted in the equilibrium equation, integration can be carried out. As expected, all terms varying with radius disappear, leaving

$$D_1 = F_1 R^3 / 4D$$

a result which is independent of the method of anchoring the centre of the plate.

Further boundary conditions can now be used. Deflection and radial gradient are zero at the edge of the clamping collars, and moment M_r is zero at the outer edge, giving the equations

$$A_1 \alpha + B_1 \alpha^{-1} + C_1 \alpha^3 + \alpha \ln \alpha = 0$$

$$A_1 + B_1 \alpha^{-2} + 3C_1 \alpha^2 + 1 + \ln \alpha = 0$$

$$2(1 - \nu) B_1 + 2(3 + \nu) C_1 + 1 + \nu = 0$$

These are satisfied by constants having the values

$$A_1 = -\tfrac{1}{2}(1 + 2 \ln \alpha + 4\alpha^2 C_1)$$

$$B_1 = \tfrac{1}{2}\alpha^2 \frac{3 + \nu - (1 + \nu)\alpha^2}{3 + \nu + (1 - \nu)\alpha^4}$$

$$C_1 = -\tfrac{1}{2} \frac{1 + \nu + (1 - \nu)\alpha^2}{3 + \nu + (1 - \nu)\alpha^4}$$

The deflection at the outer edge now becomes

$$w(1,0) = \frac{F_1 R^3}{8D} \left[\frac{-(1 + \nu) + 4(1 + \nu)\alpha^2 + (1 - 3\nu)\alpha^4}{3 + \nu + (1 - \nu)\alpha^4} - 1 - 2 \ln \alpha \right]$$

$$(3.36)$$

A similar solution has been given by Timoshenko and Woinowski-Krieger, *Theory of Plates and Shells*, p. 289.

Radial bending stress is given by $\sigma_r = -3M_r/2h^2$ where h is the half-thickness of the plate. The highest value is at the edge of the clamps,

$$\sigma_r(\alpha,0) = -\frac{3}{4}\frac{F_1}{h^2}R\frac{3+\nu-2(1+\nu)\alpha^2-(1-\nu)\alpha^4}{\alpha[3+\nu+(1-\nu)\alpha^4]}$$

3.7.2 Higher harmonics

When a line distribution of force $F = F_n \cos n\theta$ is applied to a circular plate, deflection will vary as $\cos n\theta$. For values of n equal to or greater than 2, the biharmonic equation $\nabla^4 w = 0$ is solved by the regular series for deflection

$$w = (A_n\rho^n + B_n\rho^{-n} + C_n\rho^{n+2} + D_n\rho^{-n+2})\cos n\theta \qquad (3.37)$$

where the four constants are determinate from known boundary conditions. Applied loads of this kind exert no resultant force or moment on the plate.

3.7.3 Point load at edge of disc

A thin disc rigidly clamped between central collars carries a concentrated load P at its rim, as shown in Fig. 3.18. This problem may be approached by resolving the load into a series of line-distributions applied around the rim, so that the shear force at any point is given by

$$F = F_0 + \sum F_n \cos n\theta \qquad (n = 1, 2, 3, \dots)$$

The coefficients are found by applying the Fourier theorem

$$F_n = \frac{1}{\pi}\int_0^{2\pi} F\cos n\theta\, d\theta$$

The shear force F is zero at all points on the rim except on a small arc subtending angle $\delta\theta$ at the position of the concentrated load, where the shear force per

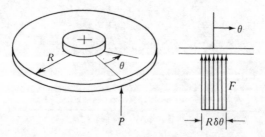

Fig. 3.18 Point load resolved into line distributions

unit length is given by $F = P/R\delta\theta$. In this region, $\cos n\theta = 1$, so the integral gives

$$F_0 = P/2\pi R \qquad F_n = P/\pi R \qquad (n = 1, 2, 3, \ldots)$$

For each harmonic distribution of load, the flexure problem may be solved by methods already described. Deflection amplitudes w_n at the rim can be specified in terms of compliance factors c_n, defined by

$$w_0 = 2c_0 F_0 R^3/D \qquad w_n = c_n F_n R^3/D \qquad (n = 1, 2, 3, \ldots)$$

Values for c_0 and c_1 can be deduced from expressions (3.30) and (3.36) for deflections. Table 3.1 gives some computed values for various values of clamping ratio α.

The total deflection of the rim at any point is obtained by summing all the harmonic deflections,

$$w = \frac{PR^2}{\pi D}(c_0 + \sum c_n \cos n\theta) \qquad (3.38)$$

The value of this expression with $\theta = 0$ gives the deflection under the load, which is expressed in terms of a numerical compliance factor C by the relation $w = CPR^2/\pi D$, where

$$C = \sum c_n \qquad (n = 0, 1, 2, \ldots)$$

Some of these values are included in Table 3.1.

From (3.38) the theoretical deflection at various angular positions around the rim may be found. In Fig. 3.19, these values are compared with some experimental measurements. Such a comparison serves to verify that the disc

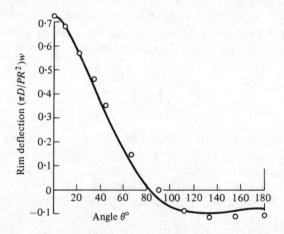

Fig. 3.19 Theoretical curve of rim deflection produced by point load P, compared with experimental values, for clamping ratio $\alpha = 0\cdot125$

is free from internal stress, as the shape of the curve may be appreciably changed by such stresses, especially if the disc is very thin.

Table 3.1 Compliance factors c_n for rim loading of disc clamped at radius ratio α

$\alpha =$	0	0·125	0·250	0·375	0·500
$n = 0$	0·159	0·126	0·085	0·050	0·025
1	∞	0·346	0·185	0·099	0·049
2	0·191	0·168	0·122	0·075	0·040
3	0·044	0·043	0·042	0·036	0·026
4	0·017	0·017	0·017	0·016	0·014
C	∞	0·729	0·476	0·300	0·175

3.7.4 Line distribution of force on a plate Consider the application of force on some contour C which lies inside the boundary of the plate. Such problems have already been considered when the deflection is axisymmetrical, but more general circumstances are now assumed. At any selected point on the contour, coordinates (n,t) may be set up, normal and tangential to the contour. On the left of the contour, let deflection be $w_1(n,t)$ and on the right $w_2(n,t)$. There can be no discontinuity in deflection across the contour, hence gradients in the direction of the contour, $\partial w_1/\partial t$ and $\partial w_2/\partial t$, must be equal at points adjacent to the contour. There is also continuity of gradients $\partial w_1/\partial n$ and $\partial w_2/\partial n$ across the contour. It follows that there must be continuity of twist across the contour C,

$$\left[\frac{\partial^2 w_1}{\partial n\,\partial t}\right]_C = \left[\frac{\partial^2 w_2}{\partial n\,\partial t}\right]_C$$

and therefore continuity of shear stresses σ_{nt}. This means that the special Kirchhoff condition (3.33) which is needed at a boundary can be simplified for loading on an interior contour. The discontinuity in shear force S_n across the contour due to applied force F per unit length can now be calculated as before, according to (3.24) and (3.25), leading to the relation

$$F = -D\left[\frac{\partial}{\partial n}\nabla^2 w_1 - \frac{\partial}{\partial n}\nabla^2 w_2\right]_C \qquad (3.39)$$

In polar coordinates, use may be made of a further simplification in dealing with a force distribution along a circle. The bending moment

$$M_r = D\left[\frac{\partial^2 w}{\partial r^2} + \nu\left(\frac{1}{r}\frac{\partial w}{\partial r} + \frac{1}{r^2}\frac{\partial^2 w}{\partial \theta^2}\right)\right]$$

must be continuous across the contour. However, the first derivative in the r-direction and all derivatives in the θ-direction are necessarily continuous across the contour. Hence the moment condition is satisfied by the additional requirement that $\partial^2 w/\partial r^2$ shall be continuous. The conditions to be used at a line distribution of force are therefore seen to be that there should be radial continuity of deflection and its first and second r-derivatives. A fourth condition (3.39) takes account of the value of the applied force distribution.

As an example, consider a force distribution $F = F_2 \cos 2\theta$ applied along a circle of radius R drawn on a solid plate of infinite extent. Following (3.37), deflections at any radius $r = \rho R$ may be expressed

$$w_1 = A_2(\rho^2 + C_2\rho^4)\cos 2\theta \qquad (\rho < 1)$$
$$w_2 = A_2(B_2\rho^{-2} + D_2)\cos 2\theta \qquad (\rho > 1)$$

Continuity of deflection and its first and second r-derivatives gives

$$B_2 = C_2 = -\tfrac{1}{3} \qquad D_2 = 1$$

while the discontinuity in shear force gives $A_2 = F_2R^3/16D$. Although there is a non-zero deflection amplitude at infinity, the total strain energy is finite.

3.7.5 Circular hole in an infinite plate   This flexure problem may be analysed by methods analogous to those used for the plane-stress problem in the previous chapter. First, consider a large plate with an axisymmetrical distribution of bending moment $M_r = M$ over its outer boundary. The condition that $M_r = 0$ at the surface of a hole of radius R is satisfied by taking the deflected shape as

$$w = \frac{MR^2}{D}\left(\frac{\ln \rho}{1 - \nu} + \frac{\rho^2}{2(1 + \nu)}\right)$$

where $\rho = r/R$. This indicates that shear force is zero everywhere, with a bending moment at the rim of the hole $M_\theta = 2M$.

A second situation can now be examined, where bending moment $M_r = M' \cos 2\theta$ is applied to the outer edge of a very large plate, producing a pure twist at large distances from the hole. The expression for deflection, following (3.37),

$$w = \frac{M'R^2}{2D}\left(\frac{\rho^2}{1 - \nu} + \frac{2 - \rho^{-2}}{3 + \nu}\right)\cos 2\theta \qquad (3.40)$$

satisfies conditions of zero bending moment M_r and zero applied transverse force V_r at the surface of the hole, as obtained by using equations (3.22) and

101

(3.34). Hoop stress at this surface is in proportion to the bending moment

$$M_\theta = -4 \frac{1+\nu}{3+\nu} M' \cos 2\theta$$

With $\nu = 0.3$, the numerical multiplier is about 1.6.

When this solution is combined with the solution for axial symmetry, the solution is obtained for a plate under pure bending with moment M_x equal to some uniform value, say M'', at the outer boundary, with $M_y = 0$. At the surface of the hole, the highest bending moment is now found to be $M_\theta = 1.8M''$.

These numerical factors become increasingly inaccurate as the hole diameter becomes equal to a small multiple of plate thickness, the factors being under-estimated by about 10% for a hole diameter of six times the plate thickness. The reason is that the simple theory of flexure assumes that transverse shear stresses are of a lower order of magnitude than bending stresses. To see whether this is so near the edge of a hole in a twisted plate, the shear force at $\rho = 1$, $\theta = 0$ may be calculated from (3.40),

$$S_r = -D \frac{\partial}{\partial r} \nabla^2 w = -\frac{8M'}{(3+\nu)R}$$

With a parabolic distribution of shear stress through the plate section, the maximum shear stress is one and a half times the mean value,

$$\sigma_{zr} = -6M'/(3+\nu)Rh$$

This may be compared with the maximum tangential bending stress at the same point,

$$\sigma_\theta = -6(1+\nu)M'/(3+\nu)h^2$$

The ratio of these stresses, $\sigma_{zr}/\sigma_\theta = h/(1+\nu)R$ increases as the hole radius R is reduced, so neglect of the deformation caused by transverse shear stress gives an increasingly large error in the calculated stress distribution. When account is taken of transverse shear stress, it is found that the stress concentration factor for a very small hole, as R/h tends to zero, approaches that for the equivalent plane-stress problem. For example, a small hole in a plate under uniform unidirectional bending gives a disturbance of stress in the surface layer which is the same as if there were no stress gradient through the thickness, the stress concentration factor then being equal to 3.0 (see Timoshenko and Woinowski-Krieger, p. 322).

3.8 Point Load at a Corner

In all the problems so far considered, the boundary has been in the form of a continuous circle. For plates with a rectangular boundary or with a boundary consisting of arcs and radial lines, it is possible that externally applied forces may be required at the right-angled corners to give equilibrium. Horizontal shear stresses act in the interior of a plate, but as previously mentioned, these must vanish at a free edge, and are replaced by vertical shear stresses. When the boundary is a continuous curve, these vertical shear stresses are balanced internally between one element of the boundary and the next. However, the element situated at a right-angled corner has to supply two sets of shear stresses acting vertically in the same direction.

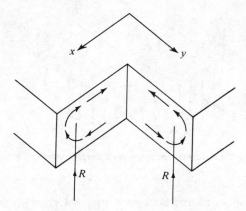

Fig. 3.20 External force to be applied at a corner

Figure 3.20 shows such a corner with a small element removed. The shear stresses shown are those produced by a positive twisting moment M_{xy}. Vertical shear stresses on each exposed face have a resultant upward force R, which was previously found to be numerically equal to the value of M_{xy} at the corner. A total force $P_c = 2R$ must therefore be applied at this corner, its value, from (3.31) being

$$P_c = 2(1 - \nu) D \frac{\partial^2 w}{\partial x \, \partial y} \tag{3.41}$$

When this force is of positive sign, it acts upwards. Boundaries of the kind considered here will have four right-angled corners. At the corner that is outermost with respect to the positively-directed coordinate axes, and also at the diagonally opposite corner, the sign of P_c is correctly given by the last equation. At the other two corners, a negative sign must be inserted.

3.8.1 Problems in rectangular coordinates As a simple illustration of this type of problem, consider a square plate having sides of length a, simply supported on all four sides. This means that the deflection is zero at all points on the boundary by reason of the constraint imposed on the boundary, but no bending moment can be resisted by the supports. It is specified that an upward pressure p acts over

103

the surface of the plate, distributed according to $p = p_0 \cos(\pi x/a) \cos(\pi y/a)$, with coordinate axes as shown in Fig. 3.21. A deflection $w = w_0 \cos(\pi x/a) \cos(\pi y/a)$ is assumed, which satisfies the conditions imposed on deflection and bending moment at the boundary. As the differential equation $\nabla^4 w = p/D$ is satisfied at

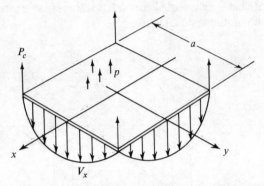

Fig. 3.21 **Boundary forces acting on a simply-supported square plate**

all points, a relation is obtained between central deflection w_0 and peak pressure p_0,

$$w_0 = p_0 a^4/4\pi^4 D$$

The total force P applied to the surface of the plate can now be obtained by integrating pressure p over the surface,

$$P = 4p_0 a^2/\pi^2$$

From the bending moment equations (3.18), stresses in the centre of the plate may be found, and by using the last two equations, these may be expressed in terms of total applied load P,

$$\sigma_x = \sigma_y = 3(1 + \nu)P/8t^2$$

where t is the plate thickness.

The transverse force V_x per unit length to be applied to the boundary which is perpendicular to the x-axis is obtained from (3.33),

$$V_x = -\frac{(3 - \nu)\pi}{16} \frac{P}{a} \cos(\pi y/a)$$

Shear forces acting on the other sides are the same. Hence the total force distributed over all four sides of the plate amount to $-(3 - \nu)P/2$.

Concentrated forces each equal to $P_c = (1 - \nu)P/8$, as obtained from (3.41), act at all four corners in the same upward direction. Physically, it can be envisaged that if these forces were not supplied, the corners would tend to lift off the line supports, by movement in a direction opposite to that of the pressure applied to the surface of the plate. Although the presence of concentrated forces at right-angled corners is a correct deduction from the simple theory of flexure of thin plates, the existence of such a force violates the assumption that transverse shear stresses are always small. A more exact analysis taking account of shear deflection shows that the concentrated forces are actually distributed over a short length of each edge adjacent to the corners.

In the problem just considered, an artificially simple pressure distribution was chosen. Any other distribution must be described by Fourier series solutions of the differential equation, and calculations tend to become lengthy. A detailed treatment of rectangular plates has been given by Timoshenko and Woinowski-Krieger, *Theory of Plates and Shells*, pp. 105-258. Rectangular plates also offer scope for variational methods, as discussed in the next chapter.

4. Strain Energy Methods

As in other branches of mechanics, an energy equation can provide an alternative way of arriving at relations otherwise derivable from conditions of equilibrium. A differential equation of equilibrium is by itself a generalized restriction on the behaviour of all individual elements, but an energy equation can give a broader view of a finite member, complete with its loads and supporting forces. Therefore, the alternative approach of formulating and equating energy quantities usually helps towards a fuller physical understanding of problems in applied elasticity. However, to a practical engineer, the important advantage of energy methods is that they can be made to yield approximate answers to everyday problems for which the rigorous solutions to differential equations often become too complicated to be of much use.

The basic assumption in the elastic analysis of structures is that the material is a linear elastic material. This is a material in which there is a linear relation between stress and strain at all times, as long as strains are infinitesimal, that is, very small in comparison with unity. Then it may be said either that a certain stress causes a proportional strain to appear, or that a certain strain is bound to be associated with the presence of a proportional stress. The idea of strain components being expressible as linear combinations of the stress components at a point is conveyed by the elasticity equations

$$
\left.
\begin{aligned}
Ee_x &= \sigma_x - \nu(\sigma_y + \sigma_z), \text{ etc.} \\
2Ge_{xy} &= \sigma_{xy}, \text{ etc.}
\end{aligned}
\right\}
\tag{4.1}
$$

In any element of a stressed body, the energy absorbed while the stress is being set up is equal to the work done by forces acting on the element. This work is given by the average force acting during the deformation, multiplied by the relative movement of the ends of the element. In an elastic body in which deformation is proportionate to applied force, the average force is equal to half the final force needed to produce the deformation. Although some structural members may respond non-linearly to applied forces by reason of their external shapes, attention is restricted here to a small element of the body. This is of material which, by definition, has a linear response. Hence the six components of stress acting on a rectangular element supply work which may be considered as strain energy stored within the material. In terms of general stress and strain components σ_{ij} and e_{ij} this energy ΔU per unit volume is given by the summation

$$\Delta U = \tfrac{1}{2}\sigma_{ij}e_{ij} \tag{4.2}$$

In equations (4.1) and (4.2) it is assumed that shear strain is defined in terms of displacements by the 'mathematical' definition.

It may be convenient to express energy in terms of stress components alone, or in terms of strain components alone. This may be done by replacing the unwanted quantities by using the elasticity relations (4.1). The resulting expressions for strain energy in the most general circumstances are

$$
\left.
\begin{aligned}
\Delta U &= \frac{1}{2E}[\sigma_x^2 + \sigma_y^2 + \sigma_z^2 - 2v(\sigma_x\,\sigma_y + \sigma_y\,\sigma_z + \sigma_z\,\sigma_x)] \\
&\quad + \frac{1}{2G}[\sigma_{xy}^2 + \sigma_{yz}^2 + \sigma_{zx}^2] \\
\Delta U &= \frac{E}{2(1+v)(1-2v)}[(1-v)(e_x^2 + e_y^2 + e_z^2) \\
&\hspace{6em} + 2v(e_x e_y + e_y e_z + e_z e_x)] \\
&\quad + 2G[e_{xy}^2 + e_{yz}^2 + e_{zx}^2]
\end{aligned}
\right\} \tag{4.3}
$$

Particular classes of elastic members may permit simplification of these expressions, if it should be known that certain components of stress or strain in these members must be zero, as will be seen later.

4.1 Linear Structure

Even when a structure is made from a linear elastic material, defined as a material which obeys the elasticity equations (4.1), the structure may or may not respond linearly to externally applied forces. Let some particular point A on a structure be displaced amount w by the application of force P at this point. Suppose it is possible to calculate the displacements of all other points in the body, and, therefore, the strains and stresses at all points. Assuming for simplicity that only one non-zero stress component σ_x is present in the structure, this stress may be written in the form of a series

$$\sigma_x = \alpha_0 + \alpha_1\,w + \alpha_2\,w^2$$

where coefficients α_0, α_1, and α_2 relate to the selected point, and higher powers of w are neglected. The conditions under which there is proportionality between load and deflection, expressed by a constant stiffness $k = P/w$, are now investigated.

Strain energy (4.3) is integrated through the total stressed volume v, and when energy due to internal stress α_0 is deducted, the remainder is equated to work done,

$$\tfrac{1}{2}Pw = \frac{1}{2E} \int [(\alpha_0 + \alpha_1 w + \alpha_2 w^2)^2 - \alpha_0{}^2] \, dv$$

Expanding this expression and dividing through by $w/2$ gives

$$P = \frac{1}{E} \int [2\alpha_0 \alpha_1 + (\alpha_1{}^2 + 2\alpha_0 \alpha_2) w + 2\alpha_1 \alpha_2 w^2 + \alpha_2{}^2 w^3] \, dv$$

If the body is initially in equilibrium, force P must be zero when w is zero, so the first term in the integral, involving internal stress α_0, can make no contribution to the integral. With this term omitted, the equation may be divided through again by w,

$$k = \frac{1}{E} \int [\alpha_1{}^2 + 2\alpha_0 \alpha_2 + \alpha_2 w(2\alpha_1 + \alpha_2 w)] \, dv \tag{4.4}$$

It is seen that stiffness k is independent of deflection w only if coefficient α_2 is zero. If this is so, it is noted that response is not affected by internal stress α_0. A structure can therefore be recognized as one which will give a linear response to loading at point A if it can be shown that strains at every point vary linearly with a displacement imposed at point A.

This is so for many structures, exceptions being those incorporating thin flexural elements such as struts or plates in which membrane stresses may be set up, as these stresses vary as the square of transverse deflection. Problems of this kind will be examined later.

4.1.1 Superposition principle   This principle states that if loads act simultaneously at two points on a linear structure, the displacement produced at any third point is equal to the sum of the displacements produced by the two loads acting separately. This can be extended to any two sets of loads, or to any number of separate loads. We first investigate whether the stiffness of a structure with respect to some load P_1 is affected by the presence of another load P_2. If load P_1 is applied by itself, either positively or negatively, displacement w_3 would be expected to vary along some line A'OA, as plotted in Fig. 4.1. With this load removed and load P_2 applied, some displacement OB is produced. Also, the structure is supposed to be stiffened by the presence of this load. Therefore, the subsequent application of P_1 will give line BC, which is somewhat steeper than line OA. Now let the loads be applied in the same order, but acting in the negative direction, so that load P_2 gives displacement OB' with a reduction in stiffness. Subsequent application of negative load P_1 will give line B'C' which is less steep than line OA'. As an alternative loading procedure, let the loads be maintained in constant ratio while they are applied together. The crooked line

C'OC will now be followed, which violates the assumption that the structure is linear.

Hence it is demonstrated that the stiffness with respect to load P_1 cannot vary linearly with load P_2. By similar arguments, it can be shown that this stiffness

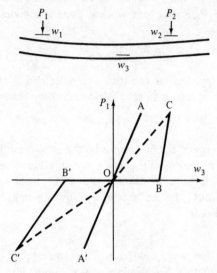

Fig. 4.1 Superposition principle

cannot vary as P_2 squared, or in any other way, without violating the premise that the structure is linear. Therefore, all the loading lines in Fig. 4.1 must be straight and parallel, which means that the displacement due to both loads acting together is equal to the sum of the displacements of each acting separately. It follows that stresses or strains at any point in the body may be similarly summed. However, it should be noted that the strain energy set up by two loads acting together is not the same as the sum of the energies for each load acting separately. This is because the strain energy in an element found from the square of the total stress, which is not the same as the sum of the squares.

4.1.2 Reciprocal theorem   In discussing the displacements at various points produced by loads acting at other points, it is useful to introduce inverse stiffness coefficients, also called compliance or influence coefficients. These are written with a double subscript, a_{mn}, the first indicating the site where displacement is measured and the second indicating the site of the load responsible for it. Thus the displacement w_1 at position (1) produced by load P_2 acting by itself at position (2) is given by $w_1 = a_{12}P_2$. If other loads continue to act while load P_2 is applied, the displacement w_1 will be an incremental displacement.

For the present purpose, displacements are measured at two points along the lines of action of loads that can be applied at these points. Suppose that load P_1

is first applied, and produces displacement $a_{11}P_1$ at its point of application. With this load remaining in place, let a second load P_2 be applied at the second point. This will produce displacement $a_{22}P_2$ at its point of application. However, the second load will also produce an additional displacement $a_{12}P_2$ at point (1). While this occurs, load P_1 remains at its full value, and therefore supplies additional work $a_{12}P_2P_1$. So the total work W done in applying loads in this order is given by

$$W = \tfrac{1}{2}a_{11}P_1^2 + \tfrac{1}{2}a_{22}P_2^2 + a_{12}P_2P_1$$

If this process of applying the two loads is now carried out by applying load P_2 first with load P_1 following, an alternative expression for work is obtained,

$$W = \tfrac{1}{2}a_{22}P_2^2 + \tfrac{1}{2}a_{11}P_1^2 + a_{21}P_1P_2$$

As the final strain energy of the body must be the same regardless of the order of application of loads, it is deduced that $a_{21} = a_{12}$. In other words, the displacement produced at point (2) by some load acting at point (1) is the same as the displacement produced at point (1) by the same load acting at point (2), always assuming that the structure is linear.

4.1.3 Displacements due to multiple loads Let loads $P_1, P_2, P_3, \ldots$, act at various points on a structure. The displacement w_1 under load P_1 will be the sum of the contributions from all the loads,

$$w_1 = a_{11}P_1 + a_{12}P_2 + a_{13}P_3 + \cdots \tag{4.5}$$

Similar expressions can be written for w_2, w_3, etc. Supposing that all loads are applied simultaneously, the total work supplied is given by

$$W = \tfrac{1}{2}P_1 w_1 + \tfrac{1}{2}P_2 w_2 + \tfrac{1}{2}P_3 w_3 + \cdots$$

Displacements are now eliminated using (4.5), and the work W is identified with strain energy U stored in the structure,

$$U = \tfrac{1}{2}P_1(a_{11}P_1 + a_{12}P_2 + a_{13}P_3 + \cdots)$$
$$+ \tfrac{1}{2}P_2(a_{21}P_1 + a_{22}P_2 + a_{23}P_3 + \cdots)$$
$$+ \tfrac{1}{2}P_3(a_{31}P_1 + a_{32}P_2 + a_{33}P_3 + \cdots)$$

Noting that P_1 appears in the first row and first column of this array, we differentiate to get

$$\frac{\partial U}{\partial P_1} = \tfrac{1}{2}(a_{11}P_1 + a_{12}P_2 + a_{13}P_3 + \cdots)$$
$$+ \tfrac{1}{2}(a_{11}P_1 + a_{21}P_2 + a_{31}P_3 + \cdots)$$

However, by the reciprocal theorem, these compliance coefficients are related by $a_{21} = a_{12}, a_{31} = a_{13}$, etc. Hence, by referring back to (4.5), it is seen that

$$\frac{\partial U}{\partial P_1} = w_1 \qquad (4.6)$$

Similarly, the displacement at any point may be found by differentiating U with respect to the load at that point, provided that energy U is expressed in terms of loads only, without reference to displacements.

The converse argument for deriving loads from known displacements can be followed through in a similar way. Here, it is convenient to use stiffness coefficients k_{mn}, the first subscript indicating the position of the load, and the second showing where displacement is measured. If a number of points on a structure are given known displacements $w_1, w_2, w_3, \ldots$, the value of load at point (1) may be written

$$P_1 = k_{11} w_1 + k_{12} w_2 + k_{13} w_3 + \cdots$$

with similar expressions for the other loads. Strain energy, now expressed in terms of displacements alone, can be differentiated to give

$$\frac{\partial U}{\partial w_1} = P_1 \qquad (4.7)$$

with similar equations for the other loads. The results (4.6) and (4.7) are often referred to as Castigliano's theorems, and are further discussed by R. V. Southwell (1941).

4.1.4 Energy in beams and shafts Some simple flexural and torsional systems will now be analysed using strain energy. The strain energy of an element is written in terms of the local value of bending moment or torque, and these forces are in turn expressed in terms of forces applied to the whole structure. Energy can then be integrated over all the members, and the total can be differentiated to give displacements. As a first step, strain energy in an element is determined.

Consider an element of an elastic beam lying initially along the x-axis. Under the action of a positive bending moment M, the left-hand end will assume some slope ϕ, while the right-hand end will assume some larger slope $\phi + \delta x \, d\phi/dx$, as shown in Fig. 4.2(a). The net amount of work done is equal to half the moment times the change in slope. The slope ϕ is given in terms of deflection by dw/dx, so when this work is equated to increase in strain energy ΔU, we have

$$\Delta U = \tfrac{1}{2} M \frac{d^2 w}{dx^2} \delta x$$

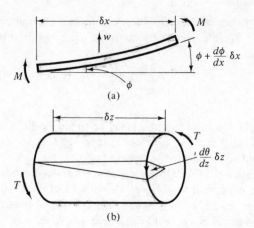

Fig. 4.2 Strain energy in an element of (a) a beam, (b) a shaft

As the equation of equilibrium is $M = EI\, d^2w/dx^2$, the total energy U can be written in alternative ways,

$$U = \tfrac{1}{2} \int EI\left(\frac{d^2 w}{dx^2}\right)^2 dx = \tfrac{1}{2}\int \frac{M^2}{EI}\, dx \qquad (4.8)$$

The first integral may be verified by using a stress $\sigma_x = -Ez\, d^2w/dx^2$ in the energy expression (4.3), z being the height of an element of the section above the centroidal axis. Other components of stress are assumed zero or negligible, in accordance with the elementary plane-stress theory of beams.

A similar argument may be used for an element of a shaft of length δz which undergoes an incremental twist $\delta z\, d\theta/dz$ under the action of torque T, as shown in Fig. 4.2(b). The strain energy supplied to the element is then

$$\Delta U = \tfrac{1}{2}T\frac{d\theta}{dz}\delta z$$

If the shaft has torsional stiffness K so that torque and twist are related by the equilibrium equation $T = GK\, d\theta/dz$, the total strain energy is obtained by integrating the whole stressed length of the shaft using either of the integrals

$$U = \tfrac{1}{2} \int GK\left(\frac{d\theta}{dz}\right)^2 dz = \tfrac{1}{2} \int \frac{T^2}{GK}\, dz \qquad (4.9)$$

4.2 Minimum Energy Principles

In the last section, energy was expressed in terms of applied forces to find displacements at various points. In the methods now examined, the deformed

112

shape of a body is expressed as some continuous function containing one or more constants which are initially unknown. Energy principles are then applied to determine these constants, and hence the deformed shape, from which stresses can be found if required. These methods are conveniently illustrated by referring to beams, even though simpler alternative methods may be available for the particular problems mentioned.

4.2.1 Deflection due to a single load   The effect of applying load P at some point on a beam is to be determined. Instead of allowing this load to act, it is supposed that the beam is deflected by a fixed amount at this point. Use can now be made of the principle that at all other points, the beam will assume a deflected shape such as to make its strain energy a minimum. When this minimum value has been found, it may be equated to work done at the single point where deflection was imposed, to give a load-deflection relationship.

To illustrate this approach, a uniform cantilever carrying a single load at its end is examined. Distance x is measured from the fixed end, and at the other end ($x = L$), a deflection w_1 is imposed. Let the deflected shape be specified in terms of constants a_n by the polynomial

$$w = w_1 \sum a_n s^n \qquad (n = 0, 1, \ldots, 4)$$

where $s = x/L$. Four boundary conditions must be satisfied. Deflection and slope are zero at $s = 0$. At $s = 1$, $w = w_1$ and no bending moment acts ($d^2 w/dx^2 = 0$). Shear forces at the ends are not initially known. Using these conditions, all but one of the constants, say a_4, can be found, giving

$$w = \tfrac{1}{2} w_1 [3s^2 - s^3 + a_4 (3s^2 - 5s^3 + 2s^4)]$$

This expression satisfies the boundary conditions for any value of a_4, and if the beam is uniform along its length, leads to strain energy

$$U = \frac{EI}{2} \int_0^L \left(\frac{d^2 w}{dx^2} \right)^2 dx = \frac{3(5 + 3a_4{}^2)}{10} \frac{EI w_1{}^2}{L^3}$$

The constant a_4 can now be given the value which makes U a minimum, and clearly, this is $a_4 = 0$. The strain energy can now be equated to work $\tfrac{1}{2} P w_1$ supplied by a force P acting at the end.

The basic limitation of this approach is that it is valid only for loading at a single point. When another load acts elsewhere, this will supply additional work as the deflected shape is varied. This means that it is no longer satisfactory to vary the deflected shape so as to minimize strain energy by itself.

4.2.2 Virtual work   A more versatile procedure is now required for recognizing a state of equilibrium of a structure carrying more than one load and giving a response that is not necessarily linear. For simplicity, a single load is considered first. Let this load P produce a displacement u measured from the no-load

113

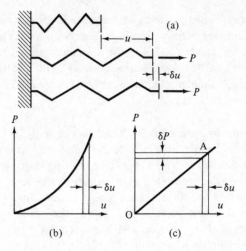

Fig. 4.3 Virtual work due to displacement from equilibrium position

position, as shown in Fig. 4.3(a). It is assumed that strain energy can be calculated in terms of displacement u from geometrical considerations without involving the load P. A small virtual displacement δu is now superimposed on the former value. This increases the strain energy by amount

$$\delta U = \frac{\partial U}{\partial u} \delta u$$

While this extra displacement is being imposed, the full load P moves through distance δu and supplies virtual work δW given by

$$\delta W = P \delta u$$

An equilibrium state is now recognized as one in which there is exact equality of increase in strain energy and work supplied by external forces during a virtual displacement, i.e., $\delta U = \delta W$. This gives the relations

$$\delta U = P \delta u \qquad P = \frac{\partial U}{\partial u} \qquad (4.10)$$

The energy supplied by an external force may be considered equivalent to the area under a curve of force P plotted vertically and displacement u plotted horizontally, as in Fig. 4.3(b). It is seen that (4.10) remains valid whether the response is linear or not. Further, the load P may be considered as having several parts P_n acting at different points, so that variations δu_n in the displacements at these points produce a change in work or strain energy given by

$$\delta W = \delta U = \sum P_n \delta u_n$$

This relation will also hold whether the response is linear or not.

114

In the special case of a structure giving displacement proportional to load, as illustrated in Fig. 4.3(c), any point A on the curve will define equal areas underneath and to the left of line OA. The increase in strain energy can then be written alternatively in terms of a variation in displacement δu or the corresponding variation in load δP,

$$\delta U = P\delta u = u\delta P \qquad (4.11)$$

It will be important to realize in subsequent discussions that the incremental work δW is taken to mean the increase produced by the full value of the external

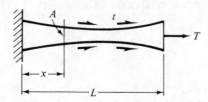

Fig. 4.4 Member carrying tensile load T and distributed tangential load t per unit length

forces moving through a variation or increment in the displacements, and is therefore given by $\delta W = P \, \delta u$. Although a force may increase from zero to value P, to give a final displacement u and a total amount of work supplied equal to $W = \frac{1}{2}Pu$ in a linear structure, it is incorrect to differentiate regarding P as a constant to obtain an increment of work $\delta W = \frac{1}{2}P \, \delta u$.

4.2.3 Variational principle   It has been seen that when a single load acts on a structure, strain energy can be found for some deflected shape which provides for some fixed displacement at the point of application of load. Minimizing strain energy with respect to the available parameters then gives the actual deflected shape, or an approximation to it. A method is now required for finding the deflected shape of a member carrying a distributed load. As additional work will be supplied by this load as the shape is varied, it is not sufficient to minimize strain energy by itself.

Perhaps the simplest example of such a system is a bar of non-uniform cross-sectional area $A(x)$ carrying tensile loads, as shown in Fig. 4.4. Displacement of any particle in the x-direction may be tentatively written as a series

$$u = a_1 x + a_2 x^2 + \cdots + a_n x^n$$

This satisfies the known condition that $u = 0$ at $x = 0$. There will also be a condition of some kind to be satisfied at end $x = L$. The first task is to ensure that these end conditions are satisfied, as these conditions define the problem to be solved. This process will eliminate some of the constants a_n, leaving others that may be varied at will without violating the end conditions of the problem. If it is assumed, for the moment, that this has been done, strain energy may be

115

calculated from the deformed shape. From the strain at any point, given by $e = du/dx$, the stress is given by $\sigma = Ee$. As it is postulated that no other component of stress acts, strain energy U is given in terms of u by

$$U = \tfrac{1}{2} \int EA \left(\frac{du}{dx} \right)^2 dx \qquad (4.12)$$

the integration being carried out over length L.

Of the parameters of u which remain available, one is selected and is varied by amount δa_n. This gives a variation in displacement

$$\delta u = \delta a_n \frac{\partial u}{\partial a_n}$$

which is a function of position x. The corresponding change in strain energy is now found to be

$$\delta U = \delta u \frac{\partial U}{\partial u} = \delta a_n \int EA \frac{du}{dx} \frac{d}{dx} \left(\frac{\partial u}{\partial a_n} \right) dx$$

Work is supplied by end force T acting through displacement $u(L)$, and also by the distributed tangential force of t per unit length which varies with x. During the variation in displacement, the work supplied is given by the full value of these loads multiplied by the distance through which they are displaced,

$$\delta W = \delta a_n \left\{ T \left[\frac{\partial u}{\partial a_n} \right]_L + \int_0^L t \frac{\partial u}{\partial a_n} dx \right\}$$

Writing $\partial u/\partial a_n = u'$, the variational relation $(\delta U - \delta W) = 0$ gives an equation

$$\int_0^L \left[EA \frac{du\, du'}{dx\, dx} - tu' \right] dx - Tu'(L) = 0 \qquad (4.13)$$

One of these equations can be set up for each of the parameters a_n that are available for variation, and from these equations, the parameters are determinate.

When end load T is either zero or is not known on account of the end condition being specified in terms of displacement, a shorter form of the equation may be derived. If the terms of the differentiation

$$\frac{d}{dx} \left(u' EA \frac{du}{dx} \right) = EA \frac{du\, du'}{dx\, dx} + u' \frac{d}{dx} \left(EA \frac{du}{dx} \right)$$

are integrated between 0 and L, the term on the left gives an integral equal to $Tu'(L)$. When this is utilized, the variational equation (4.13) becomes

$$\int_0^L \left[\frac{d}{dx}\left(EA\frac{du}{dx}\right) + t \right] u'\, dx = 0 \tag{4.14}$$

In this equation, the term in square brackets within the integral has a special significance. By considering the forces acting on an element of the bar, it can be deduced that the equation

$$\frac{d}{dx}\left(EA\frac{du}{dx}\right) + t = 0$$

expresses equilibrium of stresses. When this is satisfied, it is seen that the variational equation (4.14) is satisfied for all conceivable variations δu, and not merely by those resulting from variation of certain parameters a_n in some series representation of u.

The problem of a tensile member is used here for illustrating the variational method, and would in practice be solved by direct integration of the equilibrium equation. The variational method proves it worth in situations where solution of the differential equation of equilibrium is excessively difficult.

4.2.4 Minimum and maximum strain energy

Consider a bar of variable cross-section which is fixed at end $x = 0$ while a certain displacement $u(L)$ is imposed and maintained at end $x = L$. Suppose that $u_0(x)$ is the function giving the correct distribution of displacement, that is, the distribution which satisfies the equilibrium equation at all points. The actual distribution is now taken to be the correct one, $u_0(x)$ plus a small additional displacement $u_1(x)$, which is arranged to be zero at both ends of the bar so as to preserve the specified end values. On inserting the value $u = u_0 + u_1$ into the energy integral (4.12), it is found that

$$U = U_0 + \tfrac{1}{2}\int_0^L EA\left[2\frac{du_0}{dx}\frac{du_1}{dx} + \left(\frac{du_1}{dx}\right)^2\right] dx \tag{4.15}$$

where U_0 is the strain energy due to the presence of u_0 by itself. The first term in this integral can be written

$$EA\frac{du_0}{dx}\frac{du_1}{dx} = \frac{d}{dx}\left(EA\frac{du_0}{dx}u_1\right) - u_1\frac{d}{dx}\left(EA\frac{du_0}{dx}\right)$$

The second term on the right is zero because its bracketed part is equal to the uniform tensile force in the bar, and the first term on the right, on being inte-

117

grated, gives a zero value because u_1 is zero at both limits. Hence the strain energy becomes

$$U = U_0 + \tfrac{1}{2} \int EA \left(\frac{du_1}{dx}\right)^2 dx$$

As this integral is bound to be positive, it is seen that any distribution of displacement which departs from the correct distribution leads to an excess of strain energy. The applied force T calculated from this energy will also exceed the correct value, but by an amount proportional to the square of the error in the assumed displacement. This is a particular illustration of the Rayleigh–Ritz principle, which states that if a moderately good approximation to the deformed shape is selected, the resulting stiffness will be correct to a somewhat higher degree of accuracy. Thus, when a fixed displacement is imposed at some specified point, the correct distribution of displacements at other points is evidently the particular distribution which requires a minimum of strain energy.

Continuing with the relatively simple problem of a tensile member, the strain energy is now calculated for some fixed load distribution consisting of force $t(x)$ per unit length as shown in Fig. 4.4. A function $u = af(x)$ is selected for describing displacement, where a is a constant. It is first necessary to arrange that $f(0) = 0$ at the fixed end $x = 0$, together with $f'(L) = 0$ at the stress-free end $x = L$. The variational equation (4.14) can now be applied as if to determine the constant a. The resulting equation, on being multiplied through by a, becomes

$$\int \left[\frac{d}{dx}\left(EA\frac{du}{dx}\right) + t \right] u \, dx = 0$$

This displacement is now written $u = u_0(x) + u_1(x)$, where u_0 is the exact value which satisfies equilibrium at all points, and u_1 is a small additional value. By making use of the equilibrium equation, the last equation reduces to

$$\int_0^L (u_0 + u_1) \frac{d}{dx}\left(EA\frac{du_1}{dx}\right) dx = 0$$

This can now be rewritten

$$\int_0^L \frac{d}{dx}\left[(u_0 + u_1) EA\frac{du_1}{dx}\right] dx - \int_0^L EA\frac{du_1}{dx}\left(\frac{du_0}{dx} + \frac{du_1}{dx}\right) dx = 0$$

The first integral is zero because $u_0 = u_1 = 0$ at the fixed end $x = 0$, and $du_1/dx = 0$ at the stress-free end $x = L$. The second integral may be combined with energy

118

equation (4.15) to give the strain energy

$$U = U_0 - \tfrac{1}{2} \int EA \left(\frac{du_1}{dx}\right)^2 dx$$

This integral is necessarily of positive value.

From this equation it is seen that when a fixed loading is applied, any assumed distribution of displacement which is determined as far as possible by using the variational method will give less strain energy than the correct value. Hence the correct distribution of displacement is the one that allows the maximum amount of strain energy to be absorbed. The displacements predicted by the variational method will, under these conditions of specified loading, be generally less than the correct displacements, though not necessarily less at every point. Hence the member will appear to be stiffer than it really is, though the term stiffness has no precisely definable meaning when loads are distributed. This overestimate of stiffness is always to be expected from the use of a variational equation which allows equilibrium conditions to be slightly violated.

4.2.5 Variational equation for a beam

If a beam initially lies along the x-axis, its deflected shape is completely defined by some function $w(x)$ which gives deflection at any point. For example, this function may be a series

$$w = a_0 + a_1 x + a_2 x^2 + \cdots + a_n x^n$$

The strain energy follows from (4.8),

$$U = \tfrac{1}{2} \int EI \left(\frac{d^2 w}{dx^2}\right)^2 dx$$

If it has been arranged that certain of the coefficients in the expression for deflection may be varied without violating the known end conditions, a variation δa_n in one of these coefficients leads to a variation in strain energy given by

$$\delta U = \delta a_n \int EI \frac{d^2 w}{dx^2} \frac{d^2}{dx^2}\left(\frac{\partial w}{\partial a_n}\right) dx \tag{4.16}$$

the integral being taken along the whole length L of the beam.

While this variation in deflection occurs, the transverse loads supply an amount of work given by the full value of each load moving through a distance equal to the change in deflection. Suppose that concentrated loads P_m act at points x_m, together with some distributed load of intensity $p(x)$ per unit length. Then the variation of work W is given by

$$\delta W = \delta a_n \left\{ \sum P_m \left[\frac{\partial w}{\partial a_n}\right]_{x=x_m} + \int_0^L p \frac{\partial w}{\partial a_n} dx \right\}$$

As this virtual work must equal the change in strain energy, $\delta W = \delta U$, a variational equation follows,

$$\int \left[EI \frac{d^2 w}{dx^2} \frac{d^2 w'}{dx^2} - pw' \right] dx - \sum P_m w'(x_m) = 0 \qquad (4.17)$$

where $w' = \partial w / \partial a_n$. For each of the available coefficients a_n a different equation is obtained, so that eventually a set of simultaneous equations is obtained which is adequate for determining the coefficients.

4.2.6 Shorter form of beam equation

If any work is supplied by the two supports of the beam, this work should be included in (4.17) as an additional term. Alternatively, this work may be taken into account in the following way.

Positive shear forces S_0 and S_L and positive bending moments M_0 and M_L act in the directions shown in Fig. 4.5. These values remain unchanged as the deflection is varied, and supply an amount of work given by

$$\delta W_e = \delta a_n \left\{ \left[M_L \frac{dw'}{dx} + S_L w' \right]_{x=L} - \left[M_0 \frac{dw'}{dx} + S_0 w' \right]_{x=0} \right\}$$

As shown in equations (3.2) and (3.4) in the previous chapter, bending moments and shear forces can be expressed as derivatives of deflection w,

$$M = EI \frac{d^2 w}{dx^2} \qquad S = -\frac{d}{dx} \left(EI \frac{d^2 w}{dx} \right)$$

so that the work done can be written

$$\delta W_e = \delta a_n \left[EI \frac{d^2 w}{dx^2} \frac{dw'}{dx} - w' \frac{d}{dx} EI \frac{d^2 w}{dx^2} \right]_0^L$$

In this expression it is understood that the value of the bracketed term at point $x = 0$ is subtracted from the value at point $x = L$. The same result might be

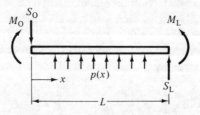

Fig. 4.5 Positive forces acting on a beam

obtained by carrying out an integration between limits $x = 0$ to $x = L$,

$$\delta W_e = \delta a_n \int_0^L \frac{d}{dx}\left[\frac{dw'}{dx}EI\frac{d^2 w}{dx^2} - w'\frac{d}{dx}EI\frac{d^2 w}{dx^2}\right]dx$$

A rearrangement of the derivatives in this expression now gives

$$\delta W_e = \delta a_n \int_0^L \left[EI\frac{d^2 w'}{dx^2}\frac{d^2 w}{dx^2} - w'\frac{d^2}{dx^2}\left(EI\frac{d^2 w}{dx^2}\right)\right]dx$$

This amount of work supplied at the ends of the beam may be added to the work done by applied loads. This total of work supplied by external forces during a variation in deflection is then equated to the change in strain energy as given by (4.16), that is, $\delta W_e + \delta W = \delta U$, giving a new variational equation

$$\int_0^L \left[\frac{d^2}{dx^2}\left(EI\frac{d^2 w}{dx^2}\right) - p\right]w'\,dx - \sum P_m w'(x_m) = 0 \qquad (4.18)$$

It can now be mentioned that over any part of the beam where concentrated loads P_m are absent, the equilibrium equation is

$$\frac{d^2}{dx^2}\left(EI\frac{d^2 w}{dx^2}\right) = p$$

as derived previously in (3.4). Hence, if the particular function $w(x)$ is chosen which satisfies the equilibrium equation at all points, the variational equation must be satisfied for any possible variation of the deflected shape. On the other hand, if the variational equation is used for determining a finite number of coefficients in an approximate expression for deflection, this expression need not precisely satisfy the equilibrium equation. It is clear that no continuous function $w(x)$ can give the correct solution when concentrated forces act within the range of integration, as the shear force must change discontinuously at the positions of these forces. Nevertheless, a continuous function can be assumed, and after its coefficients have been determined by means of (4.18), such a function will describe deflections to an accuracy depending on the number of parameters that have been varied.

A variational equation of the form of (4.18) requires an integration over a specified portion of the beam. Conditions at the limits of this integration may consist of deflection, gradient, bending moment, or shear force. Two of these are known at each end of the portion considered, and these determine four constants in a polynomial for deflection. When concentrated forces P_m act at the extreme ends of the beam, and it is desired to integrate over the whole length, there may

9

be a dilemma about whether to include these in the second term of (4.18) or whether to omit them. If the range of integration just falls short of the extreme end, this term will have to be omitted, but the shear force at the limit of integration will then reflect the presence of the concentrated load. This shear force will be taken into account as an end condition already satisfied in the polynomial which is available for variation of parameters. On the other hand, if the range of integration extends just outside of the concentrated loads, the shear force at the end must be taken as zero, so different constants will be calculated from the different end conditions. This problem does not arise if the basic form of the variational theorem (4.17) is used, as this is valid for the elastic system as a whole rather than for a specific range of integration along the beam.

4.3 Stability of Struts

A strut, that is, an elongated member transmitting a compressive load along its length, calls for rather closer attention than a tension member because of the possibility of collapse through lateral buckling. Various treatments of struts bring into prominence two or three slightly different physical ideas. Perhaps the most direct way of finding a stability criterion for a uniform strut is by examining its equilibrium in a laterally displaced configuration. The differential equation thus obtained can be solved using given conditions of end fixing. Alternatively, a strut becomes unstable when the end load becomes large enough to supply the flexural strain energy associated with a displaced configuration. There is a further way of looking at the problem. An end load, if tensile, has the effect of increasing stiffness with respect to a lateral force, while a compressive load reduces this stiffness. Instability is then identified with a condition of zero stiffness of the strut in its neutral position.

As a preliminary, the strain energy of a beam subjected to both flexure and end loading is calculated. For a straight beam lying along the x-axis and deflected upwards by amount w at any point, the bending stress in this section will be given by

$$\sigma_x = -Ez \frac{d^2 w}{dx^2}$$

where z is measured upwards from the centroidal plane. If a compressive load Q is now applied, it will set up an average stress $\sigma_x = -Q/A$ over the area of section A. The strain energy ΔU in length δx of the beam can now be found from the total stress at any point on the section,

$$\Delta U = \frac{1}{2E} \int \left(Ez \frac{d^2 w}{dx^2} + \frac{Q}{A} \right)^2 dA$$

However, as z is measured from the centroidal plane we have

$$\int z \, dA = 0 \qquad \int z^2 \, dA = I$$

where I is the second moment of area about the centroidal axis of the section. With these values inserted, the expression for total strain energy shows that this energy is simply the sum of the separate energies for flexure and end loading.

4.3.1 Change of length during bending

A strut is a member in which non-linear response must be looked for. Clearly, doubling both lateral and longitudinal loads will give deflections that are more than doubled if the strut is brought near to a state of collapse. For a constant end load, the response to a lateral load will indeed be linear, but the superposition principle need not apply for two or more separate lateral loads. This is because the longitudinal displacement of any point varies as the square of a lateral displacement imposed at any other point.

Consider any portion of a beam of length δx lying initially in position OA as shown in Fig. 4.6(a). After deflection of the beam, this element moves to position $O'A'$ inclined at angle ϕ. If the length projected on the x-axis remains constant, the true length must increase by amount δs given by $(\delta x + \delta s)/\delta x = \sec \phi$. But the slope is given by $\tan \phi = dw/dx$. Therefore, to the first approximation, the increase in length is given by

$$\delta s = \tfrac{1}{2}\left(\frac{dw}{dx}\right)^2 \delta x$$

The effect of compliance of longitudinal end constraint is now examined. Consider a strut of longitudinal stiffness k_1, loaded by means of a spring of stiffness k_2 so that the initial load is Q. To avoid introducing the idea of flexure at this stage, the strut may be assumed to have one or more joints, as shown in Fig. 4.6(b). Suppose that a small lateral deflection is accompanied by a small reduction δQ in the end load. This by itself will allow the strut to lengthen

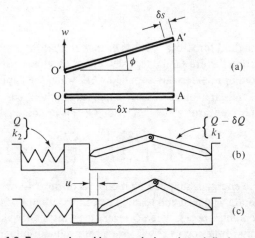

Fig. 4.6 Energy released by strut during a lateral displacement

elastically by amount $\delta Q/k_1$. However, if the ends are held a fixed distance apart, the amount of lengthening is defined by the deflected shape, as already mentioned, giving the equation

$$\frac{\delta Q}{k_1} = \frac{1}{2} \int \left(\frac{dw}{dx}\right)^2 dx$$

The initial strain energy of the strut, $U = \frac{1}{2} Q^2/k_1$, is therefore reduced by amount

$$\delta U = Q \frac{\delta Q}{k_1} = \frac{Q}{2} \int \left(\frac{dw}{dx}\right)^2 dx \tag{4.19}$$

and this energy appears as work done against the forces restraining lateral displacement of the strut. If the end constraint is now relaxed, the end will move through some small distance u, as shown in Fig. 4.6(c). Although this movement allows energy to be transferred from the spring to the strut, the energy released from the elastic system as the external constraining force is withdrawn amounts merely to $\frac{1}{2}u\delta Q$, and as this is of a lower order of magnitude than the energy given by (4.19), it can be neglected.

It appears, therefore, that the elastic energy released during a lateral deflection is not affected by stiffness of longitudinal end constraints. When the constraints are very stiff, this energy comes from a depletion of strain energy in the strut. When the stiffness tends to zero, as when a constant load is applied, the strut keeps its strain energy constant while the energy released is supplied by movement of the end load. Regardless of the degree of constraint, the energy released depends only on the initial compressive end load Q, as shown by (4.19). This energy released will count towards the energy that must be supplied by lateral forces acting on the strut in order to set up bending stresses. Hence a lesser lateral force is needed to cause a given lateral deflection, and the stiffness appears to be reduced.

4.3.2 Energy method for struts   It has been seen that the energy released by the end load during flexure can be regarded either as a loss of strain energy when the ends are fixed or as externally supplied work when the applied load remains constant. If the ends are considered to be longitudinally fixed, the strain energy which is gained by the strut while deflection $w(x)$ is imposed is given by

$$U = \frac{1}{2} \int \left[EI \left(\frac{d^2 w}{dx^2}\right)^2 - Q\left(\frac{dw}{dx}\right)^2 \right] dx \tag{4.20}$$

provided that no energy is supplied by moments or shear forces acting at the ends. For moderate values of end load Q, a positive amount of energy is absorbed in setting up bending stresses. When Q reaches the instability value, it is found that one particular function $w(x)$ will give a zero value of integral (4.20). This condition implies an exact balance between energy absorbed by bending stresses and energy released by reduction of compressive stress.

As longitudinal force Q is bound to be constant along the strut, its value at the condition of instability, that is, the condition when deflection results in no change of strain energy, can be deduced from (4.20),

$$Q = \frac{\int\limits_0^L EI\left(\frac{d^2 w}{dx^2}\right)^2 dx}{\int\limits_0^L \left(\frac{dw}{dx}\right)^2 dx}$$

For using this expression, a function for deflection $w(x)$ must be selected which satisfies restrictions placed on deflection, gradient, or bending moment at the two ends of the strut. Any remaining parameters appearing in the final expression for instability load Q can then be adjusted to give the least value of Q.

4.3.3 Variational equation for struts

When a single lateral load acts on a beam carrying end load, the shape assumed by the beam can be found by imposing a fixed deflection at the point of application of lateral load, and then finding the deflections at other points that give minimum strain energy. When several point loads or distributed loads act, a variational method is needed. The method previously used for a beam without end load can be followed, but more particular attention must be given to shear forces acting at the ends. Shear force is understood to act in a direction at right angles to the centre line of the beam. As the centre line may be inclined to the horizontal, an external end load Q applied horizontally will have a component in the direction of the shear force, and this will be additive to any external vertical force V, as shown in Fig. 4.7. Hence, shear force is given to sufficient accuracy by $S = V + Q\, dw/dx$. Shear force can now be expressed as a derivative of w,

$$S = -\frac{d}{dx}\left(EI\frac{d^2 w}{dx^2}\right)$$

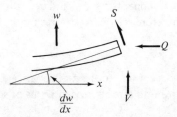

Fig. 4.7 Contribution of end load to shear force

for the purpose of finding the work supplied by vertical force V during a variation in deflection. This work is given by

$$V\delta w = -\delta a_n \left\{ \frac{d}{dx} \left(EI \frac{d^2 w}{dx^2} \right) + Q \frac{dw}{dx} \right\} w'$$

where $w' = \partial w / \partial a_n$ and a_n is a variable parameter of deflection.

This modified expression can now be inserted in the previous analysis to find the variation in external work. The variation in strain energy must now be found from (4.20), which is valid for any degree of longitudinal constraint of the ends. Hence the variational equation is found to be

$$\int \left[\frac{d^2}{dx^2} \left(EI \frac{d^2 w}{dx^2} + Qw \right) - p \right] w' \, dx = 0$$

It is noted that the equilibrium condition is expressed by a zero value of the term in square brackets.

This variational equation may be used for obtaining the value of Q for instability. With lateral pressure p made equal to zero, a function $w(x)$ in series form may be inserted. All the parameters may be determined except one, and this remains as an arbitrary multiplier, as the amount of deflection is indeterminate at the point of instability.

It is always assumed in applying these theoretical methods that the material of the strut is linear-elastic, and that the strut is so slender as to collapse by elastic instability under stresses that are low in relation to the compressive yield stress of the material. In practice, such slender struts are rarely used, and much attention has been given to the behaviour of struts in the transitional range between purely elastic instability and undirectional plastic yielding.

4.4 Flexure of Plates

First, an expression for the strain energy in an element of a plate is derived. Consider an initially flat plate lying in the (x,y) plane. If the plate is thin in relation to its dimensions measured in its plane, it can be assumed that flexure gives rise only to stress components σ_x, σ_y, and σ_{xy}, the other components being negligible. Strains are therefore related to stresses by the elasticity equations

$$Ee_x = \sigma_x - v\sigma_y \qquad Ee_y = \sigma_y - v\sigma_x \qquad 2Ge_{xy} = \sigma_{xy}$$

where e_{xy} is shear strain according to the mathematical definition. These equations can be rearranged to give stresses in terms of strains

$$\sigma_x = E'(e_x + ve_y) \qquad \sigma_y = E'(e_y + ve_x) \qquad \sigma_{xy} = (1 - v) E' e_{xy}$$

where $E' = E/(1 - v^2)$, and G is replaced by $E/2(1 + v)$. These values can now be inserted into (4.2), i.e.,

$$\Delta U = \tfrac{1}{2}[\sigma_x e_x + \sigma_y e_y + 2\sigma_{xy} e_{xy}]$$

to give strain energy ΔU per unit volume in terms of strains,

$$\Delta U = \frac{E'}{2} [e_x^2 + e_y^2 + 2v e_x e_y + 2(1 - v) e_{xy}^2]$$

Expressions (3.11) and (3.12) giving strains in terms of deflection derivatives are now introduced,

$$e_x = -z \frac{\partial^2 w}{\partial x^2} \qquad e_y = -z \frac{\partial^2 w}{\partial y^2} \qquad e_{xy} = -z \frac{\partial^2 w}{\partial x \, \partial y}$$

When strain energy per unit volume is now integrated through the plate thickness t, that is, from $z = -t/2$ to $z = t/2$, the total strain energy U becomes

$$U = \frac{D}{2} \int \left[\left(\frac{\partial^2 w}{\partial x^2} \right)^2 + \left(\frac{\partial^2 w}{\partial y^2} \right)^2 + 2v \frac{\partial^2 w}{\partial x^2} \frac{\partial^2 w}{\partial y^2} + 2(1 - v) \left(\frac{\partial^2 w}{\partial x \, \partial y} \right)^2 \right] dA$$

the integral being taken over total area A of the plate, which is taken to have uniform thickness, and a flexural stiffness constant $D = E' t^3 / 12$.

4.4.1 Variational equation for plates   The last expression, giving strain in a plate, can be slightly rearranged,

$$U = \frac{D}{2} \int \int \left\{ (\nabla^2 w)^2 - 2(1 - v) \left[\frac{\partial^2 w}{\partial x^2} \frac{\partial^2 w}{\partial y^2} - \left(\frac{\partial^2 w}{\partial x \, \partial y} \right)^2 \right] \right\} dx \, dy \quad (4.21)$$

As would be expected on physical grounds, this integrand consists of two invariant functions. If deflection w, expressed in terms of parameters a_n, is now varied by amount $\delta w = w' \, \delta a_n$, differentiation of (4.21) with respect to a_n gives

$$U' = D \int \int \nabla^2 w' \nabla^2 w \, dx \, dy$$

$$-(1 - v) D \int \int \left[\frac{\partial^2 w'}{\partial x^2} \frac{\partial^2 w}{\partial y^2} + \frac{\partial^2 w}{\partial x^2} \frac{\partial^2 w'}{\partial y^2} - 2 \frac{\partial^2 w'}{\partial x \, \partial y} \frac{\partial^2 w}{\partial x \, \partial y} \right] dx \, dy \quad (4.22)$$

This equation may be used as it stands for solving problems of plate flexure, but it is possible to reduce it to a shorter form, as was done with the equation for beams.

Work done at the boundary of the plate is now considered. A boundary is chosen so that it lies well within the outer edges of the plate, so that special conditions prevailing at these edges need not be considered at this stage. Using coordinates (n,t) normal and tangential to the chosen boundary at any point, the bending moment, twisting moment, and shear force at any point on the boundary are written M_{nn}, M_{nt}, and S_n. With these forces remaining constant, a variation is imposed on deflection w, so that the additional work done by these forces is obtained by integrating around the boundary

$$W' \, \delta a_n = \delta a_n \int \left[M_{nn} \frac{\partial w'}{\partial n} + M_{nt} \frac{\partial w'}{\partial t} + S_n w' \right] dt \qquad (4.23)$$

Forces acting on the boundary are now expressed as derivatives of w, in forms following (3.18),

$$M_{nn} = D \left[\nabla^2 w - (1 - \nu) \frac{\partial^2 w}{\partial t^2} \right] \qquad M_{nt} = (1 - \nu) D \frac{\partial^2 w}{\partial n \, \partial t}$$

$$S_n = - D \frac{\partial}{\partial n} \nabla^2 w$$

Parts of integral (4.23) that do not contain the factor $(1 - \nu)$ are taken first, the corresponding work being denoted W'_1,

$$W'_1 = D \int \left(\nabla^2 w \frac{\partial w'}{\partial n} - w' \frac{\partial}{\partial n} \nabla^2 w \right) \, dt$$

Derivatives taken in direction n normal to the boundary can be considered as vector components, so the divergence theorem can be applied to change this line integral into the surface integral

$$W'_1 = D \iint \left[\frac{\partial}{\partial x} \left(\nabla^2 w \frac{\partial w'}{\partial x} - w' \frac{\partial}{\partial x} \nabla^2 w \right) + \frac{\partial}{\partial y} \left(\nabla^2 w \frac{\partial w'}{\partial y} - w' \frac{\partial}{\partial y} \nabla^2 w \right) \right] dx \, dy$$

$$= D \iint (\nabla^2 w' \nabla^2 w - w' \nabla^4 w) \, dx \, dy \qquad (4.24)$$

Parts of integral (4.23) containing the factor $(1 - \nu)$ are now collected. This part of the work done at the boundary is denoted W'_2, and is obtained by integrating around the boundary,

$$W'_2 = - (1 - \nu) D \int \left(\frac{\partial^2 w}{\partial t^2} \frac{\partial w'}{\partial n} - \frac{\partial^2 w}{\partial n \partial t} \frac{\partial w'}{\partial t} \right) dt$$

This is rearranged in the form

$$W'_2 = -(1 - \nu) D \int \left[\nabla^2 w \frac{\partial w'}{\partial n} - \left(\frac{\partial^2 w}{\partial n^2} \frac{\partial w'}{\partial n} + \frac{\partial^2 w}{\partial n \partial t} \frac{\partial w'}{\partial t} \right) \right] dt$$

The two parts of this integral can be regarded as components of a vector measured in the n-direction, so the divergence theorem may be used to give the surface integral

$$W_2' = -(1 - v)D \iint \left[\frac{\partial}{\partial x}\left(\nabla^2 w \frac{\partial w'}{\partial x} - \frac{\partial^2 w}{\partial x^2}\frac{\partial w'}{\partial x} - \frac{\partial^2 w}{\partial x \partial y}\frac{\partial w'}{\partial y} \right) \right.$$

$$\left. + \frac{\partial}{\partial y}\left(\nabla^2 w \frac{\partial w'}{\partial y} - \frac{\partial^2 w}{\partial y^2}\frac{\partial w'}{\partial y} - \frac{\partial^2 w}{\partial x \partial y}\frac{\partial w'}{\partial x} \right) \right] dx\,dy$$

$$= -(1 - v)D \iint \left[\frac{\partial^2 w}{\partial x^2}\frac{\partial^2 w'}{\partial y^2} + \frac{\partial^2 w}{\partial y^2}\frac{\partial^2 w'}{\partial x^2} - 2\frac{\partial^2 w}{\partial x \partial y}\frac{\partial^2 w'}{\partial x \partial y} \right] dx\,dy \quad (4.25)$$

Finally, the work supplied by distributed pressure p and point loads P_m may be written as W_3',

$$W_3' = \iint pw'\,dx\,dy + \sum P_m w_m'$$

This is now added to other work quantities given in (4.24) and (4.25), to give the total work $W' = W_1' + W_2' + W_3'$ and this may be combined with the variation in strain energy (4.22) to give the variational equation

$$U' - W' = \iint (D\nabla^4 w - p)w'\,dx\,dy - \sum P_m w_m' = 0 \quad (4.26)$$

the surface integral being taken over the area enclosed by the chosen contour.

If the chosen contour excludes points where concentrated forces P_m act, the integral must be zero for any variation of deflection, which means that the equation $D\nabla^4 w = p$ must hold good at all points within the contour. Alternatively, in the absence of distributed pressure p, the equation $\nabla^4 w = 0$ must be satisfied inside any contour such as that marked C in Fig. 4.8, which excludes both isolated point loads such as P_1 and P_2 and line distributions of load F. Inside this contour the deflection w which solves this differential equation will be a con-

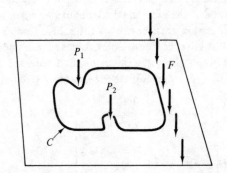

Fig. 4.8 Range of validity of differential equation

tinuous function, and its derivatives up to the third will be continuous. As the contour can be of any shape, it appears that the differential equation virtually holds over the whole of the plate. However, the solutions will necessarily have third derivatives which reflect the infinite shear force per unit length underneath a concentrated load, or the discontinuity of shear force across a line distribution of applied force. In practice, this means that convenient coordinates must be selected for describing the positions of these infinite point values. Also, when a line distribution of force extends completely across the plate, or when such a line forms a closed loop, different solutions must be selected for regions on either side of the line. These solutions must then be matched to give continuity of deflection, slope, and bending moment across the line, with the required discontinuity in shear force.

The advantage of the variational method is that it avoids these difficulties associated with exact solutions of the differential equation, as a single continuous function is selected to describe deflections over the whole plate. When this function has been adjusted to fit conditions of flexural constraint at the boundaries, the remaining parameters can be determined by consecutive applications of (4.26). In general, this solution will not precisely satisfy conditions of internal equilibrium at all points on the plate. It should be mentioned that while a fairly accurate expression for deflection can often be obtained by varying only one or two parameters, the derivatives of this expression will be subject to larger deviations from the precisely correct solution. Hence the values of stress will not be quite so accurate as values of deflection. This is obviously so at the positions of point loads, where the exact solution gives infinite stress, but a continuous function for deflection found by the variational method gives a finite stress.

4.4.2 Rectangular plate The variational equation (4.26) was derived by considering work done at a boundary of general shape lying within the extreme edges of a plate. It may be asked whether the same result is obtained by using a boundary coinciding with the outer edges. At the edges of a plate it must be assumed that shear stress σ_{nt} disappears, even though it may have a non-zero value at a distance inside of the edge equal to one or two thicknesses of the plate. Hence, a modified expression must be used for the shear force V_n which is externally applied at the edge, and also, point forces P_c will appear at right-angled corners, as previously discussed in chapter 3. The forces acting on elements of the edge of a rectangular plate are therefore given in terms of w by

$$M_{nn} = D\left[\nabla^2 w - (1-\nu)\frac{\partial^2 w}{\partial t^2}\right] \qquad P_c = 2(1-\nu)\,D\,\frac{\partial^2 w}{\partial x\,\partial y}$$

$$V_n = -D\left[\frac{\partial}{\partial n}\nabla^2 w + (1-\nu)\frac{\partial}{\partial t}\frac{\partial^2 w}{\partial n\,\partial t}\right]$$

A variation $w'\delta a_n$ imposed on the deflected shape now produces a change of work supplied at the boundaries which may be expressed in two parts, W_1' and W_2',

$$W_1' = D \int \left(\nabla^2 w \frac{\partial w'}{\partial n} - w' \frac{\partial}{\partial n} \nabla^2 w \right) dt$$

$$W_2' = -(1-v) D \int \left(\frac{\partial^2 w}{\partial t^2} \frac{\partial w'}{\partial n} + w' \frac{\partial}{\partial t} \frac{\partial^2 w}{\partial n \partial t} \right) dt + 2(1-v) D \sum w_c' \left[\frac{\partial^2 w}{\partial x \partial y} \right]_c$$

where subscript c denotes values at the corners of the plate. The first part W_1' gives the result previously expressed in (4.24), so attention may be given to the second part W_2'. If the coordinates of the corners are x_1, x_2, y_1, and y_2, this part can be written in the form of a definite integral

$$W_2' = -(1-v) D \left\{ \int \left[\frac{\partial^2 w}{\partial y^2} \frac{\partial w'}{\partial x} + w' \frac{\partial}{\partial y} \frac{\partial^2 w}{\partial x \partial y} \right]_{x_1}^{x_2} dy \right.$$

$$\left. + \int \left[\frac{\partial^2 w}{\partial x^2} \frac{\partial w'}{\partial y} + w' \frac{\partial}{\partial x} \frac{\partial^2 w}{\partial x \partial y} \right]_{y_1}^{y_2} dx - 2 \left[\left[w' \frac{\partial^2 w}{\partial x \partial y} \right]_{x_1}^{x_2} \right]_{y_1}^{y_2} \right\}$$

This is evidently a partly integrated form of the surface integral

$$W_2' = -(1-v) D \int \int \left[\frac{\partial}{\partial x} \left(\frac{\partial^2 w}{\partial y^2} \frac{\partial w'}{\partial x} + w' \frac{\partial}{\partial y} \frac{\partial^2 w}{\partial x \partial y} \right) \right.$$

$$\left. + \frac{\partial}{\partial y} \left(\frac{\partial^2 w}{\partial x^2} \frac{\partial w'}{\partial y} + w' \frac{\partial}{\partial x} \frac{\partial^2 w}{\partial x \partial y} \right) - 2 \frac{\partial^2}{\partial x \partial y} \left(w' \frac{\partial^2 w}{\partial x \partial y} \right) \right] dx \, dy$$

which on rearrangement is found to be identical with the expression in (4.25). As this analysis is not different from the last in any other respect, the resulting variational equation is exactly the same as that already found, i.e., (4.26). Hence the same conclusions may be drawn from this equation when the chosen contour coincides with the outer edges of a rectangular plate. Incidentally, this argument confirms the Kirchhoff boundary condition (3.33) for the edge of a plate.

4.4.3 Simply supported rectangular plate It is stipulated here that deflection is zero at all points on the boundary. Bending moment acting parallel to the boundary must also be zero. However, if deflection w is zero, the curvature in a direction parallel to the boundary must be zero. Hence the bending moment condition is satisfied by making the curvature normal to the boundary zero. Two simple conditions are therefore imposed on any suggested function for w. Shear forces are not statically determinate, so they are not taken into account.

While polynomial series may be used, Fourier series have the advantages that boundary conditions are easily satisfied, and that the integration process is simple. Coordinates x and y are measured from the centre of the rectangle parallel to the sides of length a and b. To consider a specific problem, a uniform pressure p is applied, giving a deflected shape which is expected to be symmetrical about the coordinate axes. Such a shape is expressed by the series

$$w = \frac{pa^4}{D} \sum \sum A_{mn} \cos m\pi \frac{x}{a} \cos n\pi \frac{y}{b} \quad \begin{pmatrix} m = 1, 3, 5, \ldots \\ n = 1, 3, 5, \ldots \end{pmatrix}$$

Assuming that $b > a$, sufficient accuracy for most purposes is obtained by retaining the first term only in x and the first two terms in y,

$$w = \frac{pa^4}{D} \cos \pi \frac{x}{a} \left(A_1 \cos \pi \frac{y}{b} + A_3 \cos 3\pi \frac{y}{b} \right)$$

The best values of A_1 and A_3 may be found by applying a variational equation of the form of (4.26),

$$\iint (D\nabla^4 w - p) \frac{\partial w}{\partial A_n} \, dx \, dy = 0 \quad (n = 1, 3)$$

Utilizing the fact that the definite integral

$$\int_{-a/2}^{a/2} \cos k\pi \frac{x}{a} \cos n\pi \frac{x}{a} \, dx \quad (n = 1, 3, \ldots)$$

has value $a/2$ for $k = n$, and value zero for $k \neq n$, the following values of the constants may be found

$$A_1 = \frac{16}{\pi^6} \Bigg/ \left(1 + \frac{a^2}{b^2} \right)^2 \qquad A_3 = -\frac{16}{3\pi^6} \Bigg/ \left(1 + 9 \frac{a^2}{b^2} \right)^2$$

The deflection when dimension b is infinitely long is best found by repeating the calculation with an expression for w which is not a function of y,

$$w = A_1 \frac{pa^4}{D} \cos \frac{x}{a}$$

which leads to a value $A_1 = 4/\pi^5$.

Expressing the maximum deflection, which occurs at the centre of the plate, in the form

$$w(0, 0) = \alpha pa^4 / D$$

the following numerical values of coefficient α are obtained

b/a	1·0	1·5	2·0	3·0	4·0	∞
α	0·0041	0·0077	0·0101	0·0121	0·0125	0·0131

These values agree to within about 2% with values found by more exact series solutions of the biharmonic equation. The maximum bending stress, which occurs in the middle of the plate is given by

$$\sigma_x = \tfrac{1}{2}E' t \left(\frac{\partial^2 w}{\partial x^2} + \nu \frac{\partial^2 w}{\partial y^2}\right)$$

where t is the plate thickness.

4.4.4 Rectangular plate with built-in edges
As before, an expression for deflection is selected containing the first two terms in y, as the longer side of the plate of length b is parallel to this direction:

$$w = \frac{pa^4}{D}\left[1 + \cos 2\pi \frac{x}{a}\right]\left[A_2\left(1 + \cos 2\pi \frac{y}{b}\right) + A_4\left(1 - \cos 4\pi \frac{y}{b}\right)\right]$$

This expression gives zero deflection at all points on the boundary, and also zero gradient normal to the boundary. When the variational equation is applied for variations in coefficients A_2 and A_4, simultaneous equations are obtained,

$$\left(3 + 2\frac{a^2}{b^2} + 3\frac{a^4}{b^4}\right)A_2 + 2A_4 = \frac{1}{4\pi^4}$$

$$2A_2 + \left(3 + 8\frac{a^2}{b^2} + 48\frac{a^4}{b^4}\right)A_4 = \frac{1}{4\pi^4}$$

For any given ratio b/a, these can be solved to obtain maximum deflection in terms of numerical coefficient α as before. Some computed values are shown below.

b/a	1·0	1·2	1·5	2·0	∞
α	0·00125	0·00169	0·00213	0·00242	0·00257

Of these values, that for $b/a = 2 \cdot 0$ is least accurate, being about 5% below the correct value. Values obtained from series solutions of the differential equation are given by Timoshenko and Woinowski-Krieger, *Theory of Plates and Shells*, p. 202. It is seen from the table that for values of $b/a > 2$, the value of α may be taken as that corresponding to an infinite ratio b/a.

An advantage of this variational method is that it can deal with irregular load distributions. By using relatively simple expressions for deflected shape and by carrying out numerical integrations if necessary, numerical results can be found which might be difficult to find by other means.

4.4.5 Circular plate

When the boundary on which loading is specified is of a circular shape, polar coordinates (r,θ) are required. To obtain an expression for strain energy, the x and y derivatives in the general expression (4.21) can be replaced by r and θ derivatives, as discussed in chapter 1.

If the total strain energy U is split into two parts U_1 and U_2, these may be written

$$U_1 = \frac{D}{2} \int\int (\nabla^2 w)^2 \, r \, dr \, d\theta$$

$$U_2 = -(1-v) D \int\int \left[\frac{\partial^2 w}{\partial r^2} \left(\frac{1}{r^2} \frac{\partial w}{\partial r} + \frac{1}{r^2} \frac{\partial^2 w}{\partial \theta^2} \right) - \left(\frac{\partial^2 w}{\partial r \, \partial \theta \, r} \right)^2 \right] r \, dr \, d\theta \quad (4.27)$$

For some variation in deflection $\delta w = w' \delta a_n$, these expressions may be differentiated, and for convenience in the subsequent calculation, the second expression is considerably rearranged in the following manner,

$$U_2' = -(1-v) D \int\int \left\{ \frac{\partial}{\partial r} \left[\frac{\partial w'}{\partial r} \left(\frac{\partial w}{\partial r} + \frac{1}{r} \frac{\partial^2 w}{\partial \theta^2} \right) + w' \frac{\partial}{\partial \theta} \frac{\partial^2 w}{\partial r \, \partial \theta \, r} \right] \right.$$

$$\left. + \frac{\partial}{\partial \theta} \left[\frac{1}{r} \frac{\partial w'}{\partial \theta} \frac{\partial^2 w}{\partial r^2} + w' \frac{\partial}{\partial r} \frac{\partial^2 w}{\partial r \, \partial \theta \, r} \right] - 2 \frac{\partial^2}{\partial r \, \partial \theta} \left[w' \frac{\partial^2 w}{\partial r \, \partial \theta \, r} \right] \right\} dr \, d\theta \quad (4.28)$$

In the following derivation of a variational equation, a general result is obtained by examining a partial annulus subtending some angle less than 2π. Positive bending moments M_{rr} and $M_{\theta\theta}$ and positive applied shear forces V_r and V_θ are shown in Fig. 4.9. Also shown are the positive concentrated forces P_c that may be present at the right-angled corners. These forces are given in terms of derivatives by the equations

$$M_{rr} = D\left[\nabla^2 w - (1-\nu)\left(\frac{1}{r}\frac{\partial w}{\partial r} + \frac{1}{r^2}\frac{\partial^2 w}{\partial \theta^2}\right)\right]$$

$$M_{\theta\theta} = D\left[\nabla^2 w - (1-\nu)\frac{\partial^2 w}{\partial r^2}\right]$$

$$V_r = -D\left[\frac{\partial}{\partial r}\nabla^2 w + (1-\nu)\frac{1}{r}\frac{\partial}{\partial \theta}\frac{\partial^2}{\partial r\partial\theta}\frac{w}{r}\right] \qquad (4.29)$$

$$V_\theta = -D\left[\frac{1}{r}\frac{\partial}{\partial \theta}\nabla^2 w + (1-\nu)\frac{\partial}{\partial r}\frac{\partial^2}{\partial r\partial\theta}\frac{w}{r}\right]$$

$$P_c = 2(1-\nu)D\frac{\partial^2}{\partial r\partial\theta}\frac{w}{r}$$

At the curved boundaries of radii r_1 and r_2 and at the radial boundaries at angles θ_1 and θ_2, a variation in deflection produces a variation in work represented by

$$W' = \int\left[r\left(\frac{\partial w'}{\partial r}M_{rr} + w'V_r\right)\right]_{r_1}^{r_2} d\theta + \int\left[\frac{1}{r}\frac{\partial w'}{\partial \theta}M_{\theta\theta} + w'V_\theta\right]_{\theta_1}^{\theta_2} dr + [[w'P_c]_{r_1}^{r_2}]_{\theta_1}^{\theta_2}$$

When these boundary forces are replaced by the derivatives of w already specified, this expression is split into two parts W'_1 and W'_2, the first being independent of the factor $(1-\nu)$. The first reduces to the surface integral

$$W'_1 = D\int\int\left\{\frac{\partial}{\partial r}\left[r\frac{\partial w'}{\partial r}\nabla^2 w - rw'\frac{\partial}{\partial r}\nabla^2 w\right] + \frac{\partial}{\partial \theta}\left[\frac{1}{r}\frac{\partial w'}{\partial \theta}\nabla^2 w - w'\frac{\partial}{\partial r}\nabla^2 w\right]\right\} dr\,d\theta$$

$$= D\int\int(\nabla^2 w'\,\nabla^2 w - w'\nabla^4 w)r\,dr\,d\theta$$

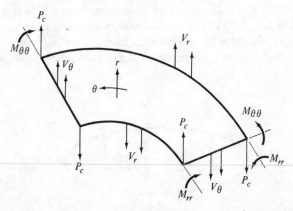

Fig. 4.9 Positive boundary forces acting on a sector of an annulus

By a similar procedure, the second part W_2' made up of terms containing factor $(1 - \nu)$ may be collected and expressed as a surface integral. This gives an expression identical with (4.28), which shows that $U_2' - W_2' = 0$.

When total values of strain energy and work are taken, they give the variational equation

$$U' - W' = \iint (D\nabla^4 w - p) w' \, r \, dr \, d\theta - \sum P_m w_m' = 0 \qquad (4.30)$$

This equation could have been deduced from the cartesian form (4.26) simply by changing coordinates. For a complete annulus, the derivation can be simplified by using the condition that all quantities must be single-valued with respect to θ. Since (4.30) is derived by eliminating work done at the boundary, this equation will not solve problems in which forces act only at the boundary, when it may be necessary to use an equation containing general expressions for energy and work.

As an application of this method, the deflection of a plate under four point loads is considered. The loads act at radius γR on two diameters at right angles, the radius of the plate being R, as shown in Fig. 4.10. An expression for deflection

$$w = (a_2 \rho^2 + a_3 \rho^3 + a_4 \rho^4) \cos 2\theta$$

is chosen, where ρ is radius ratio r/R. Expressions for bending moment M_{rr} and applied transverse force V_r as in (4.29) are used to make these values zero at the outer edge, so that two of the constants are expressed in terms of the third, i.e., $a_3 = -0.577 a_2$ and $a_4 = 0.157 a_2$. A value $\nu = 0.3$ is assumed. With these values, the expression for deflection is inserted into (4.30) to obtain

$$a_2 = 0.667 \frac{PR^2}{D} (\gamma^2 - 0.577 \gamma^3 + 0.157 \gamma^4)$$

In the particular case of loading at the outer edge, i.e., $\gamma = 1$, a deflection amplitude $w(R, 0) = 0.224 \, PR^2/D$ is found. The numerical coefficient may be compared with a value 0.25 found from a solution of the differential equation of flexure. The calculation described is obviously rough, as it does not allow the deflected shape of a radial line to change as radius of loading γ is increased from 0 to 1. For better accuracy, further terms in the polynomial for deflection are needed.

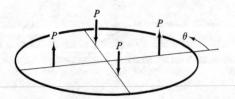

Fig. 4.10 Four-point loading of a disc

A calculation of this kind is very much simpler than one in which the equation $\nabla^4 w = 0$ is satisfied. The latter method requires separate solutions for the regions $\rho < \gamma$ and $\rho > \gamma$. These have to be matched at $\rho = \gamma$ to give continuity of bending moment, with a specified discontinuity in shear force which is expressed as a Fourier series, as mentioned in the previous chapter.

4.5 Stability of Plates

In the last section, attention was restricted to bending stresses, which vary from a positive value on one surface of a plate to a numerically equal negative value on the opposite surface. Strain energy was calculated from these stresses only. Attention must now be given to stresses which are constant through the thickness, denoted σ_x^*, σ_y^*, and σ_{xy}^*. An additional amount of strain energy is associated with these stresses while the plate is in a perfectly flat condition. It may be calculated independently of bending stress energy, the total being simply the sum of the strain energies due to the bending stress system and the plane stress system. However, the initial value of strain energy associated with the plane stress system is not of primary importance here, as the effects to be studied are produced by changes in this stored energy which occur as the plate is deflected from its initial state of flatness. Regardless of whether the plane stress system consists of internal stresses or stresses imposed by external loads acting on the edges of the plate, work will be done in overcoming the resistance of these stresses as elements of the plate stretch or contract during flexure.

4.5.1 Interaction energy
Considering an element of length δx in the x-direction, flexure will produce some gradient $\partial w / \partial x$ in this direction. The amount of stretch δs, as shown in Fig. 4.6(a) was previously found to be $\delta s = \frac{1}{2}(\partial w / \partial x)^2 \, \delta x$. An additional strain e_x' is therefore set up during flexure, given by

$$e_x' = \frac{1}{2}\left(\frac{\partial w}{\partial x}\right)^2 \tag{4.31}$$

The additional strain in the y-direction is given by a similar expression, with y written in place of x.

To find the additional shear strain produced, suppose that an element is first rotated about the x-axis through angle $\partial w / \partial y$, as shown in Fig. 4.11. This raises edge AB to a position A$'$B$'$ at a height δw above the (x,y) plane, given by $\delta w = \delta y(\partial w / \partial y)$, where δy is the side of the element measured in the y-direction. When the element is subsequently rotated about the y-axis through angle $\partial w / \partial x$, the edge A$'$B$'$ moves to position A$''$B$''$, and all points on this edge move distance δu in the negative x-direction given by $\delta u = \delta w(\partial w / \partial x)$. During these movements, the element has been regarded as a rigid body, but as part of a plate, the element will not have this freedom of movement. To restore its position so that line OA$''$

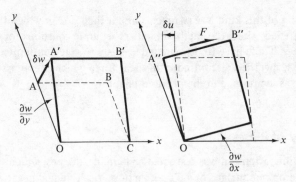

Fig. 4.11 Second-order shear strain

lies vertically over the y-axis, the movement δu must be reversed, setting up a positive shear strain $e'_{xy} = \frac{1}{2}\delta u/\delta y$, given by

$$e'_{xy} = \tfrac{1}{2}\frac{\partial w}{\partial x}\frac{\partial w}{\partial y} \tag{4.32}$$

These strains caused by flexure, which are uniform through the plate thickness t, are set up while coplanar stresses act at their full values, so the strain energy of a volume element is increased by amount

$$\varDelta U_I = \sigma_x^* e'_x + \sigma_y^* e'_y + 2\sigma_{xy}^* e'_{xy}$$

For the whole plate, this interaction energy is obtained by inserting strains from (4.31) and (4.32) and integrating through the whole volume,

$$U_I = \frac{t}{2}\int\int\left[\sigma_x^*\left(\frac{\partial w}{\partial x}\right)^2 + \sigma_y^*\left(\frac{\partial w}{\partial y}\right)^2 + 2\sigma_{xy}^*\frac{\partial w}{\partial x}\frac{\partial w}{\partial y}\right]dx\,dy \tag{4.33}$$

This energy must be supplied in addition to energy required for setting up bending stresses. When the direct stresses are negative, i.e., compressive, giving negative energy U_I, this means that less energy need be supplied by lateral loads to produce a given deflection. When compressive stresses become high enough, a state of instability will arise in which the energy U_B required to set up bending stresses is exactly supplied by the release of interaction energy, i.e., $U_B + U_I = 0$.

4.5.2 Stability of circular plates When an annulus is complete and contains no radial cuts, continuity in the θ-direction leads to a simplification of the strain energy expression. A particular product in the energy expression (4.27) can be written

$$\frac{\partial^2 w}{\partial r^2}\frac{\partial^2 w}{\partial \theta^2} = \frac{\partial}{\partial \theta}\left(\frac{\partial w}{\partial \theta}\frac{\partial^2 w}{dr^2}\right) - \frac{\partial w}{\partial \theta}\frac{\partial^3 w}{\partial r^2\,\partial \theta}$$

As the first term on the right integrates to zero, the second term may be substituted in the energy integral, which can then be partly integrated from inner radius r_1 to outer radius r_2, so that the total strain energy due to bending becomes

$$U_B = \frac{D}{2} \int_{r_1}^{r_2} \int_{0}^{2\pi} (\nabla^2 w)^2 \, r \, dr \, d\theta \; -(1-\nu)\frac{D}{2}\int_{0}^{2\pi} \left[\left(\frac{\partial w}{\partial r} \right)^2 - \frac{\partial}{\partial r}\left\{ \frac{1}{r}\left(\frac{\partial w}{\partial \theta} \right)^2 \right\} \right]_{r_1}^{r_2} d\theta \tag{4.34}$$

For problems of axial symmetry, this equation and equation (4.33) for interaction energy can be further simplified, so that the criterion for instability, i.e., $U_B + U_I = 0$, takes the form

$$\int_{r_1}^{r_2} \left\{ \left[\frac{1}{r}\frac{d}{dr}\left(r\frac{dw}{dr} \right) \right]^2 + \frac{t}{D}\sigma_r^*\left(\frac{dw}{dr} \right)^2 \right\} r \, dr - (1-\nu)\left[\left(\frac{dw}{dr} \right)^2 \right]_{r_1}^{r_2} = 0 \tag{4.35}$$

If, therefore, the gradient dw/dr is known as a function of r, and an assumption is made about the way in which stress σ_r^* is distributed radially, this equation will give the numerical value of this stress required to cause instability. When the deflected shape is not known initially, a variational form of this equation must be used for determining deflection parameters. However, for the moment, the procedure used is to guess a deflected shape in series form which satisfies any known boundary conditions, and then to insert it into (4.35) to determine the critical level of coplanar stress.

As an example, consider a uniform disc of outside radius R and thickness t, simply supported at its outer edge so that no bending moment is transmitted. A uniform compressive stress p acts radially on the outer edge, and it is required to find the value of this stress that causes buckling. As indicated by Lamé's equations, the radial stress will be uniform throughout the disc, and of value $\sigma_r^* = -p$. An axisymmetrical deflected shape is now selected,

$$w = a_0 + a_2 \rho^2 + a_4 \rho^4$$

where ρ is radius ratio r/R. With $\nu = 0.3$, the condition of zero moment at the edges,

$$\frac{d^2 w}{dr^2} + \frac{\nu}{r}\frac{dw}{dr} = 0$$

eliminates one of the constants, $a_4 = -0.197a_2$. When the expression for gradient is inserted into (4.35) and integration is carried out, the critical compressive stress is found to be $p = 4.2D/R^2 t$. As the rigidity constant is $D = Et^3/12(1 - \nu^2)$, this simplifies to $p = 0.38 \, Et^2/R^2$.

This value is virtually the same as that obtained from the Bessel function solution of the equilibrium equation, described, for example, by F. E. Relton

(1946). This solution gives a radial gradient equal to a constant times $J_1(2 \cdot 05\rho)$. The advantage of the energy method is that it can be applied when geometrical conditions are less elementary. For example, when a disc is clamped over a fraction of its radius between rigid concentric collars, solution of the differential equation becomes difficult, while the energy method remains straightforward.

As a further example, consider a thin solid disc in which sufficient internal stress has been induced to give zero stiffness in axisymmetrical deflection. A deflected shape is assumed consisting of terms in ρ^2, ρ^3, and ρ^4. Bending moment is zero at the outer edge, and also, in this situation, shear force is zero at the outer edge, i.e., $d(\nabla^2 w)/dr = 0$. These conditions lead to an expression for deflection

$$w = a_2(\rho^2 - 0 \cdot 82\,\rho^3 + 0 \cdot 23\rho^4)$$

In principle, any distribution of radial stress may be assumed, unless the plastic strain history of the disc indicates some particular distribution, the only restriction being that this stress must vanish at the outer edge. For simplicity, it is assumed that $\sigma_r^* = -p(1 - \rho)$. Equilibrium then indicates a hoop stress at the rim of value p. When these expressions for stress and deflection are inserted into the instability equation (4.35), the critical level of internal stress is found to be $p = 17D/R^2 t = 1 \cdot 5Et^2/R^2$.

When a more complicated distribution of internal stress is assumed, or when measured values are used, it is a simple matter to evaluate the integral in (4.35) numerically. From the solution of the last problem, it is noted that a given value of internal stress p will affect stiffness to a degree depending on the ratio R^2/t^2, that is, on the relative thinness of the plate. The internal stress cannot, of course, exceed the plastic yield stress of the material. Taking for an example a soft steel sheet with a limiting value of p/E of about $0 \cdot 0015$, it appears that axisymmetric instability cannot result from internal stresses if the diameter-to-thickness ratio is less than about 60. It is noted in passing that when internal stresses of opposite sign are induced in a disc, so that radial stress in the middle is tensile and hoop stress at the rim is compressive, instability appears as a buckling of the rim to a deflection equal to some constant times $\cos 2\theta$.

It is convenient to mention here a method for estimating the effect of small internal stresses on the stiffness of plates, which is useful for either circular or square plates. The deflected shape under a specified type of loading can be first calculated using the differential equation of flexure, and it is assumed that although deflections may be larger or smaller when internal stress is present, the distribution of deflection remains the same as for an initially stress-free plate. In such a plate, let the specified loading produce deflection w_n at a specified point. Stiffness coefficient k_n' can be defined in terms of the strain energy due to bending alone, by the relation

$$U_B = \tfrac{1}{2}k_n' w_n^2$$

For a given set of values of internal stress, the interaction energy may be calculated using the deflected shape for a stress-free plate, and will take the form $U_I = C w_n{}^2$, where C is a calculated coefficient. With internal stress present, the work done by the imposed loads will be given by $W = \frac{1}{2} k_n w_n{}^2$, where k_n is the actual stiffness under these conditions. This stiffness can now be found using the energy equation $W = U_B + U_I$, which gives $k_n = k'_n + 2C$. While this method is exactly correct while internal stresses are small, it becomes less correct with larger stresses and overestimates the stress level that causes instability, that is, the condition of zero stiffness k_n. The obvious reason for this is that the distribution of deflection changes when this condition is approached.

4.5.3 Thin-walled beams

As the primary function of a beam is to sustain a bending moment, the cross-section should be such as to do this with economy of material. According to the elementary theory of bending, a beam made from material of tensile-compressive yield stress Y can sustain a limiting moment $M = YI/h$. In this formula, I is the second moment of area of the section and h is the greatest vertical distance from the centroidal axis to the edge of the section, assuming the moment acts about a horizontal axis. It can be seen that the section should be made as deep as possible, with thin walls, so as to obtain the largest I/h value for a given area of section. A limit is reached in this direction when the section becomes unstable, and buckling occurs.

As a rather over-simplified example, consider a channel section having a base of width $2a$ and limbs of height a as shown in Fig. 4.12, the wall thickness being t. The applied moment M gives compressive bending stress in the free edges of the section. Buckling would be expected to occur in such a way as to tend to relieve this compressive stress, which means that the free edges will tend to follow a wave of some wave-length L. Taking coordinates as shown, with the lateral deflection of the limb denoted u, a very simple deflected shape may be written

$$u = u_0 \frac{y^2}{a^2} \cos 2\pi \frac{z}{L}$$

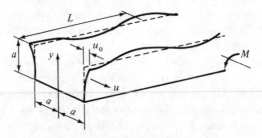

Fig. 4.12 Instability of thin-walled beam

141

This expression does not ensure complete absence of bending moment and shear force along the free edge ($y = a$), and to arrange for these conditions, a more elaborate expression would be needed.

A possible method of analysing this problem will be outlined here, though calculations will not be followed through in detail. Bending strain energy U_B, as given by (4.21), may be calculated for the assumed deflected shape. As this shape is imposed, energy is released due to the presence of bending stress σ_z, and follows from (4.33),

$$U_I = \frac{t}{2} \int\limits_0^L \int\limits_0^a \sigma_z \left(\frac{\partial u}{\partial z} \right)^2 dy\, dz$$

In this problem, it is not difficult to find bending stress in terms of bending moment,

$$\sigma_z = 3M(a - 4y)/5a^3\, t$$

which gives a negative value of U_I. When bending moment M increases to a critical value M_c so as to give $U_B + U_I = 0$, buckling may be expected. This moment may be minimized with respect to wave-length L, which indicates a value of approximately $L = 3a$. The moment can then be expressed as $M_c = CEt^3$, where C is a numerical coefficient of the order of unity.

It is of some practical interest to design a section which will fail simultaneously by buckling and plastic yielding. This condition is expressed by

$$M_c = CEt^3 = BYa^2\, t$$

where the numerical constant B for this particular problem is found to be 5/9. Hence the thickness t, given by

$$\frac{t^2}{a^2} = \frac{B}{C} \frac{Y}{E}$$

is the minimum value to which wall thickness can be profitably reduced. With a ratio $E/Y = 500$, which is representative for steel, this calculation indicates a value a/t very roughly equal to 30, but as the model chosen for the buckling deformation overestimates buckling strength, the limiting thickness will be somewhat larger. For light alloys with relatively low elastic modulus, the limiting thickness will be larger.

When the possibility of buckling arises, the beam should be turned so that the free edges of the section carry tensile rather than compressive bending stress. If the free edges of a section are necessarily exposed to compressive stress, they are sometimes stiffened either by bending over the extreme edge at a right angle, or by welding transverse spacers between the limbs of the section at intervals along the beam.

4.5.4 Variational equation A shorter form of the variational equation will now be derived for a plate subject to flexure while coplanar stresses are acting. This is the two-dimensional form of an equation previously derived for struts. There is a slight advantage in expressing coplanar stresses σ_{ij}^* as force N_{ij} per unit length of a line segment drawn on the plate, so that when the plate is of thickness t, we have

$$N_{ij} = t\sigma_{ij}^*$$

this force being positive when tensile. It is also more concise to use the summation convention for repeated suffices, which allows interaction energy (4.33) to be written

$$U_{\mathrm{I}} = \tfrac{1}{2} \iint N_{ij}\, w,_i\, w,_j\, dx\, dy$$

where a comma indicates differentiation with respect to coordinate x_i and x_i may be put equal to x or y successively. This energy may be regarded as part of the strain energy, and may be added to the flexural strain energy. For a variation δa_n in a parameter of deflection w, an additional change in energy $\delta U_{\mathrm{I}} = U_{\mathrm{I}}'\delta a_n$ is represented by

$$U_{\mathrm{I}}' = \iint N_{ij}\, w',_i\, w,_j\, dx\, dy \qquad (4.36)$$

Attention must now be given to the work done along a contour during a variation in displacement at the contour. As forces N_{ij} are considered to act in the central surface of the plate, they will have upward components due to the inclination of this surface to the (x,y) plane, as shown in Fig. 4.13. The total upward force on an element of edge of tangential length δt is given by

$$R = \left(S_{\mathrm{n}} + N_{\mathrm{nn}}\frac{\partial w}{\partial \mathrm{n}} + N_{\mathrm{nt}}\frac{\partial w}{\partial \mathrm{t}} \right)\delta t$$

where the shear force is given by

$$S_{\mathrm{n}} = -D\frac{\partial}{\partial \mathrm{n}}\nabla^2 w$$

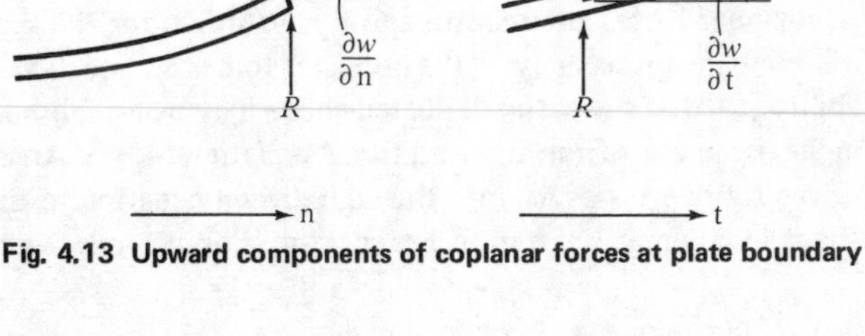

Fig. 4.13 Upward components of coplanar forces at plate boundary

The extra work δW_c supplied by the coplanar forces around the whole contour is therefore given, for variation $w'\,\delta a_n$, by

$$W_c'\,\delta a_n = \delta a_n \int \left(N_{nn}\frac{\partial w}{\partial n} + N_{nt}\frac{\partial w}{\partial t} \right) w'\, dt$$

As the expression to be integrated is in the form of the component of a vector resolved in the n-direction, the divergence theorem can be applied to give the surface integral

$$W_c' = \iint (w'\, N_{ij}\, w,_j),_i \, dx\, dy$$

The expression to be integrated may be expanded thus,

$$(w'\, N_{ij}\, w,_j),_i = w',_i\, N_{ij}\, w,_j + w'\, N_{ij},_i\, w,_j + w'\, N_{ij}\, w,_{ij}$$

As equilibrium of any element requires that $N_{ij,\,i} = 0$, the second term in the last expression is zero, leaving

$$W_c' = \iint (w',_i\, N_{ij}\, w,_j + w'\, N_{ij}\, w,_{ij})\, dx\, dy$$

When this value, and also U_I' from (4.36), representing additional terms due to the presence of coplanar forces, are added to the quantities W' and U' formerly derived for a plate free from coplanar forces, it is seen that the equation for plate flexure, (4.26), needs to be modified by the addition of the term

$$U_I' - W_c' = -\iint w'\, N_{ij}\, w,_{ij}\, dx\, dy$$

giving the new equation

$$\iint (D\nabla^4 w - N_{ij}\, w,_{ij} - p)\, w'\, dx\, dy = 0 \qquad (4.37)$$

If this is to be true for all possible variations, we must have

$$D\nabla^4 w - N_{ij}\, w,_{ij} - p = 0$$

This is the equilibrium equation for a plate containing coplanar stresses, and may be more directly obtained by examining the equilibrium of an element under its vertical forces and moments.

Clearly, a solution of the equation $D\nabla^4 w - p = 0$ will no longer be appropriate when coplanar forces are present, and the error in using the simpler equation will increase appreciably as the coplanar forces approach values that cause instability. At that stage, the deflected shape may be established by determining its parameters from the variational equation (4.37). An alternative is to insert a given set of stresses N_{ij} into the equilibrium equation to find the deflected shape by a numerical iterative procedure. For example, such a

144

procedure can be worked out for dealing with axisymmetric deflections of discs.

4.6 Non-linear Deflection of Plates

When an initially flat plate is deflected, all components of bending stress vary linearly through the thickness of the plate, and have zero value at the middle surface. Additional stresses which are constant through the thickness may arise when deflections become large. This cannot occur in a beam, for if no longitudinal load is applied at its ends, there can be no average longitudinal stress at any section. However, in a plate, it can be envisaged that certain strips which deflect to a large curvature may have their ends more or less anchored by adjacent strips which have a smaller deflection. Hence a deflected plate may contain membrane stresses in its middle plane even though the outer edges may be free from forces acting in the plane of the plate. In this case, the membrane stresses constitute an internal stress system which is brought into existence as a deflection is imposed. If the boundaries are fixed so that they cannot move in the plane of the plate, these membrane stresses may be larger, and will require forces to be applied at the boundaries to maintain the boundaries fixed. As strain energy depends on stress squared, any additional stresses, regardless of whether they are positive or negative, will add to the strain energy. Hence membrane stresses invariably have the effect of increasing the apparent stiffness of a plate.

As membrane stresses vary with deflection squared, they have the effect of making response to loading non-linear as deflections become large, although the response remains linear if values of deflections are small in relation to plate thickness. This behaviour is in contrast to that of a beam with no longitudinal constraint, which responds linearly until deflections become appreciable in relation to the length of the beam, provided the beam remains elastic. When a plate initially contains an internal stress system, the membrane stresses may interact with the internal stresses to give a changed stiffness, even for infinitesimal deflections. This effect has already been discussed using the idea of interaction energy. A general method is now outlined for calculating the membrane stresses set up in an initially flat plate during flexure.

It is common experience that a thin strip can be held at one end and deflected at the other through a fairly large distance without suffering any ill effects, whereas if the strip should be stretched longitudinally through the same distance, permanent extension or fracture would result. This suggests that in a plate, deflections w may be of a larger order of magnitude than extensions u and v measured in the (x,y) plane in which the plate lies, while strains within the material remain restricted to small values. This means that although squares of derivatives of u and v are neglected in the linear theory of elasticity, because they are of a secondary order of smallness, squares of derivatives of w cannot be similarly discarded. The components of strain produced by deflection alone are proportional to deflection squared, and have already been found in equations

(4.31) and (4.32). These may be of the same order of magnitude as the strains which vary linearly with displacements u and v, and which were derived in chapter 1. The two parts of each strain component are simply added to give total values

$$
\left.
\begin{aligned}
e_x &= \frac{\partial u}{\partial x} + \tfrac{1}{2}\left(\frac{\partial w}{\partial x}\right)^2 \qquad e_y = \frac{\partial v}{\partial y} + \tfrac{1}{2}\left(\frac{\partial w}{\partial y}\right)^2 \\
e_{xy} &= \tfrac{1}{2}\left(\frac{\partial u}{\partial y} + \frac{\partial v}{\partial x}\right) + \tfrac{1}{2}\frac{\partial w}{\partial x}\frac{\partial w}{\partial y}
\end{aligned}
\right\}
\tag{4.38}
$$

Although these equations may bear a resemblance to those used in theories of finite strain, it is emphasized that in this situation, the strains remain infinitesimal, and the linear elasticity equations still apply.

An equation of strain compatibility is obtained by differentiating these expressions,

$$
\frac{\partial^2 e_y}{\partial x^2} + \frac{\partial^2 e_x}{\partial y^2} - 2\frac{\partial^2 e_{xy}}{\partial x\,\partial y} = \left(\frac{\partial^2 w}{\partial x\,\partial y}\right)^2 - \frac{\partial^2 w}{\partial x^2}\frac{\partial^2 w}{\partial y^2}
$$

It follows from the discussion of stress functions in chapter 2 that the combination of strains on the left of this equation is equal to $\nabla^4 M/E$, where M is the stress function which gives stresses that are elastically consistent with these strains. As the strains are set up by stretching of the middle surface of the plate, the stresses corresponding to these strains are the membrane stresses. Hence the stress function M is given by

$$
\nabla^4 M = -E\left[\frac{\partial^2 w}{\partial x^2}\frac{\partial^2 w}{\partial y^2} - \left(\frac{\partial^2 w}{\partial x\,\partial y}\right)^2\right]
\tag{4.39}
$$

The membrane stresses, written m_{ij}, are given, as usual, by the derivatives

$$
m_x = \frac{\partial^2 M}{\partial y^2} \qquad m_y = \frac{\partial^2 M}{\partial y^2} \qquad m_{xy} = -\frac{\partial^2 M}{\partial x\,\partial y}
$$

These membrane stresses are average stresses through the thickness of the plate, and have no direct connection with bending stresses. It may be noticed that since displacements u and v vanish in this analysis, the result should not be affected by superimposing a set of displacements produced by some set of coplanar forces acting around the boundary. This is so, since the stresses produced by such forces are bound to be derivable from some stress function F satisfying the equation $\nabla^4 F = 0$. It appears therefore that the membrane stresses derived from (4.39) are arbitrary to the extent of possible contributions made

146

by coplanar boundary forces. Hence the stresses at the boundary must be specified before those in the interior can be found.

It is seen that the only kind of surface into which an initially flat sheet can be bent without setting up membrane stresses is one which gives a zero value of the right-hand side of equation (4.39). This invariant function of curvatures can be zero only if one of the principal curvatures is zero at every point. This is so if the surface is generated by the movement of a straight line, so as to form the surface of a cylinder or cone having any shape of cross-section. Such a surface is said to be developable, that is, it can be developed from a flat sheet or returned to a state of flatness without setting up membrane stresses. However, these stresses will be set up when any other deflected shape is imposed.

4.6.1 Circular plates   Membrane stresses in circular plates can be calculated fairly simply, especially when the deflected shape is symmetrical about the axis of the plate. For such problems, (4.39) can be expressed in polar coordinates and partly integrated to give

$$\frac{d}{dr}\left[\frac{1}{r}\frac{d}{dr}\left(r\frac{dM}{dr}\right)\right] = -\frac{E}{2r}\left(\frac{dw}{dr}\right)^2 \tag{4.40}$$

Only the very simple example of spherical curvature will be worked through here, in which deflection is given by $w = w_1\rho^2$, where ρ is the ratio of radius r to outside radius R and w_1 is deflection at the rim. With this expression inserted, (4.40) can be integrated twice to give radial stress

$$m_r = \frac{1}{r}\frac{dM}{dr} = \frac{Ew_1^2}{4R^2}(A - \rho^2) + \frac{B}{r^2}$$

The constant of integration B may be put equal to zero as the disc has no central hole. If no radial stress is applied at the rim, constant A takes a value of unity. Hoop stress is now found to be

$$m_\theta = \frac{d^2M}{dr^2} = \frac{Ew_1^2}{4R^2}(1 - 3\rho^2)$$

If is seen that hoop stress is compressive at the rim ($\rho = 1$). This might be expected, as points on the rim approach nearer to the axis of the disc during flexure, resulting in a shortening of the circumference.

The strain energy U_M due to membrane stress is independent of bending strain energy and may be written, for the present purpose,

$$U_M = \frac{\pi t}{E}\int_0^R (m_r^2 + m_\theta^2 - 2\nu m_r m_\theta)r\, dr$$

where t is the uniform thickness of the disc. By noting that the hoop stress satisfies the equilibrium equation $m_\theta = d(rm_r)/dr$, it is found that when the disc

147

has a stress-free rim, the cross-product of stresses in the integral gives no contribution, the remaining terms give strain energy

$$U_M = \pi E t w_1{}^4 / 24\,R^2$$

The bending strain energy from (4.34) is

$$U_B = \pi E t^3\, w_1{}^2 / 3(1-\nu)\,R^2$$

To set up spherical curvature in an initially flat plate, only a moment M_R acting at the rim is required, and this acts over the whole circumference of length $2\pi R$. The rim deflects through angle $2w_1/R$, but as this angle will not increase linearly from zero as the bending moment is applied, the work W supplied by the bending moment must be written as an integral

$$W = 4\pi \int\limits_0^{w_1} M_R\, dw_1$$

in which M_R is regarded as varying in some way with deflection. An energy equation $W = U_B + U_M$ can now be formed. On differentiating through with respect to w_1, any particular value of M_R is determined as a function w_1,

$$M_R = \frac{E t^3 w_1}{6(1-\nu)\,R^2}\left(1 + \alpha\,\frac{w_1{}^2}{t^2}\right)$$

where the numerical coefficient $\alpha = (1-\nu)/4$. With $\nu = 0.3$, this constant becomes $\alpha = 0.175$. As an alternative to expressing work W as an integral, the variational procedure could be applied to the expressions for energy and work to give the same result.

This result means that when deflection w_1 becomes equal to plate thickness t, the moment that must be applied to the rim is larger by a factor 1.175 than the moment expected from the linear theory. Naturally, this factor will be different in each different geometrical situation. When the edges of the plate are prevented from moving radially, the factor α will be larger. For example, consider a diaphragm rigidly clamped around its edge to suppress both flexure and radial movement at the edge, while a uniform pressure is applied to the surface of the diaphragm. The solution for this problem was obtained in chapter 3, i.e., $w = w_0(1-\rho^2)^2$, where w_0 is the central deflection. Zero radial displacement at the edge requires constants of integration to be adjusted so that membrane stresses satisfy the relation $m_\theta - \nu m_r = 0$ at the edge. The above method of analysis then gives a value $\alpha = 0.49$, which means that when deflection becomes equal to the plate thickness, the applied pressure is larger by a factor 1.49 than that predicted by the linear theory. In these examples, the deflected shape appropriate for infinitesimal deflections has been used. This will be unsatisfactory for very large deflections in which the membrane stress energy predominates, and a variational procedure for establishing the changed deflected shape will then be needed.

Due to the non-linearity that develops with increasing deflection, deflections must be limited to one-half or even one-quarter of the plate thickness when stiffnesses are determined experimentally. Further, a nominally flat plate, before it is loaded, should not depart from flatness at any point by more than about 10% of its thickness if it is to give satisfactory experimental results. In a plate that is not initially flat, a component of membrane stress may be introduced which varies directly with deflection rather than with deflection squared, giving a change in stiffness even for infinitesimal deflections.

4.7 Variational Method for Torsion

In order to have a sound basis for applying energy methods to the torsion of bars of solid section, the problem must be carefully defined. A parallel-sided bar of uniform isotropic material is considered, which is acted on only by couples applied at its ends, all other external forces being absent.

A section at right angles to the axis is selected at some distance away from the ends. By St. Venant's principle, the stress distribution over this section will depend only on the shape of the section and the couple applied, and not on any irregularities of stress distribution over the extreme ends of the bar. The z-axis is taken to lie along the axis of the bar, so that the section considered lies in the (x,y) plane. It can now be said that any two straight lines drawn at right angles on this section will remain straight and at right angles as the bar is twisted. This follows from the symmetry of the situation by the following argument. Suppose that, for some reason, an initially straight line drawn out-wards from the centre of the section becomes convex in the direction of the torque applied to the upper part of the bar. On viewing the section from the other side, and applying the same argument, it would be decided that the line should curve in the opposite way. Hence such an argument must be invalid. The same reasoning may be applied to an increase or decrease in the initial angle between any two lines drawn on the section. It is concluded that the section must retain its original shape, and as rectangular elements drawn on the section must remain rectangular, shear strain e_{xy} must be everywhere zero.

This argument does not apply to longitudinal displacement at any selected point, the direction of which can be quite logically associated with the direction of the torque vector. Generally, warping of the section will occur. However, all sections taken at various points along the bar would be expected to deviate from their original state of flatness by the same amount, since the bar may be of any desired length, and all sections are identically loaded. Therefore, after twisting, the axial spacing of corresponding points on any two sections must be the same regardless of where the point is located on the section. Hence the portion of the bar between the two chosen sections must stretch or contract uniformly. However, it may be argued that an elastic bar will accommodate a given amount of twist per unit length while absorbing a minimum of strain energy. Direct strains in either the axial or transverse directions will merely add to the strain energy without interacting with the externally applied torque, so it is deduced that such strains are everywhere zero.

4.7.1 Shear stress and torque   The only strains remaining are shear strains, which have corresponding stress components in an elastic bar. The stress equilibrium equations (1.7) are now applied to these stresses, and the restriction imposed is found to be

$$\frac{\partial \sigma_{zx}}{\partial x} + \frac{\partial \sigma_{zy}}{\partial y} = 0$$

This will always be satisfied if stresses are derived from a stress function ϕ,

$$\sigma_{zx} = \frac{G\theta}{L} \frac{\partial \phi}{\partial y} \qquad \sigma_{zy} = -\frac{G\theta}{L} \frac{\partial \phi}{\partial x} \qquad (4.41)$$

where G is the shear modulus of the material, and θ is the angle of twist in a bar of length L. These factors are often omitted in defining stresses in terms of ϕ, but their inclusion helps to simplify the equations to be derived here.

The stresses are evidently components of a vector. It is required that at points on the boundary of the section, the stress vector must be parallel to the boundary, since there must be no complimentary shear stress acting on the lateral sides of the bar. This means that ϕ must have zero gradient in a direction tangential to the boundary, which is arranged by making ϕ zero at all points on the boundary of the section.

Positive shear stresses are shown in Fig. 4.14, in which the z-axis is directed upwards. The anticlockwise torque T acting on the section is now

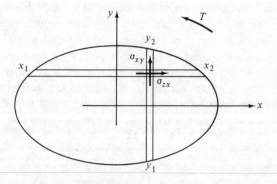

Fig. 4.14 Shear stresses and torque

obtained by integrating moments of shear stresses over the section,

$$T = \int\int (x\sigma_{zy} - y\sigma_{zx})\, dx\, dy$$

By replacing stresses by derivatives from (4.41), this integral can be written in the form

$$T = -\frac{G\theta}{L} \int\int \left[\frac{\partial}{\partial x}(x\phi) + \frac{\partial}{\partial y}(y\phi) - 2\phi \right] dx\, dy$$

$$= -\frac{G\theta}{L} \left\{ \int [x\phi]_{x_1}^{x_2}\, dy + \int [y\phi]_{y_1}^{y_2}\, dx - \int\int 2\phi\, dx\, dy \right\}$$

At the ends of any vertical or horizontal strip, as shown in Fig. 4.14, the stress function is zero, so the first two integrals are of zero value, leaving an expression for torque

$$T = \frac{G\theta}{L} \int\int 2\phi\, dx\, dy \qquad (4.42)$$

As yet, no restriction has been placed on the function $\phi(x,y)$ except that it is zero on the boundary.

4.7.2 Strain energy in torsion   From the general expression (4.3) for strain energy, this can be written, for a bar of length L,

$$U = \frac{L}{2G} \int\int (\sigma_{zx}^2 + \sigma_{zy}^2)\, dx\, dy$$

On inserting values in terms of ϕ from (4.41), this becomes

$$U = \frac{G\theta^2}{2L} \int\int \left[\left(\frac{\partial\phi}{\partial x}\right)^2 + \left(\frac{\partial\phi}{\partial y}\right)^2 \right] dx\, dy$$

It is now supposed that the stress function ϕ is expressed in terms of parameters a_n, any one of which can be varied. However, such a variation $\delta\phi = \phi'\, \delta a_n$ must be such that the function ϕ continues to be zero at the boundary, which means that ϕ' has to be zero at the boundary. The corresponding variation in strain energy follows by differentiating U,

$$U' = \frac{G\theta^2}{L} \int\int \left(\frac{\partial\phi'}{\partial x}\frac{\partial\phi}{\partial x} + \frac{\partial\phi'}{\partial y}\frac{\partial\phi}{\partial y} \right) dx\, dy$$

The first term in this integral can be re-written

$$\frac{\partial\phi'}{\partial x}\frac{\partial\phi}{\partial x} = \frac{\partial}{\partial x}\left(\phi'\frac{\partial\phi}{\partial x} \right) - \phi'\frac{\partial^2\phi}{\partial x^2}$$

which, on being integrated, gives

$$\iint \frac{\partial \phi'}{\partial x} \frac{\partial \phi}{\partial x}\, dx\, dy = \int \left[\phi' \frac{\partial \phi}{\partial x} \right]_{x_1}^{x_2} dy - \int \phi' \frac{\partial^2 \phi}{\partial x^2}\, dx\, dy$$

As ϕ' is zero at the ends $x = x_1$ and $x = x_2$ of any horizontal strip drawn on the section, the first term on the right is zero. The same argument may be applied to the second term of the integral, leading to an alternative integral

$$U' = -\frac{G\theta^2}{L} \iint \phi' \nabla^2 \phi\, dx\, dy$$

Considering the work done by external forces, the torque has already been calculated in equation (4.42).

According to question (4.11) derived for a linear elastic body, a variation δT in the torque produces a variation δW in work supplied given by $\delta W = \theta\, \delta T$, which can be written, for a variation δa_n in ϕ,

$$W' \delta a_n = \delta a_n \frac{G\theta^2}{L} \iint 2\phi'\, dx\, dy$$

Equilibrium with respect to this variation now requires that $U' - W' = 0$, giving the variational equation

$$\int (\nabla^2 \phi + 2) \phi'\, dA = 0 \tag{4.43}$$

with integration over cross-sectional area A.

4.7.3 Relationship with exact solution   The variational equation will be satisfied for all possible variations only if the differential equation $\nabla^2 \phi = -2$ is satisfied at all points over the area of the section. However, an approximate solution may be found which need not satisfy the differential equation exactly. It must be possible to vary one or more of the parameters of ϕ without violating the condition that ϕ is zero at the boundary. Practical success with this method clearly depends on being able to find a coordinate system which permits this boundary condition to be satisfied.

One might enquire how the solution thus found differs from the exactly correct solution which satisfies the differential equation. It may be recalled that in the derivation of the variational equation, no mention was made of compatibility of strains. Referring to the equations of compatibility (1.11) derived in chapter 1, the two non-zero components of strain in a twisted bar are, strictly speaking, required to satisfy the equations.

$$\frac{\partial}{\partial x}\left(\frac{\partial e_{xz}}{\partial y} - \frac{\partial e_{yz}}{\partial x} \right) = 0 \qquad \frac{\partial}{\partial y}\left(\frac{\partial e_{yz}}{\partial x} - \frac{\partial e_{xz}}{\partial y} \right) = 0$$

which means that

$$\frac{\partial e_{xz}}{\partial y} - \frac{\partial e_{yz}}{\partial x} = C$$

where C is some constant which is unvarying over the whole section. When strains are replaced by stresses, and stresses are again replaced by derivatives from (4.41), this restriction is equivalent to saying that $\nabla^2 \phi$ must be constant over the section. If this is so, the variational equation will ensure that this constant value is equal to (-2). On the other hand, the solution that will generally emerge from the variational method will not be such that $\nabla^2 \phi$ is constant over the whole section, and then, compatibility of strains will be violated.

This draws attention to a distinctive feature of the variational method described here for torsion problems. In the method previously used for beams and plates, the error involved in the variational method was due to violating the conditions of stress equilibrium, the error being such that the apparent stiffness was higher than the correct value. In using the method for torsion, equilibrium is ensured, but the compatibility relations are violated. The stiffness is now less than the correct value, which is proved as follows.

Suppose that some function $\phi(x,y) = af(x,y)$ is chosen, which is slightly different from the exact solution. When the variational equation (4.43) is applied for determining the constant a, one obtains the form

$$\int (\nabla^2 \phi + 2) \phi \, dA = 0$$

where the integration is taken over area A of the section. This function is now supposed to consist of the exact solution ϕ_0 plus an additional part ϕ_1. Bearing in mind that $\nabla^2 \phi_0 + 2 = 0$, the last integral becomes

$$\int (\phi_0 + \phi_1) \nabla^2 \phi_1 \, dA = 0 \qquad (4.44)$$

Now consider the differentiation

$$\frac{\partial}{\partial x} \left(\phi_0 \frac{\partial \phi_1}{\partial x} \right) = \frac{\partial \phi_0}{\partial x} \frac{\partial \phi_1}{\partial x} + \phi_0 \frac{\partial^2 \phi_1}{\partial x^2}$$

On integrating each term over the whole area, the term on the left gives zero on account of the zero value of ϕ_0 on the boundary. With similar expressions having y written in place of x, one obtains the relation

$$\int \phi_0 \nabla^2 \phi_1 \, dA = - \int \int \left(\frac{\partial \phi_0}{\partial x} \frac{\partial \phi_1}{\partial x} + \frac{\partial \phi_0}{\partial y} \frac{\partial \phi_1}{\partial y} \right) dx \, dy \qquad (4.45)$$

Due to the symmetry of the right-hand side, subscripts of ϕ can be interchanged on the left. With the equality thus obtained, the condition (4.44) imposed by the variational equation can be written

$$\int \phi_1 (\nabla^2 \phi_0 + \nabla^2 \phi_1)\, dA = 0$$

Again, using the fact that $\nabla^2 \phi_0 = -2$ and using (4.45) with both subscripts equal to (1), the last condition is finally expressed as

$$\int 2\phi_1\, dA = - \int\int \left[\left(\frac{\partial \phi_1}{\partial x} \right)^2 + \left(\frac{\partial \phi_1}{\partial y} \right)^2 \right] dx\, dy$$

It is seen that this equation gives the additional torque due to the superposition of ϕ_1 as indicated by (4.42). As the sum of the squares in the integral is bound to give a positive integral, the additional torque is shown to be negative. Hence the variational method in the form in which it has been discussed here will always underestimate torsional stiffness, unless the exact stress function happens to be used.

5. Numerical Methods

A number of problems in elasticity can be reduced to the solution of equations of the type,

$$\nabla^2 \phi = f(x, y) \quad \text{or} \quad \nabla^4 \phi = f(x, y)$$

where ϕ or its derivatives may be defined in terms of the stresses, strains, or displacements and $f(x,y)$ is a function of the position which can be equal to zero or to a non-zero constant. Once a function ϕ that satisfies the governing equation and the appropriate boundary conditions is found, it becomes possible to deduce the stress distribution throughout the whole solid. It is clear that this method of analysis is of general validity. Unfortunately, its application is restricted to relatively simple configurations, failing to provide a satisfactory solution for the complex shapes and forms of loading that are often encountered in practice. In those cases, alternative methods of solution, valid only for the particular problem in hand, have to be used. These methods do not seek to find an analytical expression for the stress distribution, but only to determine the actual numerical values of the stress at a few selected points, assuming a known value of the load to be applied to the solid, whose dimensions are also known. The problem has to be solved again in its entirety if one of the dimensions or the form in which the load is applied are changed.

The most direct numerical method consists in the replacement of the governing equation by the corresponding finite difference equations. This reduces the problem of finding ϕ to the solution of a system of simultaneous linear equations in which the unknowns are the values of ϕ at the selected points or nodes. In another method, the complete solid under consideration is assumed to consist of a number of small elements, whose behaviour can be readily formulated. The whole assembly is then analysed by considering the equilibrium and continuity of displacements at the interfaces between adjoining elements. The first, or *finite difference method* has a purely mathematical nature; once the governing equation has been derived, little or no further reference is made to the actual physical problem. The second, for which the name *finite element method* has recently been coined, has the advantage of retaining the physical characteristics of the problem throughout the process of its solution, but it also leads to a large number of simultaneous linear equations that have to be solved in order to find the values of the stress resultants at the edges of each element. In either case the practical application of numerical analysis depends on the difficulties associated with the solution of a large number of simultaneous linear equations. It was therefore only with the introduction of the *relaxation technique* by Southwell that numerical analysis became a powerful tool when dealing with elasticity problems. Relaxation was developed primarily as a hand computation technique and as a result some of its importance has been lost with the widespread use of high speed electronic computers. However, this chapter would not be complete without a general description of the relaxation technique which still provides an excellent introduction to the two general methods of finite differences and finite elements.

5.1 The Relaxation Technique

5.1.1 Torsion of uniform shafts
This problem can be used as a simple introduction to the finite difference method. The equations required have been outlined in the previous chapter. With a slightly different notation, the expressions for torque T acting on area of cross-section A, and for shear stresses acting on the section may be written

$$\sigma_{xz} = \frac{\partial \phi}{\partial y} \qquad \sigma_{yz} = -\frac{\partial \phi}{\partial x} \qquad T = \int 2\phi \, dA$$

The stress function ϕ must satisfy the differential equation

$$\nabla^2 \phi = -2G\theta \tag{5.1}$$

The general procedure is to find the solution $\phi(x,y)$ in terms of twist θ per unit length of bar. By integrating over the section, torque T is found in terms of θ. By differentiating at any desired point, shear stress can be found in terms of θ. By eliminating θ between these expressions, shear stress is obtained in terms of torque.

The first step is to express the differential equation in finite difference form. Instead of dealing with a physical quantity such as a stress function, the solution can be equally well calculated by considering some general numerical quantity $S(x,y)$ having no particular physical associations. By setting up coordinates (x,y) at some chosen point on the section, function S may be expanded about this point as a Taylor series with coefficients a_{ij},

$$S = S_0 + a_{10} x + a_{01} y + a_{20} x^2 + a_{11} xy + a_{02} y^2 + \cdots$$

Now consider the (x,y) plane to be covered by a square net of mesh size d, as shown in Fig. 5.1. Values of S at each of the four points shown may be added together to obtain

$$S_1 + S_2 + S_3 + S_4 = 4S_0 + 2(a_{20} + a_{02})d^2$$

156

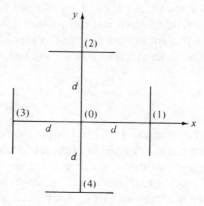

Fig. 5.1 Nodal points

However, for small distances x and y, differentiation gives

$$\nabla^2 S = 2(a_{20} + a_{02})$$

Hence, by combining the last two equations, we obtain

$$\nabla^2 S = -N/d^2$$

where

$$N = 4S_0 - (S_1 + S_2 + S_3 + S_4) \tag{5.2}$$

The value of N may be taken as any convenient integer. A satisfactory numerical solution will then consist of a series of numbers S at the nodal points of a net, such that (5.2) is satisfied at every point, together with any boundary conditions that may be imposed. When this distribution has been determined, the integral V over the section and the gradient g in any direction n can be found numerically,

$$V = \int S \, dA \qquad g = \frac{\partial S}{\partial n}$$

The torsion problem can now be solved by simply scaling the results already found. Comparison of the differential equations

$$\nabla^2 \phi = -2G\theta \qquad \nabla^2 S = -N/d^2$$

suggests a proportionality $\phi = 2G\theta d^2 S/N$. Hence torque and shear stress are given by

$$T = 4G\theta d^2 \, V/N \qquad \sigma_{zn} = 2G\theta d^2 \, g/N$$

For torsion of a solid section, the differential equation governing stress function ϕ and the boundary condition $\phi = 0$ are together sufficient for establishing $\phi(x,y)$ at all points on the section. When an internal boundary is present, this is not so. It is known that ϕ must have a constant value around such a boundary, because no shear stress acts on the internal surface, but this value of ϕ is not initially known. As discussed in other texts, for instance, Dugdale's *Elements of Elasticity*, the necessary condition to be satisfied is

$$\int_C \frac{\partial \phi}{\partial n}\, ds = 2G\theta A_1$$

where the integral is taken along the line s of perimeter C of the internal cavity of area A_1, and normal n is directed from the solid material towards the cavity. It was previously seen that the torsion problem could be examined by using any numerical quantity $S(x,y)$ satisfying the finite-difference equation (5.2) at all points. When an internal boundary is present, such a generalized solution must satisfy the relation

$$\int_C \frac{\partial S}{\partial n}\, ds = \frac{NA_1}{d^2} \tag{5.3}$$

Multiplication of all S-values by any common factor will not affect the validity of such a solution, since constant N will also be increased in the same proportion. The torque and shear stress can be deduced from the generalized solution $S(x,y)$ in the manner already described. When a set of numbers at nodal points is given, equations (5.2) and (5.3) can be applied to test whether these numbers represent a correct solution to the torsion problem.

However, the treatment given so far tends to lack a direct appeal to the physical senses. Although it is possible to proceed to a purely numerical solution of the finite-difference equation, it is often considered preferable to invest the method with some readily acceptable mechanical ideas by introducing the membrane analogy.

5.1.2 Membrane analogy for torsion   A flexible membrane is considered to be stretched across an aperture having the same outline as the cross-section of a uniform shaft. The membrane is stretched so that it has a uniform surface tension f per unit length in all directions. When pressure p is applied to the under side of the membrane, an out-of-plane displacement $Z(x,y)$ is produced. As discussed in more detail in *Theory of Elasticity* by Timoshenko and Goodier, the surface of the displaced membrane satisfies the equation

$$\nabla^2 Z = -p/f \tag{5.4}$$

It can be seen that a solution to this equation can be considered equivalent to a solution of the equation (5.1) which governs the stress function ϕ, provided that the quantities $2G\theta$ and p/f are regarded as being interchangeable.

Consider now a string net, as shown in Fig. 5.2, in which all strings are stretched to tension fd, where d is the mesh size of the net. The net is loaded

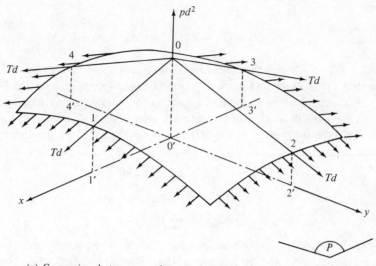

(a) Comparison between membrane and equivalent string element.

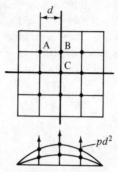

(b) Equivalent string net showing nodal points.

Fig. 5.2 String net approximation to membrane model

by forces equal to pd^2 acting at the nodal points in a direction normal to the plane originally occupied by the unloaded net. The deflection of the net resembles that of a continuous membrane, the resemblance becoming more exact as the mesh size is made smaller. The equilibrium equation (5.4) for a membrane can be adapted to express equilibrium of a net by considering four strings of length d joining a central node (0) to four adjacent nodes, as shown in Fig. 5.2(a). Taking as the reference plane the plane defined by the unloaded net, we can write

$$fd\left(\frac{Z_0-Z_1}{d}+\frac{Z_0-Z_2}{d}+\frac{Z_0-Z_3}{d}+\frac{Z_0-Z_4}{d}\right)=pd^2$$

or,

$$4Z_0 - (Z_1 + Z_2 + Z_3 + Z_4) = \frac{pd^2}{f} \tag{5.5}$$

where $Z_0, \ldots, Z_4$ are the deflections of nodes $0, \ldots, 4$. Similar equations may be written for all the nodes, so that the problem of solving equation (5.4) or (5.1) is reduced to the solution of a system of simultaneous linear equations with the nodal deflections as the only unknowns. Thus, for the net of Fig. 5.2(b), given the symmetry, the deflections at A, B, and C can be obtained from,

$$4Z_C - 4Z_B = Pd^2/f$$

$$4Z_B - 2Z_A - Z_C = pd^2/f$$

$$4Z_A - 2Z_B = pd^2/f$$

since the deflections at the boundary are all zero, corresponding to the boundary condition $\phi = 0$ for the stress function. Having found Z_A, Z_B, and Z_C the shape of the membrane can be estimated.

In general, it is found that a much finer mesh is required in order to achieve a reasonably accurate result. The number of equations to be handled may then become so large that their direct solution is no longer practicable. The relaxation technique that will now be described was introduced to deal with this situation. The first step is to write equation (5.5) in a dimensionless form, by the definition of new variables,

$$z = \frac{fZN}{pd^2} = \frac{ZN}{2G\theta d^2} \tag{5.6}$$

where N is any convenient scaling factor, say 100. The governing equations are then replaced by,

$$4z(0,0) = [z(d,0) + z(-d,0) + z(0,d) + z(0,-d)] + N \tag{5.7}$$

which applies at any point $(0,0)$ selected within the boundary. It does not apply at points on the boundary where values are assumed to be known and in the torsion problem are zero. The solution can be found by writing down any set of values at the nodal points, for example, $z = 0$ everywhere. Equation (5.7) is then applied at all nodes to derive a second set of values. Then, using the new values, the equation can be applied again to get a third set of values. If this process is continued for a sufficient number of times, the values will converge to the correct ones, which precisely satisfy equation (5.7). However, the rate of convergence is slow and the work to be carried out excessive since a new calculation is required at each point on each occasion. More systematic techniques will be described later, the most popular being the *relaxation method* first developed by Southwell. This method is based on the physical relaxation of restraints imposed on the deformation of the string net itself rather than on an abstract juggling with numbers or expressions. Consider again the string net and assume that a number of jacks are used to support the net at the nodes, preventing their free displacement. If the nodes are fixed, as in Fig. 5.3(a), the load is entirely taken by the jacks. If now we let the jack at node (0,0) move by $\zeta(0,0)$, we see that a certain proportion of the load is taken by the net itself, so that the load taken by the jack is relieved by

$$4(Sd)(\zeta(0,0)/d) = 4S\zeta(0,0)$$

as illustrated in Fig. 5.3(b). On the other hand, if, as shown in Fig. 5.3(c), the jack at node $(d,0)$ is allowed to move by $\zeta(d,0)$, the load taken by the jack at $(0,0)$ is increased by $S\zeta(d,0)$. In general, the total force supported by the jack at $(0,0)$ is,

$$F(0,0) = pd^2 - S\{4\zeta(0,0) - [\zeta(d,0) + \zeta(-d,0) + \zeta(0,d) + \zeta(0,-d)]\}$$

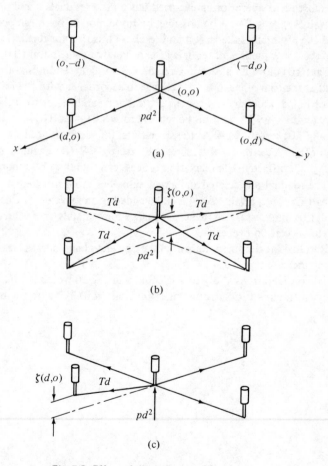

Fig. 5.3 Effect of displacements of nodes

161

In the relaxation technique, the jacks are allowed to move by a certain amount one at a time, until the forces $F(0,0)$ all become zero. The membrane is then supporting the entire external load and the total movement of the jacks is equal to the corresponding deflection Z. The preceding equation may be called the *residual formula* and may be expressed in the non-dimensional form,

$$\rho(0,0) = N - 4\eta(0,0) + [\eta(d,0) + \eta(-d,0) + \eta(0,d) + \eta(0,-d)] \qquad (5.8)$$

where $\eta = f\zeta n/pd^2$ and ρ is a residual, equal to zero when the net is in equilibrium without the help of the jacks. When that happens, the displacements η are equal to the deflections ζ.

The procedure for solving a problem is as follows. With a convenient mesh the known values are written down at the boundary points, with guessed values at the interior points. The residual formula is then applied to derive residuals at all the interior points, which are also written down. This need be done only once as subsequent operations are carried out on the residuals ρ and not on the original values η. It will now be possible, by inspection of the mesh points, to see which of the residuals is the largest and to chose this first for relaxation. We now examine the consequences of reducing the residual at some point $(0,0)$ by letting the jack at $(0,0)$ move by a certain amount, say +1. The residuals at the node itself and at the four adjacent nodes change in accordance with the relaxation pattern of Fig. 5.4, i.e., by -4 for the central node and +1 for the other four nodes. A residual $\rho(0,0)$ can thus be relaxed to $\rho(0,0) - \Delta\rho(0,0)$ by changing $\eta(0,0)$ to $\eta(0,0) + \Delta\rho(0,0)/4$. At the same time, the residuals at all adjacent nodes, $\rho(d,0), \ldots, \rho(0,-d)$ will increase by $\Delta\rho(0,0)/4$. There will be no change in $\eta(d,0), \ldots, \eta(0,-d)$. The relaxation process can therefore be summarized:

(1) The residual ρ is deleted and a new value $\rho - \Delta\rho$ is inserted.

(2) At this point, a value $\Delta\rho/4$ is to be added to the η value. If this is recorded, it is unnecessary to make the change immediately as η values are not used in the relaxation process.

(3) Residuals at the surrounding points are deleted and new values inserted, each being increased by $\Delta\rho/4$.

Special considerations apply to a selected point $(0,0)$ which is adjacent to a boundary, with one of its surrounding points, say $(d,0)$, lying on the boundary.

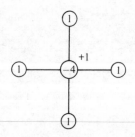

Fig. 5.4 Relaxation pattern

A change in $\rho(0,0)$ is then carried over to only three of the surrounding points in the usual way but no incremental value is carried over to the point on the boundary as no residual can exist at this point. As the residual formula does not apply at points on the boundary, it is not violated by this procedure.

(4) Having found solutions for η that cancel all residuals, the mesh size is subdivided by 2, 4, . . . in order to obtain a more accurate representation of the membrane. The corresponding values of (pd^2/f) would then have to be subdivided by 4, 16, ... Alternatively, the values calculated for η in the coarse net, multiplied by 4, 16, . . . can be used for the finer net with the same value of (pd^2/f) and of the scaling factor N.

The problem is completed by correlating the value of the torque applied to the shaft to the volume under the membrane and the slope to the shear stress, finding the angle of twist per unit length θ.

As a typical example, consider the shaft of square cross-section of Fig. 5.5 only one quadrant of which is shown. Taking $N = 100$, we first assign the values 125, 150, and 175 to η_P, η_Q, and η_R and find residuals -100, -75, and 0 respectively. Taking P and its symmetrical as pivotal points, we relax by -50. The residuals then become 100 at P and -175 at Q. Q is now taken as a pivot and relaxed by -60, the residuals are then -20, 65, and -240 at P, Q, and R respectively. The next pivot is R, which is relaxed by -60 and so on, following

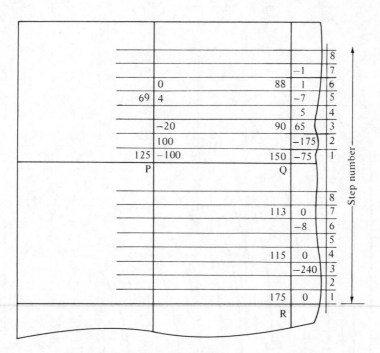

Fig. 5.5 Relaxation process: shaft of square cross-section

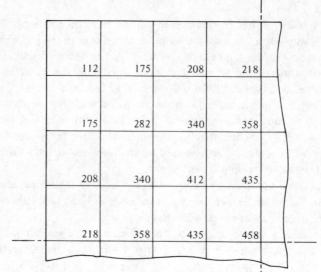

Fig. 5.6 Results of relaxation with a finer mesh

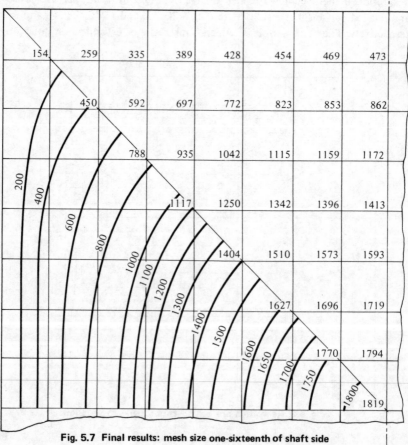

Fig. 5.7 Final results: mesh size one-sixteenth of shaft side

the sequence of Fig. 5.5. The final results are 69, 88, and 113 for P, Q, and R respectively, with an error of less than 1/4.

The coarse mesh is now subdivided as shown in Fig. 5.6 and initial values are guessed for the intermediate nodes. By relaxing all supports we find the values shown in the figure. A further reduction in the mesh size yields the results shown in Fig. 5.7, in which contour lines have also been drawn.

It is sometimes desirable to relax a block of points simultaneously as illustrated in Fig. 5.8. In that case, the total changes in the residuals at the points within the block are -2η and at the points directly connected to it, η.

For triangular nets–Fig. 5.9(a)–the following relaxation formula is obtained:

$$\rho_0 = n - 6\eta_0 + \eta_1 + \eta_2 + \eta_3 + \eta_4 + \eta_5 + \eta_6$$

and for hexagonal nets–Fig. 5.9(b),

$$\rho_0 = n - 3\eta_0 + \eta_1 + \eta_2 + \eta_3$$

Very often, the net can not be drawn to fit exactly over the whole section, and while it is possible to derive residual formulae by considering the equilibrium of elements formed by strings of different lengths, it is easier to follow another approach, based on the series expansion of equation (5.4). Writing,

$$Z \approx Z_0 + \left(\frac{\partial Z}{\partial x}\right)_0 (x - x_0) + \tfrac{1}{2}\left(\frac{\partial^2 Z}{\partial x^2}\right)_0 (x - x_0)^2 + \cdots$$

$$+ \left(\frac{\partial Z}{\partial y}\right)_0 (y - y_0) + \tfrac{1}{2}\left(\frac{\partial^2 Z}{\partial y^2}\right)_0 (\lambda - y_0)^2 + \cdots$$

we find that the deflections at points $(x_0 + m_1 d, y_0)$, $(x_0, y_0 + m_2 d)$, $(x_0 - d, y_0)$, and $(x_0, y_0 - d)$, illustrated in Fig. 5.10, are,

$$Z_A = Z_0 + \left(\frac{\partial Z}{\partial x}\right)_0 m_1 d + \tfrac{1}{2}\left(\frac{\partial^2 Z}{\partial x^2}\right)_0 m_1{}^2 d^2$$

$$Z_B = Z_0 + \left(\frac{\partial Z}{\partial y}\right)_0 m_2 d + \tfrac{1}{2}\left(\frac{\partial^2 Z}{\partial y^2}\right)_0 m_2{}^2 d^2$$

$$Z_C = Z_0 - \left(\frac{\partial Z}{\partial x}\right)_0 d + \tfrac{1}{2}\left(\frac{\partial^2 Z}{\partial x^2}\right)_0 d^2$$

$$Z_D = Z_0 - \left(\frac{\partial Z}{\partial y}\right)_0 d + \tfrac{1}{2}\left(\frac{\partial^2 Z}{\partial y^2}\right)_0 d^2$$

165

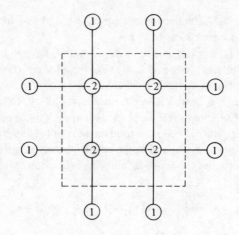

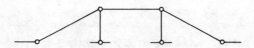

Fig. 5.8 Relaxation pattern: block relaxation

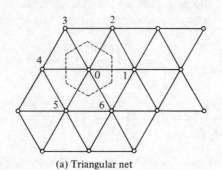

(a) Triangular net

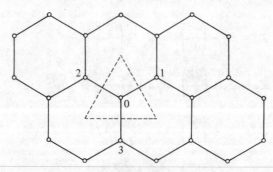

(b) Hexagonal net

Fig. 5.9 Triangular and hexagonal nets

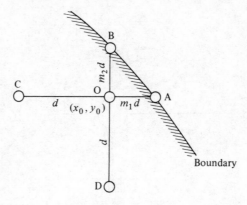

Fig. 5.10 Boundary intersecting two sides of a net element

from which we obtain,

$$d^2\left[\left(\frac{\partial^2 Z}{\partial x^2}\right)_0 + \left(\frac{\partial^2 Z}{\partial y^2}\right)_0\right] = d^2\ [(\nabla^2 Z)_0] = -Z_0\left(\frac{2}{m_1} + \frac{2}{m_2}\right)$$

$$+ \frac{2}{m_1(1+m_1)}\ Z_A + \frac{2}{m_2(1+m_2)}Z_B + \frac{2}{1+m_1}\ Z_C + \frac{2}{1+m_2}Z_D$$

The residual equation is therefore,

$$\rho(0,0) = n - \left(\frac{2}{m_1} + \frac{2}{m_2}\right)\eta(0,0) + \frac{2}{m_1(1+m_1)}\ \eta(m_1 d, 0)$$

$$+ \frac{2}{1+m_1}\ \eta(0,-d) + \frac{2}{m_2(1+m_2)}\ \eta(0, m_2\ d) + \frac{2}{1+m_2}\ \eta(0,-d) \tag{5.9}$$

A displacement of +1 at the central node brings about the change in residuals given by the relaxation pattern of Fig. 5.11. Usually it will be found that the boundary intersects at the most two of the strings, since if this is not so it becomes necessary to choose a finer mesh. Also, from equation (5.9) it is apparent that a displacement +1 in node $(-d,0)$ would cause a change in the residual at $(0,0)$ of $2/(1 + m_1)$. Similarly, a displacement +1 of node $(0,-d)$ causes a change in the residual at $(0,0)$ equal to $2/(1 + m_2)$.

5.1.3 Torsion of hollow sections   When studying the torsion of a shaft of hollow cross-section it is found that the boundary condition $Z = 0$ only applies to the external boundary, while for every internal boundary it is necessary to satisfy the following condition:

$$Z = C_j \oint_j \frac{\partial Z}{\partial n} ds = -p\ \frac{A_j}{S} \tag{5.10}$$

167

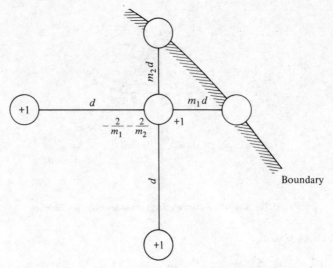

Fig. 5.11 Relaxation pattern: boundary intersecting two sides of a net element

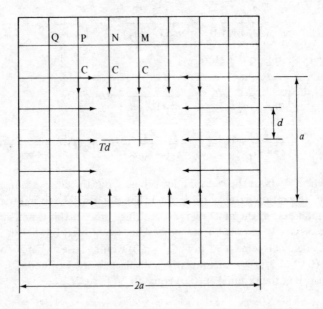

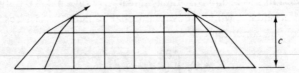

Fig. 5.12 Net used for the analysis of a hollow shaft

where the subindex j refers to the boundary considered, A_j is the area it encloses, and C_j a constant. (See for example Timoshenko and Goodier.) The internal boundary condition may be expressed in the form,

(Average slope of Z normal to boundary) × (Perimeter) = (Area) × $(-p/t)$

Taking the case of a square shaft with a central square hole it is easy to see that the above condition is equivalent to,

$$\sum (Z - C) + p\frac{A}{f} = 0 \qquad (5.11)$$

where the summation extends over the whole inside perimeter. Thus, in Fig. 5.12,

$$4Z_M + 8Z_N + 4Z_P - 16C + 16\,\frac{Pd^2}{f} = 0$$

$$4z_M + 8z_N + 4z_p - 16c + 16n = 0 \qquad (5.12)$$

The values of z at the various nodes as well as the value of the constant c may be found by relaxation, noting that all points on the internal boundary should be displaced by +1, the change in the residuals at M, N, and P is +1, while the residual of equation (5.12) is −16. Equation (5.12) also shows that a displacement of points M, N, or P of +1 causes a change in its residual of +4, +8, and +4 respectively. The final values are, with $n = 100$,

$$z_M = 189 \qquad z_N = 185 \qquad z_P = 170$$
$$z_Q = 110 \qquad c = 282$$

Before proceeding to the treatment of curved boundaries, consider a cluster of nodes adjacent to an internal boundary as in Fig. 5.13(a) and assume that the central node is displaced by +1. From the residual formula, the residuals at the internal nodes are as shown. The residual of equation (5.10) will be +1 since there is a change in slope $(1/d)$ over a length d. On the other hand, if the whole boundary, previously maintained fixed, is displaced by +1, the residuals at all the internal nodes directly connected to the boundary change by +1 while the residual of equation (5.10) changes by −(1 × number of nodes on boundary). If now we consider the configuration of Fig. 5.13(b), in which the boundary intersects one of the strings, and we displace the central node by +1, a fictitious node P, just outside the boundary, will be displaced by $-(1 - m)/m$, since the string is pinned down at the boundary. From the relaxation formula applied to the three internal nodes and to the fictitious external node P we find that the effect of the combined displacement +1 and $-(1 - m)/m$ on the residuals is as shown in Fig. 5.13(b). At the same time, the residual of equation (5.10) changes by $+1/m$ due to the change in slope of the string between P and the central node.

Giving a displacement +1 to the whole boundary results in a displacement $+1/m$ of point P, and hence a change in the residual at the central node of the same amount, while the average change in slope results in a residual $\Sigma(-1/m)$ where the sum is extended over the whole boundary.

The relaxation pattern of Fig. 5.13 may be extended to the more general case where two strings are intersected by the boundary.

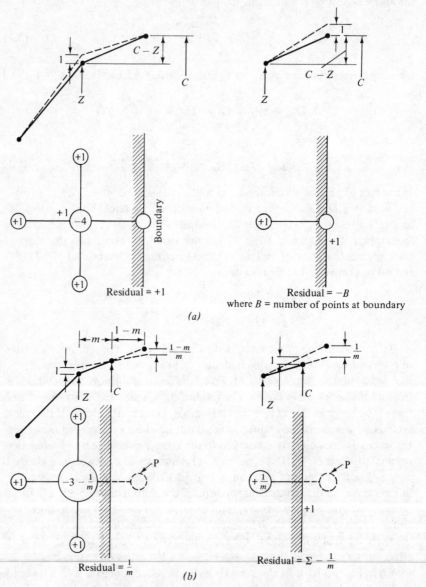

Fig. 5.13 Relaxation pattern for internal boundaries

5.2 Finite Difference Approximations

5.2.1 Approximate solution of the equations of Laplace and Poisson
In the absence of body forces, it can be shown that

$$\nabla^2(\sigma_x + \sigma_y) = 0$$

an equation that expresses, in restricted form, the conditions of compatibility, equilibrium, and elasticity. This is the Laplace equation found in field problems such as electric potential or heat conduction. While a knowledge of $(\sigma_x + \sigma_y)$ is not by itself of any great value in the determination of the stress distribution, it is however useful when $(\sigma_x - \sigma_y)$ has been determined by some other method, for example, by photoelasticity (see chapter 7). Given the similarity between Laplace's equation and that governing the deflection of a membrane, also known as Poisson's equation, it is clear that the same relaxation technique may be used for both. In fact, the computation is made easier since the term $(-p/f)$ is now set equal to zero. The string net provides, in this way, a useful basis for the determination of the stress invariant $(\sigma_x + \sigma_y)$.

An alternative formulation of the residual formulae, without making any reference to the string net, is to express the governing equation by means of finite differences. We can, in fact, write, with the notation of Fig. 5.2(a),

$$\left(\frac{\partial Z}{\partial x}\right)_0 \approx \frac{Z_0 - Z_3}{d} \qquad \left(\frac{\partial Z}{\partial x}\right)_1 \approx \frac{Z_1 - Z_0}{d}$$

$$\left(\frac{\partial^2 Z}{\partial x^2}\right)_0 \approx \frac{1}{d}\left[\left(\frac{\partial Z}{\partial x}\right)_1 - \left(\frac{\partial Z}{\partial x}\right)_0\right] = \frac{Z_1 - 2Z_0 + Z_3}{d^2}$$

$$\left(\frac{\partial Z}{\partial y}\right)_0 \approx \frac{Z_0 - Z_4}{d} \left(\frac{\partial Z}{\partial y}\right)_2 \approx \frac{Z_2 - Z_0}{d} \tag{5.13}$$

$$\left(\frac{\partial^2 Z}{\partial y^2}\right)_0 \approx \frac{1}{d}\left[\left(\frac{\partial Z}{\partial y}\right)_2 - \left(\frac{\partial Z}{\partial x}\right)_0\right] = \frac{Z_2 - 2Z_0 + Z_4}{d^2}$$

the substitution of these approximate correlations in Poisson's equation reduces the exact expression to the same residual formula as previously found from the consideration of the equilibrium of the string net. In this particular case the use of finite differences does not add anything to what has been done so far. There are however many other cases where the finite difference approximation facilitates the numerical solution of the problem.

5.2.2 The generalized Poisson's equation. Torsion of stepped shaft of circular cross-section
In a shaft of circular cross-section of variable diameter, the governing equation is

$$\frac{\partial^2 \phi}{\partial r^2} - \frac{3}{r}\frac{\partial \phi}{\partial r} + \frac{\partial^2 \phi}{\partial z^2} = 0 \tag{5.14}$$

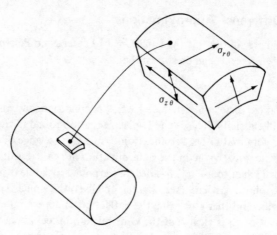

Fig. 5.14 Stresses in circular shaft

(See Timoshenko and Goodier, loc. cit., p. 307.) The shear stresses, illustrated in Fig. 5.14 are derived from the stress function ϕ through,

$$\sigma_{z\theta} = \frac{1}{r^2}\frac{\partial \phi}{\partial r} \qquad \sigma_{r\theta} = -\frac{1}{r^2}\frac{\partial \phi}{\partial z}$$

The boundary conditions require the stress function ϕ to be zero at the centre line and constant at the surface. The value of this constant can be found by considering a circular shaft of uniform cross-section of radius a in which the shear stress at a distance r from the centre line is,

$$\sigma_{z\theta} = \frac{2Tr}{\pi a^4} \qquad \sigma_{r\theta} = 0$$

where T is the torque. The shear stresses are acting in the directions defined in Fig. 5.14. From the definition of the stress function we find that,

$$\phi = \frac{T}{2\pi}\left(\frac{r}{a}\right)^4$$

and at the surface $r = a$, $\phi = (T/2\pi)$, an expression which is independent of the shaft diameter. For ease of computation $T/2\pi$ may be taken to be equal to some convenient value, say 1000. Once the corresponding values of ϕ within the shaft have been found, the true values for a given torque are calculated by the application of a scaling factor equal to $T/2000\pi$. Equation (5.14) can be expressed in finite difference form taking

$$\left(\frac{\partial \phi}{\partial r}\right)_0 \approx \tfrac{1}{2}\frac{\phi_1 - \phi_0}{d}$$

with the remaining derivatives defined by equations (5.13). The governing equation is therefore equivalent to the following residual formula,

$$\rho = \phi_1 + \phi_2 + \phi_3 + \phi_4 - 4\phi_0 - \frac{3d}{r}\frac{\phi - \phi_3}{2}$$

valid for the points defined in Fig. 5.15. This implies that the same relaxation technique that was previously used for the solution of Poisson's equation can be used here. The relaxation pattern follows immediately from the residual formula.

Curved boundaries are treated in the same way as was done for non-circular shafts of uniform cross-section, the only difference being that the term $(\partial\phi/\partial r)_0$ previously defined from ϕ_1 and ϕ_3 is now taken to be approximately equal to $(\phi_0 - \phi_3)/d$. Thus if points 1 and 2 are on the curved boundary, point 3 is still within the shaft. The residual formula is then,

$$\rho = \frac{2}{m_2(m_2 + 1)}\phi_1 + \frac{2}{1 + m_1}\phi_2 + \frac{2}{1 + m_2}\phi_3 + \frac{2}{m_1(1 + m_1)}\phi_4$$

$$- \left(\frac{2}{m_1} + \frac{2}{m_2}\right)\phi_0 - \frac{3d}{r}(\phi_0 - \phi_3)$$

5.2.3 The biharmonic equation
An important group of problems in elasticity is governed by the biharmonic equation,

$$\nabla^4 \phi = f(x, y)$$

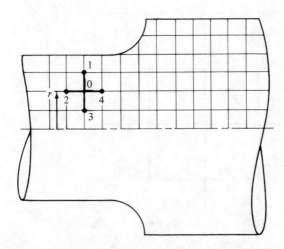

Fig. 5.15 Net used for the analysis of stepped circular shaft

In particular, the stress distribution in a solid under plane stress or strain conditions is found to be derived from the Airy stress function ϕ such that

$$\sigma_{xx} = \frac{\partial^2 \phi}{\partial y^2} \qquad \sigma_{yy} = \frac{\partial^2 \phi}{\partial x^2} \qquad \sigma_{xy} = -\frac{\partial^2 \phi}{\partial x \, \partial y} \tag{5.15}$$

and

$$\nabla^4 \phi = \frac{\partial^4 \phi}{\partial x^4} + 2 \frac{\partial^4 \phi}{\partial x^2 \, \partial y^2} + \frac{\partial^4 \phi}{\partial y^4} = 0 \tag{5.16}$$

in the absence of body forces. Since it may not be possible to find a satisfactory solution for this problem, a numerical method is often used in order to determine values of ϕ at specific points within the solid. Once these values are found the stresses are derived by means of

$$(\sigma_{xx})_0 = \left(\frac{\partial^2 \phi}{\partial y^2}\right)_0 = \frac{\phi(0, d) - 2\phi(0, 0) + \phi(0, -d)}{d^2}$$

$$(\sigma_{yy})_0 = \left(\frac{\partial^2 \phi}{\partial x^2}\right)_0 = \frac{\phi(d, 0) - 2\phi(0, 0) + \phi(-d, 0)}{d^2}$$

$$(\sigma_{xy})_0 = -\left(\frac{\partial^2 \phi}{\partial x \, \partial y}\right)_0 = \frac{[\phi(d, -d) - \phi(-d, -d)] - [\phi(d, d) - \phi(-d, d)]}{4d^2}$$

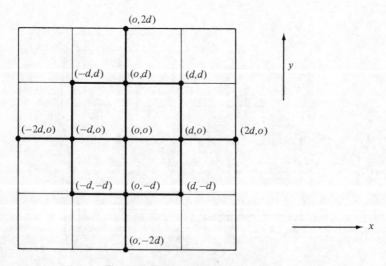

Fig. 5.16 Net for biharmonic equation

174

with the notation of Fig. 5.16. With the same notation, we can write the following expressions:

$$\left(\frac{\partial^4 \phi}{\partial x^4}\right)_0 = \left[\frac{\partial^2}{\partial x^2}\left(\frac{\partial^2 \phi}{\partial x^2}\right)\right]_0 \approx \frac{1}{d^2}\left[\left(\frac{\partial^2 \phi}{\partial x^2}\right)_{d,0} - 2\left(\frac{\partial^2 \phi}{\partial x^2}\right)_0 + \left(\frac{\partial^2 \phi}{\partial x^2}\right)_{-d,0}\right]$$

$$= \frac{1}{d^4}[\{\phi(2d,0) - 2\phi(d,0) + \phi(0,0)\} - 2\{\phi(d,0) - 2\phi(0,0) + \phi(-d,0)\}$$

$$+ \{\phi(0,0) - 2\phi(-d,0) + \phi(-2d,0)\}]$$

$$= \frac{1}{d^4}[\phi(2d,0) - 4\phi(d,0) + 6\phi(0,0) - 4\phi(-d,0) + \phi(-2d,0)]$$

$$\left(\frac{\partial^4 \phi}{\partial x^2 \partial y^2}\right)_0 = \left[\frac{\partial^2}{\partial y^2}\left(\frac{\partial^2 \phi}{\partial x^2}\right)\right]_0 \approx \frac{1}{d^2}\left[\left(\frac{\partial^2 \phi}{\partial x^2}\right)_{0,d} - 2\left(\frac{\partial^2 \phi}{\partial x^2}\right)_0 + \left(\frac{\partial^2 \phi}{\partial x^2}\right)_{0,-d}\right]$$

$$= \frac{1}{d^4}[\phi(d,d) + \phi(-d,d) + \phi(d,-d) + \phi(-d,-d) - 2\phi(d,0) - 2\phi(0,d)$$

$$- 2\phi(0,-d) - 2\phi(-d,0) + 4\phi(0,0)]$$

$$\left(\frac{\partial^4 \phi}{\partial y^4}\right)_0 \approx \frac{1}{d^4}[\phi(0,2d) - 4\phi(0,d) + 6\phi(0,0) - 4\phi(0,-d) + \phi(0,-d)]$$

so that the governing equation can be replaced by the set of simultaneous linear equations,

$$d^4(\nabla^4 \phi)_0 \approx 20\phi(0,0) - 8[\phi(d,0) + \phi(-d,0) + \phi(0,d) + \phi(0,-d)]$$

$$+ 2[\phi(d,d) + \phi(-d,d) + \phi(-d,-d) + \phi(d,-d)]$$

$$+ \phi(2d,0) + \phi(-2d,0) + \phi(0,2d) + \phi(0,-2d)$$

$$= -\rho_0 = 0 \tag{5.17}$$

The form of this equation is very similar to that of the residual formula previously derived and the same relaxation technique could be used. A relaxation pattern can easily be drawn by noting that a displacement +1 of the central node causes its residual to change by -20, while the residuals at the remaining nodes are,

Node	Change in residual due to +1 displacement at (0,0)
(d,0); (0,d); (−d,0); (0,−d)	+8
(d,d); (−d,d); (−d,−d); (d,−d)	−2
(2d,0); (−2d,0); (0,2d); (0,−2d)	−1

Unfortunately, the relaxation pattern in this case is such that it makes the relaxation process extremely tedious, not only because of the number of points affected by a displacement of the central node, but also because the total number of relaxations required to reduce all residuals to zero is increased. The slow convergence of the biharmonic relaxation makes a point-by-point manual process of the type previously used for the solution of the torsion problems quite unsuitable when dealing with the biharmonic equation. Other techniques, specially developed for the solution of simultaneous linear equations with high speed digital computers are preferred. Some of these techniques are briefly described in a separate section.

As shown in Fig. 5.16, the minimum size of the grid element required for the evaluation of $(\nabla^4 \phi)$ by means of equation (5.17) is $(4d \times 4d)$. This would preclude the use of that equation for all those internal points situated at a depth d from the boundary. In order to overcome this difficulty, it is assumed that the grid extends outside the boundary, providing enough additional external points to enable the application of the finite difference equation to all internal points, excluding those situated just on the boundary. The values of ϕ at the external points are chosen in such a way that the boundary conditions are satisfied. To do this, take the small element of the solid shown in Fig. 5.17 and express the equilibrium between the external force of components X and Y per unit length and the stresses,

$$\sigma_{xx} \cos \alpha + \sigma_{xy} \sin \alpha = X$$
$$\sigma_{yy} \sin \alpha + \sigma_{xx} \cos \alpha = Y$$

(5.18)

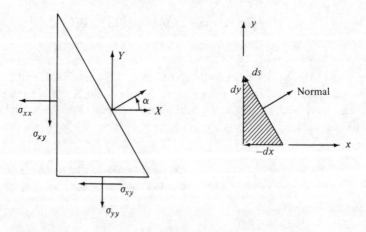

Fig. 5.17 Boundary stress conditions

176

since $\sin \alpha = -dx/ds$ and $\cos \alpha = dy/ds$ the equilibrium equations can also be written in the form

$$\frac{\partial^2 \phi}{\partial y^2} dy + \frac{\partial^2 \phi}{\partial x \partial y} dx = X ds$$

$$\frac{\partial^2 \phi}{\partial x^2} dx + \frac{\partial^2 \phi}{\partial x \partial y} dy = -Y ds$$

equivalent to,

$$\frac{\partial}{\partial y}\left(\frac{\partial \phi}{\partial y} dy + \frac{\partial \phi}{\partial x} dx\right) = \frac{\partial}{\partial y}(d\phi) = X ds$$

$$\frac{\partial}{\partial x}\left(\frac{\partial \phi}{\partial x} dx + \frac{\partial \phi}{\partial y} dy\right) = \frac{\partial}{\partial x}(d\phi) = -Y ds$$

therefore the boundary conditions become,

$$\frac{\partial \phi}{\partial y} = \int X ds \qquad \frac{\partial \phi}{\partial x} = -\int Y ds \qquad (5.19)$$

It is desirable to visualize the meaning of these equations by applying them to a few simple cases frequently found in practice. Assume first that the boundary is stress free. In that case, both equations are equivalent to,

$$\frac{\partial \phi}{\partial x} = \frac{\partial \phi}{\partial y} = 0 \quad \text{or} \quad \frac{\partial \phi}{\partial s} = \frac{\partial \phi}{\partial n} = 0$$

where s and n refer to the directions along the boundary and normal to it respectively. It follows that ϕ must be constant along the boundary and it must approach the boundary with zero slope, the shape of a cross section through the ϕ surface being as shown in Fig. 5.18. At all external boundaries we may take $\phi = 0$, while at internal boundaries, ϕ = constant. In order to find the finite difference expression of $\nabla^4 \phi$ at point (0,0), we need an additional point (2d,0), one mesh size outside the boundary. The value of ϕ at this external point must be such that

$$\left(\frac{\partial \phi}{\partial n}\right)_P = 0 \approx \frac{\phi(2d, 0) - \phi(0, 0)}{2d}$$

hence

$$\phi(2d, 0) = \phi(0, 0)$$

a conclusion that remains valid for any pair of points situated symmetrically with respect to the boundary. In the case of Fig. 5.18, we can therefore calculate

177

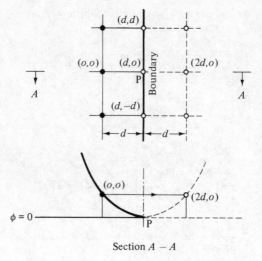

Section $A - A$

Fig. 5.18 Extrapolation of stress function at the boundary

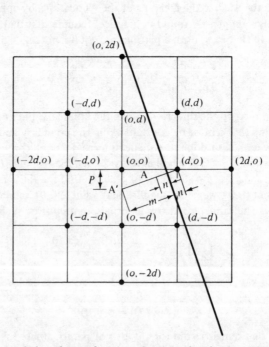

Fig. 5.19 Extrapolation of stress function when boundary intersects net element

178

$\nabla^4 \phi$ at $(0,0)$ by replacing $\phi(2d,0)$ in equation (5.17) by $\phi(0,0)$ with $\phi(d,d) = \phi(d,0) = \phi(d,-d) = 0$.

In general it is not possible to draw the grid to conform with the boundary. More often than not, a situation such as that of Fig. 5.19 will arise, in which the value of ϕ at $(d,0)$ amongst other external points has to be extrapolated from the values of ϕ at the internal points in order to evaluate $\nabla^4 \phi$ at $(0,0)$. In the first place, $\phi(d,0) = \phi_A$ where A is symmetrical to $(d,0)$ with respect to the boundary. Assuming that the variation of the stress function when moving along a normal to the boundary is parabolic,

$$\phi_A = \phi_{A'} \left(\frac{n}{m}\right)^2 \qquad (5.20)$$

ϕ_A can in turn be interpolated between the values of ϕ at $(0,0)$ and $(0,-d)$ either linearly or by means of a more accurate rule. Assuming again a parabolic variation of ϕ along the line $(0,-2d) - (0,0)$, it is easy to show that,

$$\phi_{A'} \approx \phi(0,0) + [2\phi(0,-d) - \tfrac{3}{2}\phi(0,0) - \tfrac{1}{2}\phi(0,-2d)]\frac{p}{d}$$

$$+ [\phi(0,-2d) - 2\phi(0,-d) + \phi(0,0)]\frac{p^2}{2d^2} \qquad (5.21)$$

The values of ϕ at all the remaining external points can be expressed in terms of the corresponding values at internal points in exactly the same way. Having done this, it then becomes possible to formulate $\nabla^4 \phi$ for all the internal points in terms of the ϕ values within the boundary. The resulting set of simultaneous linear equations incorporates the boundary conditions.

A parabolic approximation can also be used to extrapolate the stress function outside the boundary when one or both of the components of the external force X or Y are different from zero. Thus, in Fig. 5.18,

$$\left(\frac{\partial \phi}{\partial y}\right)_P \approx \frac{\phi(d,d) - \phi(d,-d)}{2d} = (Xd)_P \qquad (5.22)$$

where $(Xd)_P$ is the resultant of the distributed force X over a length $d/2$ at either side of P. Also,

$$\left(\frac{\partial \phi}{\partial x}\right)_P \approx -(Yd)_P$$

Assuming that the variation of ϕ when moving along the x direction is given by,

$$\phi = ax^2 + bx + c$$

where a, b, and c are parameters and taking P as the origin,

$$\phi(2d,0) = \phi(0,0) - 2(Yd)_P \, d$$

179

Similar equations can be derived when the boundary is along the x-direction or, indeed, for any other configuration.

The preceding considerations will now be illustrated by means of a few examples.

Example 1 It is proposed to find the maximum deflection of a square plate with built-in edges under uniformly distributed pressure. The plate deflection w must satisfy the biharmonic equation,

$$\nabla^4 w = \frac{p}{D}$$

where p is the pressure and D the flexural rigidity. The boundary conditions at the built-in edges are,

$$w = 0 \qquad \frac{\partial w}{\partial n} = 0$$

Before expressing the governing equation in finite difference form we define a new, dimensionless variable,

$$\phi = w \frac{D}{pd^4}$$

and the governing equation becomes,

$$d^4 \nabla^4 \phi = 1$$

which reduces to,

$$20\phi(0,0) - 8[\phi(d,0) + \phi(-d,0) + \phi(0,d) + \phi(0,-d)]$$
$$+ 2[\phi(d,d) + \phi(-d,d) + \phi(d,-d) + \phi(-d,-d)] + \phi(2d,0)$$
$$+ \phi(-2d,0) + \phi(0,2d) + \phi(0,-2d) = 1 \qquad (5.23)$$

As a first approximation, we draw the very coarse grid of Fig. 5.20 and write equation (5.23) for the only node,

$$20\phi_A + 4\phi_A = 1$$

$$w_A = \frac{p}{D}(\phi_A d^4) = 0.0026\frac{pL^4}{D}$$

which compared with the known exact solution,

$$W_A = 0.00105\frac{pL^4}{D}$$

shows that a much finer mesh is required. The next step is to draw the grid of Fig. 5.21, for which the finite difference formulation of the governing equation

180

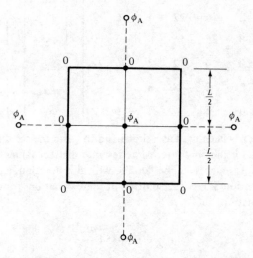

Fig. 5.20 Plate with built-in edges: Preliminary analysis

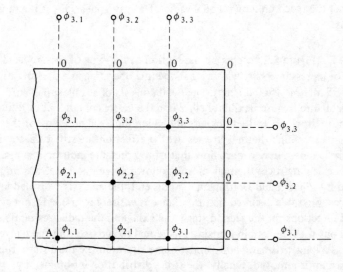

Fig. 5.21 Plate with built-in edges: use of finer mesh

is given by

$$\begin{vmatrix} 20 & -32 & 8 & 4 & 0 & 0 \\ 2 & -16 & 22 & 4 & -16 & 2 \\ -8 & 25 & -16 & -8 & 6 & 0 \\ 0 & 0 & 2 & 2 & -16 & 22 \\ 0 & 3 & -8 & -8 & 24 & -8 \\ 0 & -8 & 4 & 21 & -16 & 2 \end{vmatrix} \times \begin{vmatrix} \phi_{1,1} \\ \phi_{2,1} \\ \phi_{2,2} \\ \phi_{3,1} \\ \phi_{3,2} \\ \phi_{3,3} \end{vmatrix} = \begin{vmatrix} 1 \\ 1 \\ 1 \\ 1 \\ 1 \\ 1 \end{vmatrix}$$

Matrix notation has been used as a shorthand to write the six simultaneous linear equations for the only internal nodes with distinct values of ϕ. While it is assumed that most readers will be familiar with this notation, a brief description is included in section 5.3.2. The solution of these equations provides a slightly better approximation,

$$W_A = 0.00154 \frac{pL^4}{D}$$

Improved accuracy is possible by subdividing the grid. If, for example, a maximum error of 5% is specified, it becomes necessary to use a mesh size of about $L/40$, resulting in some 200 simultaneous linear equations whose solution is simplified by the fact that each unknown nodal value of ϕ only appears in, at the most, 12 equations.

Example 2 Figure 5.22 shows a cross-section of the type of closure head needed for a pressure vessel with a large opening operating at a pressure of the order of 35 MN/m^2 (500 lbf/in^2). The head consists of a substantial, stiff flange welded to a more flexible shell. When the vessel contains a hot fluid, the temperature drop across the thickness and along the axial direction of the flange may cause high thermal stresses, whose effect on the structural stability of the whole design may overshadow that of the pressure-induced stresses. We shall now calculate these thermal stresses following a method that was originally proposed by Bijlaard and Dohrmann. To this end, the analysis is divided into four stages. First it is assumed that the flange is entirely restrained from expanding in all directions. In the second stage, the axial and the radial restraints are removed but the hoop strain is maintained equal to zero, so that the flange behaves as a long, prismatic body under plane strain. In the third stage, this hoop restraint is removed. Finally, the stress distribution is obtained by summation of the stresses determined in the first three stages. The temperature distribution is assumed to be known. It is also assumed that the effect of the shell on the flange can be introduced at the end of the calculation by setting up equations correlating the edge forces and moments at the flange–shell intersection with the deformations they produce. This part of the problem will not be discussed here, since its treatment follows the lines discussed in chapter 6.

In the first stage, we take

$$e_{zz} = e_{rr} = e_{\theta\theta} = 0$$

Since

$$e_{zz} = \frac{1}{E}[\sigma_{zz} - \nu(\sigma_{rr} + \sigma_{\theta\theta})] + \alpha T$$

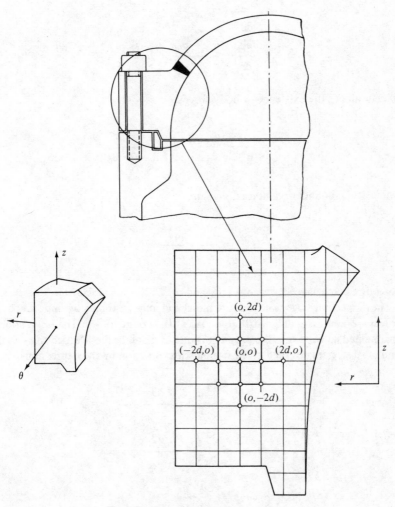

Fig. 5.22 Closure head in pressure vessel

183

and the companion equations, the stresses in this first stage are,

$$(\sigma_{zz})_1 = (\sigma_{rr})_1 = (\sigma_{\theta\theta})_1 = -\frac{E\alpha T}{1-2\nu} \tag{5.24}$$

where the temperature T is a function of the position. The coefficient of thermal expansion, like E and ν, may be assumed to remain constant throughout the flange, although this need not impose any restriction on the generality of the method. The calculated stresses must satisfy the equilibrium conditions,

$$\frac{\partial \sigma_{zz}}{\partial z} + \bar{Z} = 0$$

$$\frac{\partial \sigma_{rr}}{\partial r} + \frac{\sigma_{rr} - \sigma_{\theta\theta}}{r} + \bar{R} = 0$$

which implies the existence of body forces,

$$\bar{Z} = \frac{E\alpha}{1-2\nu}\frac{\partial T}{\partial z} \qquad \bar{R} = \frac{E\alpha}{1-2\nu}\frac{\partial T}{\partial r}$$

while at the boundary, a normal pressure

$$p = -\frac{E\alpha T}{1-2\nu}$$

is required to balance the stresses σ_{zz} and σ_{rr}.

In the second stage, we relax the radial and axial restraints by imposing body forces $-\bar{Z}$ and $-\bar{R}$ and a distributed load at the boundary $-p$. The hoop strain is maintained zero, so that the problem is one of plane strain in which the radial and axial stresses can be derived from a stress function by the expressions,

$$\sigma_{zz} = \frac{\partial^2 \phi}{\partial r^2} + \frac{E\alpha T}{1-2\nu}$$

$$\sigma_{rr} = \frac{\partial^2 \phi}{\partial x^2} + \frac{E\alpha T}{1-2\nu}$$

$$\sigma_{rz} = -\frac{\partial^2 \phi}{\partial r \, \partial x}$$

The equilibrium conditions,

$$\frac{\partial \sigma_{zz}}{\partial z} + \frac{\partial \sigma_{rz}}{\partial r} + (-\bar{Z}) = 0$$

$$\frac{\partial \sigma_{rr}}{\partial r} + \frac{\partial \sigma_{rz}}{\partial z} + (-\bar{R}) = 0$$

are then satisfied and the compatibility conditions lead to,

$$\nabla^4 \phi = -\frac{E\alpha}{1-\nu} \nabla^2 T \tag{5.25}$$

It will be noted that in the absence of any heat generation within the flange,

$$\nabla^2 T = k \frac{\partial T}{\partial t} \quad (= 0 \text{ under steady state conditions})$$

Equation (5.25) can be expressed in the same form as equation (5.17) taking,

$$\rho_0 = \frac{d^2 E\alpha}{1-\nu} [T(d, 0) + T(-d, 0) + T(0, d) + T(0, -d) - 4T(0, 0)]$$

in general, different from zero. $T(d,0), \ldots, T(0,0)$ are all supposed to be known. As in the previous example, the problem is reduced to the solution of a set of simultaneous linear equations, entirely defined as soon as the boundary conditions are included in their formulation. In deriving these conditions we follow the procedure that led to equations (5.18). Taking $X = p \cos \alpha$ and $Y = -p \sin \alpha$,

$$\sigma_{zz} \cos \alpha + \sigma_{rz} \sin \alpha = -p \cos \alpha$$

$$\sigma_{rr} \sin \alpha + \sigma_{rz} \cos \alpha = -p \sin \alpha$$

which lead to $\partial\phi/\partial r$ = constant, $\partial\phi/\partial z$ = constant, conditions that are satisfied by taking $\partial\phi/\partial s = \partial\phi/\partial n = 0$. The boundary conditions are then those previously found when $X = Y = 0$. The function ϕ can be extrapolated outside the boundary and the values at all external nodes expressed in terms of the internal values, thus defining a set of simultaneous linear equations, as many as internal nodes, with the same number of unknowns. Once ϕ values are obtained, the stresses for this second stage are, at a node (0,0),

$$(\sigma_{zz})_2 = \frac{1}{d^2}[\phi(-d,0) - 2\phi(0,0) + \phi(d,0)] + \frac{E\alpha}{1-2\nu}T(0,0)$$

$$(\sigma_{rr})_2 = \frac{1}{d^2}[\phi(0,-d) - 2\phi(0,0) + \phi(0,d)] + \frac{E\alpha}{1-2\nu}T(0,0)$$

$$(\sigma_{rz})_2 = \frac{1}{d^2}[\phi(d,-d) + \phi(-d,d) - \phi(d,d) - \phi(-d,-d)]$$

$$(\sigma_{\theta\theta})_2 = \nu[(\sigma_{xx})_2 + (\sigma_{rr})_2] \tag{5.26}$$

In order to maintain zero hoop strain, it has been necessary to introduce a stress,

$$\sigma_{\theta\theta} = (\sigma_{\theta\theta})_1 + (\sigma_{\theta\theta})_2 =$$

$$\frac{\nu}{d^2}[\phi(d,0) + \phi(-d,0) + \phi(0,d) + \phi(0,-d) - 4\phi(0,0)] - E\alpha T(0,0)$$

at point (0,0). This stress has a resultant, over the whole cross-section,

$$F = \int_{\text{Area}} \sigma_{\theta\theta}\, dA$$

and moment resultant

$$M = \int_{\text{Area}} \sigma_{\theta\theta}\, x dA$$

where x is the distance to the neutral axis of the flange cross-section. It is clear that a force $-F$ and moment $-M$ must now be introduced so that the hoop stress is finally,

$$(\sigma_{\theta\theta})_{\text{Total}} = \frac{\nu}{d^2}[\phi(d,0) + \phi(-d,0) + \phi(0,d) + \phi(0,-d) - 4\phi(0,0)]$$

$$- E\alpha T(0,0) - \frac{F}{A} - \frac{M_x}{I}$$

where A is the cross-sectional area and I is the second moment of area with respect to the neutral axis. The total stresses in the radial and axial direction are,

$$(\sigma_{zz})_{\text{Total}} = (\sigma_{zz})_1 + (\sigma_{zz})_2 = \frac{1}{d^2}[\phi(-d,0) - 2\phi(0,0) + \phi(d,0)]$$

$$(\sigma_{rr})_{\text{Total}} = (\sigma_{rr})_1 + (\sigma_{rr})_2 = \frac{1}{d^2}[\phi(0,-d) - 2\phi(0,0) + \phi(0,d)]$$

and the shear stress,

$$(\sigma_{rx})_{\text{Total}} = (\sigma_{rx})_2$$

The radial displacement is,

$$u = \frac{\sigma_{\theta\theta}\, r}{E}$$

with a rotation equal to (du/dx). As a check, when there is no heat generation and in the steady state condition, assuming the temperature to vary linearly through the thickness between T_0 and T_i on the outside and inside faces respectively, it is found that in a flange of rectangular cross-section,

$$(\sigma_{\theta\theta})_{\text{Total outside}} = E\alpha \frac{T_i - T_0}{2}$$

$$(\sigma_{\theta\theta})_{\text{Total inside}} = E\alpha \frac{T_0 - T_i}{2}$$

all other stresses being zero. In these equations h is the flange thickness, measured in the radial direction. The results may be compared to those obtained through the application of equations (6.9), developed in chapter 6 for thin discs.

Example 3   In chapter 6 the stress distribution in a thin rotating disc is found from the consideration of the general elastic equations in a solid with rotational symmetry, taking all axial stress components equal to zero. This reduction of the problem to a simple plane stress case is not possible when dealing with a thick, tapering disc in which the hub may be several times thicker than the rim. Referring to equations (6.1) and (6.2), and eliminating the displacements, the following compatibility conditions are obtained,

$$\frac{\partial e_{\theta\theta}}{\partial r} = \frac{1}{r}(e_{rr} - e_{\theta\theta})$$

$$2\frac{\partial e_{zr}}{\partial z} = r\frac{\partial^2 e_{\theta\theta}}{\partial z^2} + \frac{\partial e_{zz}}{\partial r}$$

with the notation of Figs. 6.1 and 6.2, the force F being equal to $m\omega^2 r$. A number of stress functions have been used for the solution of these problems. (See, for instance R. V. Southwell, *Relaxation Methods in Engineering Science*, or P. P. Benham and R. D. Hoyle, *Thermal Stress*.) The stress function that will be used here is defined by,

$$\sigma_{rr} = \nu\nabla^2\,\phi - \frac{\partial^2\,\phi}{\partial r^2} - \frac{1}{8}\frac{3-2\nu}{1-\nu}m\omega^2\,r^2 - E\alpha T$$

$$\sigma_{\theta\theta} = \nu\nabla^2\,\phi - \frac{1}{r}\frac{\partial\phi}{\partial r} - \frac{1}{8}\frac{1+2\nu}{1-\nu}m\omega^2\,r^2 - E\alpha T$$

$$\sigma_{zz} = (2-\nu)\,\nabla^2\,\phi - \frac{\partial^2\,\phi}{\partial z^2} - \frac{1}{2}\frac{\nu}{1-\nu}m\omega^2\,r^2 + E\alpha T \tag{5.27}$$

$$\frac{\partial\sigma_{zr}}{\partial z} = \frac{\partial}{\partial r}\left[(1-\nu)\,\nabla^2\,\phi - \frac{\partial^2\,\phi}{\partial z^2} + E\alpha T\right]$$

It can easily be shown that equilibrium and compatibility conditions are satisfied if,

$$\nabla^2(\nabla^2\,\phi) = \nabla^4\,\phi = 0 \tag{5.28}$$

where

$$\nabla^2 = \frac{\partial^2}{\partial r^2} + \frac{1}{r}\frac{\partial}{\partial r} + \frac{\partial^2}{\partial z^2}$$

The order of the governing equation can be reduced by defining two new stress functions χ and ψ,

$$\frac{\partial\chi}{\partial r} = -r\nabla^2\,\phi \quad\text{and}\quad \frac{\psi}{r} + \frac{1-\nu}{r}\chi = \frac{\partial\phi}{\partial r} \tag{5.29}$$

Equation (5.28) then becomes,

$$\nabla^4\,\phi = \nabla^2\left(-\frac{1}{r}\frac{\partial\chi}{\partial r}\right) = -\frac{1}{r}\frac{\partial}{\partial r}\left(\frac{\partial^2\chi}{\partial z^2} - \frac{1}{r}\frac{\partial\chi}{\partial r} + \frac{\partial^2\chi}{\partial r^2}\right) = 0$$

and defining the operator,

$$I^2 = \frac{\partial^2}{\partial z^2} - \frac{1}{r}\frac{\partial}{\partial r} + \frac{\partial^2}{\partial r^2} \tag{5.30}$$

the governing equation is reduced to,

$$I^2\chi = 0 \tag{5.31}$$

Eliminating ϕ between equations (5.29) and (5.31),

$$\frac{\partial}{\partial r}\nabla^2\,\phi = -\frac{1}{r}\frac{\partial^2\chi}{\partial r^2} + \frac{1}{r^2}\frac{\partial\chi}{\partial r} = \frac{\partial^3\phi}{\partial r\partial z^2} + \frac{1}{r}\frac{\partial^2\phi}{\partial r^2} - \frac{1}{r^2}\frac{\partial\phi}{\partial r} + \frac{\partial^3\phi}{\partial r^3}$$

$$= \left(\frac{\partial^2}{\partial z^2} + \frac{1}{r}\frac{\partial}{\partial r} - \frac{1}{r^2} + \frac{\partial^2}{\partial r^2}\right)\left(\frac{\psi + (1-\nu)\chi}{r}\right)$$

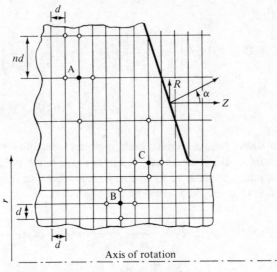

(a) Grid used for analysis

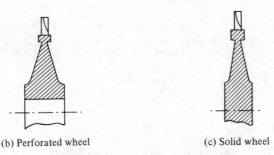

(b) Perforated wheel   (c) Solid wheel

Fig. 5.23 Stress analysis of turbine wheel

which simplifies to

$$I^2 \psi = \frac{\partial^2 \chi}{\partial z^2} \tag{5.32}$$

The governing biharmonic equation has thus been reduced to the two simultaneous differential equations (5.31) and (5.32). Both equations can be approximated by means of their finite difference expressions. Taking, for example a grid of rectangular mesh as in Fig. 5.23 for a nodal point A, equation (5.31) is,

$$\chi(d,0) + \chi(-d,0) + \chi(0,nd)\left[\frac{1}{n^2} - \frac{d}{2nr}\right] + \chi(0,-nd)\left[\frac{1}{n^2} + \frac{d}{2nr}\right]$$

$$- 2\chi(0,0)\left[1 + \frac{1}{n^2}\right] = 0 \tag{5.33}$$

and equation (5.32),

$$\psi(d,0) + \psi(-d,0) + \psi(0,nd)\left[\frac{1}{n^2} - \frac{d}{2nr}\right] + \chi(0,-nd)\left[\frac{1}{n^2} + \frac{d}{2nr}\right] - 2\psi(0,0)\left[1 + \frac{1}{n}\right.$$

$$= \chi(d,0) - 2\chi(0,0) + \chi(-d,0) \quad (5.34)$$

Very often it is advisable to use mixed grids, one part being of rectangular mesh and the rest square. In the square mesh region, at a point such as B, the finite difference equations are similar to those that have just been derived, with $n = 1$. At a point such as C, situated between the two regions,

$$\chi(d,0) + \chi(-d,0) + \chi(0,nd)\left[\frac{2}{n(n+1)} - \frac{d}{2nr}\right] + \chi(0,-d)\left[\frac{2}{n+1} - \frac{d}{2r}\right]$$

$$- \chi(0,0)\left[2 + \frac{2}{n} - \frac{(1-n)d}{2nr}\right] = 0 \quad (5.35)$$

$$\psi(d,0) + \psi(-d,0) + \psi(0,nd)\left[\frac{2}{n(n+1)} - \frac{d}{2nr}\right] + \psi(0,-d)\left[\frac{2}{n+1} - \frac{d}{2r}\right]$$

$$- \psi(0,0)\left[2 + \frac{2}{n} - \frac{(1-n)d}{2nr}\right] = \chi(d,0) - 2\chi(0,0) + \chi(-d,0) \quad (5.36)$$

The stresses are found from,

$$\sigma_{rr} = \frac{1}{r^2}[\psi + (1-\nu)\chi] - \frac{1}{r}\frac{\partial}{\partial r}[\psi + \chi] - \frac{1}{8}\frac{3-2\nu}{1-\nu}m\omega^2 r^2 - E\alpha T$$

$$\sigma_{\theta\theta} = -\frac{\nu}{r}\frac{\partial\chi}{\partial r} - \frac{1}{r^2}[\psi + (1-\nu)\chi] - \frac{1}{8}\frac{1+2\nu}{1-\nu}m\omega^2 r^2 - E\alpha T$$

$$\sigma_{zz} = \frac{1}{r}\frac{\partial\psi}{\partial r} - \frac{1}{2}\frac{\nu}{1-\nu}m\omega^2 r^2 + E\alpha T \quad (5.37)$$

Also,

$$\frac{\partial\sigma_{zr}}{\partial z} = \frac{\partial}{\partial r}\left[-\frac{1-\nu}{r}\frac{\partial\chi}{\partial r} - \frac{\partial^2\phi}{\partial z^2} + E\alpha T\right] = -\frac{1-\nu}{r}\left[\frac{\partial^2\chi}{\partial r^2} - \frac{1}{r}\frac{\partial\chi}{\partial r}\right]$$

$$+ E\alpha\frac{\partial T}{\partial r} - \frac{\partial^2}{\partial z^2}\left[\frac{\partial\phi}{\partial r}\right] = -\frac{1-\nu}{r}\left[l^2\chi - \frac{\partial^2\chi}{\partial z^2}\right] - \frac{\partial^2}{\partial z^2}\left[\frac{\psi}{r} + \frac{1-\nu}{r}\chi\right]$$

$$+ E\alpha\frac{\partial T}{\partial r}$$

190

and taking into account equation (5.31),

$$\sigma_{zr} = -\frac{1}{r}\frac{\partial \psi}{\partial z} + E\alpha \int_0^z \frac{\partial T}{\partial r} dz \qquad (5.38)$$

Before proceeding to the numerical solution of the simultaneous linear equations equivalent to the governing equation, the boundary conditions must be determined. Taking the angle between the positive axial direction and the outward normal to the boundary to be β (measured in the anticlockwise direction),

$$\sigma_{zz}\cos\beta + \sigma_{zr}\sin\beta = Z$$
$$\sigma_{rr}\sin\beta + \sigma_{zr}\cos\beta = R \qquad (5.39)$$

where Z and R are the axial and radial components of distributed loads on the boundary. From equations (5.37) to (5.39),

$$\frac{1}{r}\frac{\partial \psi}{\partial s} = Z + \frac{1}{2}\frac{\nu}{1-\nu}m\omega^2 r^2\cos\beta + E\alpha\left[T\cos\beta - \sin\beta\int\frac{\partial T}{\partial r}dz\right]$$

$$\left\{\frac{1}{r^2}[\psi + (1-\nu)\chi] - \frac{1}{r}\frac{\partial \chi}{\partial r}\right\}\sin\beta - \frac{1}{r}\frac{\partial \psi}{\partial n} = R + \frac{1}{8}\frac{3-2\nu}{0-\nu}m\omega^2 r^2\sin\beta$$

$$+ E\alpha\left[T\sin\beta - \cos\beta\int\frac{\partial T}{\partial r}dz\right] \qquad (5.40)$$

where $(\partial/\partial s)$ and $(\partial/\partial n)$ denote derivatives with respect to the tangential direction and to the normal direction to the boundary respectively. Alternatively the boundary equations may be expressed in the form,

$$-\frac{\partial \psi}{\partial z}\sin\beta + \frac{\partial \psi}{\partial r}\cos\beta = rZ + \frac{1}{2}\frac{\nu}{1-\nu}m\omega^2 r^3\cos\beta + E\alpha r\left[T\cos\beta - \sin\beta\int\frac{\partial T}{\partial r}dz\right]$$

$$\left[\frac{\psi + (1-\nu)\chi}{r} - \frac{\partial \chi}{\partial r}\right]\sin\beta - \frac{\partial \psi}{\partial z}\cos\beta - \frac{\partial \psi}{\partial r}\sin\beta$$

$$= Rr + \frac{1}{8}\frac{3-2\nu}{1-\nu}m\omega^2 r^3\sin\beta + E\alpha r\left[T\sin\beta - \cos\beta\int\frac{\partial T}{\partial r}dz\right] \qquad (5.41)$$

In turbine wheels, of the types illustrated in Fig. 5.23, the distributed external load is zero everywhere with the exception of the rim, where it equals the pull exerted by the blades, and the hub, where it equals any pressure stress due to a possible shrink fit on the shaft. At the rim,

$$\frac{\partial \psi}{\partial s} = -E\alpha\int\frac{\partial T}{\partial r}dz$$

and, if there is no heat transfer, $\partial\psi/\partial s = 0$, ψ = constant along the rim. Also,

$$\frac{\psi + (1-\nu)\chi}{r} - \frac{\partial\chi}{\partial r} - \frac{\partial\psi}{\partial r} = Rr + \frac{1}{8}\frac{3-2\nu}{1-\nu}m\omega^2 r^3 + E\alpha rT \quad (5.42)$$

where R equals the force exerted by the blades (per unit area). A similar condition holds at the hub, while, in the case of solid wheels the symmetry with respect to the centre line implies $\sigma_{zr} = 0$, i.e., ψ = constant along centre line since $(\partial T/\partial r)$ must be zero. In addition, since the stresses must have some finite value, $\psi = \chi = 0$. The derivatives of both functions with respect to the radial direction are also zero at the centre line, given the symmetry of the configuration.

To represent the boundary conditions in finite difference form, the reference grid is extended and the additional external nodes are included in the computation. Consider for example the turbine wheel with a solid shaft of Fig. 5.24 and assume a constant temperature throughout. On the shaft centre line we know that $\psi = \chi = 0$. Applying the first of equations (5.40) to the boundary AB,

$$\psi_{AB} = \frac{1}{8}\frac{\nu}{1-\nu}m\omega^2 r^4 \quad (5.43)$$

and at BC,

$$\psi_{BC} = \text{constant} = \frac{1}{8}\frac{\nu}{1-\nu}m\omega^2 a^4$$

Along the sloping face CD,

$$\frac{1}{r}\frac{\Delta\psi}{d\sqrt{(1+n^2)}} = \frac{1}{2}\frac{\nu}{1-\nu}m\omega^2 r^2 \cos\beta$$

$$\Delta\psi = \frac{1}{2}\frac{\nu}{1-\nu}m\omega^2 r^3\, nd = \frac{1}{2}\frac{\nu}{1-\nu}m\omega^2 r^3\,\Delta r$$

and

$$\psi_{CD} = \psi_{BC} + \sum\Delta\psi$$

from which the value of ψ at any point on CD may be found. Obviously,

$$\psi_{CD} = \psi_{BC} + \int_a^r \frac{1}{2}\frac{\nu}{1-\nu}m\omega^2 r^3\, dr = \frac{1}{8}\frac{\nu}{1-\nu}m\omega^2 r^4$$

and it is concluded that

$$\psi_{\text{boundary}} = \frac{1}{8}\frac{\nu}{1-\nu}m\omega^2 r^4$$

For a point on AB, the second of equations (5.40) states that $\psi(d,0) = \psi(-d,0)$ hence the two nodal equations (5.33) and (5.34) become

$$\chi(d,0) + \chi(-d,0) + \chi(0,d)\left[1 - \frac{d}{2r}\right] + \chi(0,-d)\left[1 + \frac{d}{2r}\right] - 4\chi(0,0) = 0$$

$$2\psi(-d,0) + \psi(0,d)\left[1 - \frac{d}{2r}\right] + \psi(0,-d)\left[1 + \frac{d}{2r}\right] - 4\psi(0,0)$$

$$= \chi(d,0) - 2\chi(0,0) + \chi(-d,0)$$

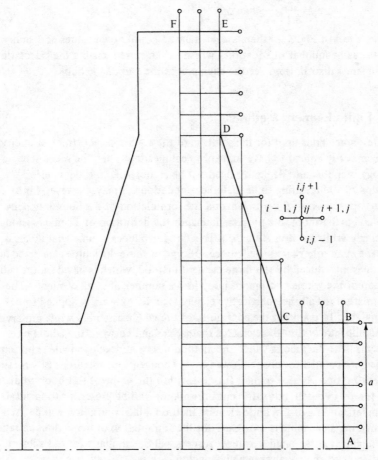

Fig. 5.24 Typical grid and wheel profile

where the values of ψ are given by equation (5.43). At a point on BC,

$$\psi(0,0) + (1-\nu)\chi(0,0) - \frac{a}{2d}[\chi(d,0) - \chi(-d,0) + \psi(d,0) - \psi(-d,0)]$$

$$= \frac{1}{8}\frac{3-2\nu}{1-\nu}m\omega^2 a^4$$

Similar equations hold for EF, adding the effect of the distributed load R, while the conditions at DE are similar to those at AB. At CD, from equation (5.41),

$$\psi(0,0) + (1-\nu)\chi(0,0) - \frac{r}{2nd}[\chi(0,nd) - \chi(0,-nd)]$$

$$- \frac{r}{2d}[\chi(d,0) - \chi(-d,0)]\cot\beta - \frac{r}{2nd}[\psi(0,nd) - \psi(0,-nd)]$$

$$= \frac{1}{8}\frac{3-2\nu}{1-\nu}m\omega^2 r^4$$

In the wheel of Fig. 5.24 there are a total of 171 unknown values of ψ and χ and the same number of equations, which, when solved, enable the calculation of the stress distribution. See the examples at the end of the book.

5.3 Finite Element Methods

The reference grids used for the solution of the equations governing elasticity problems only appear to have some physical meaning when the membrane analogy is applicable. While this need not be an obstacle to the numerical solution of the resulting finite difference equations, it may, however, hinder the interpretation of the problem and the formulation of the boundary conditions. Furthermore, the problem demands the definition of a suitable though seemingly arbitrary stress function. This fact introduces a somewhat illogical element in an otherwise logical process. Some of these difficulties are eased and new ones introduced by the finite element method, which is based on the subdivision of the whole continuous body into a number of small elements. The deformations of all the constitutive elements under externally applied forces are matched to maintain the continuity of the whole assembly while preserving the equilibrium within the elements themselves and between the adjoining elements. The lines along which the method has been developed and is presently applied are parallel to those followed by the conventional methods of structural analysis for open lattice frames. It is not within the scope of this book to discuss such frames in detail, but sufficient information will be presented to facilitate the understanding of the finite element method. Matrix notation will be used as providing a convenient shorthand for the formulation of the various equations. Although it is anticipated that most readers will be familiar with this subject, a few notes on matrix algebra will be included.

194

5.3.1 Stiffness of an elementary beam
The simplest constitutive element that can be imagined is a beam, and, although such element can only be found in open lattice frames, it is useful to consider it here as a starting point, before treating the more complex two-dimensional elements that will be required in the analysis of continua.

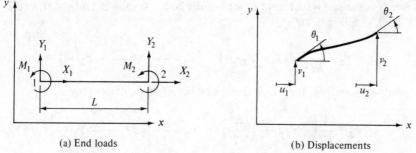

(a) End loads   (b) Displacements

Fig. 5.25 End displacements and loads in elementary beam

In the elementary beam of Fig. 5.25, under the end loads X_1, Y_1, M_1 at end 1 and X_2, Y_2, M_2 at end 2, the displacements and rotations are u_1, v_1, θ_1 and u_2, v_2, θ_2. For brevity, the array,

$$\begin{bmatrix} X_1 \\ Y_1 \\ M_1 \end{bmatrix}$$

will be referred as $\mathbf{F}_1$ and similarly,

$$\mathbf{F_2} = \begin{bmatrix} X_2 \\ Y_2 \\ M_2 \end{bmatrix}$$

$\mathbf{F}_1$ and $\mathbf{F}_2$ are called the generalised end forces. In the same way the generalised end displacements are defined by,

$$\mathbf{d}_1 = \begin{bmatrix} u_1 \\ v_1 \\ \theta_1 \end{bmatrix} \qquad \mathbf{d}_2 = \begin{bmatrix} u_2 \\ v_2 \\ \theta_2 \end{bmatrix}$$

For the beam to be in equilibrium under the action of $\mathbf{F}_1$ and $\mathbf{F}_2$, it is apparent that the following conditions must be satisfied,

$$X_1 = -X_2$$
$$Y_1 = -Y_2$$
$$M_1 = -M_2 - Y_2 L$$

Assuming first that end 1 is displaced while end 2 is fixed, as in Fig. 5.26(a), it can easily be shown by applying Castigliano's theorem that

$$\theta_1 = \frac{1}{EI}\left(M_1 L - Y_1 \frac{L^2}{2}\right) \qquad v_1 = \frac{1}{EI}\left(-M_1 \frac{L^2}{2} + Y_1 \frac{L^3}{3}\right) \qquad (5.44)$$

where I is the second moment of area of the beam. On the other hand, if end 1 is fixed, as in Fig. 5.26(b),

$$\theta_2 = \frac{1}{EI}\left(M_2 L + Y_2 \frac{L^2}{2}\right) \qquad v_2 = \frac{1}{EI}\left(M_2 \frac{L^2}{2} + Y_2 \frac{L^3}{3}\right)$$

equations that, from the equilibrium conditions, are equivalent to,

$$\theta_2 = \frac{1}{EI}\left(-M_1 L + Y_1 \frac{L^2}{2}\right) \qquad v_2 = \frac{1}{EI}\left(-M_1 \frac{L^2}{2} + Y_1 \frac{L^3}{6}\right) \qquad (5.45)$$

It it also clear that the axial force results in a change in length,

$$u_1 - u_2 = X_1 \frac{L}{EA} \qquad (5.46)$$

where A is the cross-sectional area of the beam.

Y_1 and M_1 can be expressed in terms of the end displacements and rotations v_1 and θ_1 (with $v_2 = \theta_2 = 0$) from equations (5.44). They can also be expressed in terms of v_2 and θ_2 (with $v_1 = \theta_1 = 0$) from equations (5.45), thus

$$(Y_1) \text{ due to end displacements } (1) = \frac{12EI}{L^3}v_1 + \frac{6EI}{L^2}\theta_1$$

$$(Y_1) \text{ due to end displacements } (2) = -\frac{12EI}{L^3}v_2 + \frac{6EI}{L^2}\theta_2$$

and the end loads due to the combined displacements of end 1 and 2 are

$$Y_1 = \left(\frac{12EI}{L^3}v_1 + \frac{6EI}{L^2}\theta_1\right) + \left(-\frac{12EI}{L^3}v_2 + \frac{6EI}{L^2}\theta_2\right) \qquad (5.47)$$

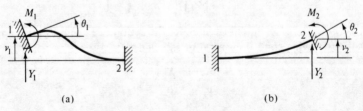

(a) (b)

Fig. 5.26 End displacements and loads in beams with one fixed end

similarly

$$M_1 = \left(\frac{6EI}{L^2}v_1 + \frac{4EI}{L}\theta_1\right) + \left(-\frac{6EI}{L^2}v_2 + \frac{2EI}{L}\theta_2\right)$$

$$X_1 = \frac{EA}{L}u_1 - \frac{EA}{L}u_2 \tag{5.48}$$

The proportionality between loads and displacements is brought out more clearly by using matrix notation, a method that has the additional advantage of providing a convenient shorthand for writing down the large number of equations with which we shall have to deal.

5.3.2 Matrix notation The generalized end forces and displacements are examples of column vectors, i.e., matrices with only one column.

A matrix

$$\mathbf{K} = \begin{bmatrix} a_1 & b_1 & c_1 \\ a_2 & b_2 & c_2 \\ a_3 & b_3 & c_3 \end{bmatrix}$$

with three rows and three columns will be referred to as a 3 x 3 square matrix. If the elements of the rows of $\mathbf{K}$ are multiplied by the corresponding elements of $\mathbf{d}$ and summed, the product $\mathbf{Kd}$ is,

$$\mathbf{Kd} = \begin{bmatrix} a_1 & b_1 & c_1 \\ a_2 & b_2 & c_2 \\ a_3 & b_3 & c_3 \end{bmatrix} \begin{bmatrix} u \\ v \\ \theta \end{bmatrix} = \begin{bmatrix} a_1 u + b_1 v + c_1 \theta \\ a_2 u + b_2 v + c_2 \theta \\ a_3 u + b_3 v + c_3 \theta \end{bmatrix}$$

which, with the addition of the appropriate subindex may refer to point 1 or 2. Note that the product is also a column vector like $\mathbf{d}$. The rules that have been used for this multiplication can also be extended to any other type of matrix with m rows and n columns. The following rules and definitions are worth remembering:

(a) Equality: Two matrices $\mathbf{A}$ and $\mathbf{B}$ are equal if they are both m x n (m rows and n columns) and their corresponding elements are equal, $a_{ij} = b_{ij}$, where $i = 1, 2, \ldots, m$ and $j = 1, 2, \ldots, n$.

(b) Addition: The sum of two matrices $\mathbf{A}$ and $\mathbf{B}$ (m x n both, otherwise addition is not possible) is another matrix $\mathbf{C}$ (also m x n) such that $c_{ij} = a_{ij} + b_{ij}$.

(c) The product of a number k and a matrix $\mathbf{A}$ is the matrix defined by the general element ka_{ij}.

(d) The product of a m x n matrix $\mathbf{A}$ and a n x p matrix $\mathbf{B}$ is the m x p matrix $\mathbf{C}$ whose elements are given by,

$$c_{ij} = \sum_{r=1}^{n} a_{ir} b_{rj}$$

197

It must be noted that the number of columns of **A** must be equal to the number of rows of **B** otherwise the multiplication is not possible. The matrices are then said to be incompatible. Taking **A** and **B** to be 3 x 3 matrices,

$$\mathbf{A} = \begin{bmatrix} a_{11} & a_{12} & a_{13} \\ a_{21} & a_{22} & a_{23} \\ a_{31} & a_{32} & a_{33} \end{bmatrix} \qquad \mathbf{B} = \begin{bmatrix} b_{11} & b_{12} & b_{13} \\ b_{21} & b_{22} & b_{23} \\ b_{31} & b_{32} & b_{33} \end{bmatrix}$$

$$\mathbf{AB} = \begin{bmatrix} (a_{11}b_{11} + a_{12}b_{21} + a_{13}b_{31})(a_{11}b_{12} + a_{12}b_{22} + a_{13}b_{32}) \\ (a_{21}b_{11} + a_{22}b_{21} + a_{23}b_{31})(a_{21}b_{12} + a_{22}b_{22} + a_{23}b_{32}) \\ (a_{31}b_{11} + a_{32}b_{21} + a_{33}b_{31})(a_{31}b_{12} + a_{32}b_{22} + a_{33}b_{32}) \end{bmatrix}$$

$$\begin{bmatrix} (a_{11}b_{13} + a_{12}b_{23} + a_{13}b_{33}) \\ (a_{21}b_{13} + a_{22}b_{23} + a_{23}b_{33}) \\ (a_{31}b_{13} + a_{32}b_{23} + a_{33}b_{33}) \end{bmatrix}$$

In general, $\mathbf{AB} \neq \mathbf{BA}$.

(e) The inverse of a matrix $\mathbf{A}, \mathbf{A}^{-1}$ is defined by

$$\mathbf{AA}^{-1} = \mathbf{A}^{-1}\mathbf{A} = \mathbf{I}$$

where **I** is a matrix in which all the elements outside the principal diagonal—running from the top left-hand corner to the botton right-hand corner—are zero while those along the principal diagonal are equal to unity.

(f) The transpose of a matrix $\mathbf{A}, \mathbf{A}^t$ is obtained by changing the rows into columns, thus if

$$\mathbf{A} = \begin{bmatrix} a_{11} & a_{12} & a_{13} \\ a_{21} & a_{22} & a_{23} \\ a_{31} & a_{32} & a_{33} \end{bmatrix} \qquad \mathbf{A}^t = \begin{bmatrix} a_{11} & a_{21} & a_{31} \\ a_{12} & a_{22} & a_{32} \\ a_{13} & a_{23} & a_{33} \end{bmatrix}$$

It follows from the above rules that

$$(\mathbf{AB})^{-1} = \mathbf{B}^{-1}\mathbf{A}^{-1}$$

$$(\mathbf{AB})^t = \mathbf{B}^t\mathbf{A}^t$$

5.3.3 Matrix representation of the force–displacement equations in the elementary beam It is now possible to write equations (5.47) and (5.48) in matrix form.

198

With the load and displacement matrices as previously defined and taking,

$$
\mathbf{K}_{11} = \begin{bmatrix} \dfrac{EA}{L} & 0 & 0 \\[2ex] 0 & \dfrac{12EI}{L^3} & \dfrac{6EI}{L^2} \\[2ex] 0 & \dfrac{6EI}{L^2} & \dfrac{4EI}{L} \end{bmatrix} \qquad
\mathbf{K}_{12} = \begin{bmatrix} \dfrac{-EA}{L} & 0 & 0 \\[2ex] 0 & \dfrac{-12EI}{L^3} & \dfrac{6EI}{L^2} \\[2ex] 0 & \dfrac{-6EI}{L^2} & \dfrac{2EI}{L} \end{bmatrix}
$$

we can write

$$
\mathbf{F}_1 = \mathbf{K}_{11}\,\mathbf{d}_1 + \mathbf{K}_{12}\mathbf{d}_2 \tag{5.49}
$$

Similarly, taking

$$
\mathbf{K}_{21} = \begin{bmatrix} \dfrac{-EA}{L} & 0 & 0 \\[2ex] 0 & \dfrac{-12EI}{L^3} & \dfrac{-6EI}{L^2} \\[2ex] 0 & \dfrac{6EI}{L^2} & \dfrac{2EI}{L} \end{bmatrix} \qquad
\mathbf{K}_{22} = \begin{bmatrix} \dfrac{EA}{L} & 0 & 0 \\[2ex] 0 & \dfrac{12EI}{L^3} & \dfrac{-6EI}{L^2} \\[2ex] 0 & \dfrac{-6EI}{L^2} & \dfrac{4EI}{L} \end{bmatrix}
$$

we find

$$
\mathbf{F}_2 = \mathbf{K}_{21}\,\mathbf{d}_1 + \mathbf{K}_{22}\,\mathbf{d}_2 \tag{5.50}
$$

Both equations (5.49) and (5.50) can be grouped together into a single one by defining a force matrix,

$$
\mathbf{F} = \begin{bmatrix} \mathbf{F}_1 \\ \mathbf{F}_2 \end{bmatrix}
$$

and a displacement matrix

$$
\mathbf{d} = \begin{bmatrix} \mathbf{d}_1 \\ \mathbf{d}_2 \end{bmatrix}
$$

together with a stiffness matrix

$$
\mathbf{K} = \begin{bmatrix} \mathbf{K}_{11} & \mathbf{K}_{12} \\ \mathbf{K}_{21} & \mathbf{K}_{22} \end{bmatrix}
$$

It will be noted that $\mathbf{F}$ and $\mathbf{d}$ are column vectors with 6 elements and $\mathbf{K}$ is a 6 x 6 matrix. Equations (5.49) and (5.50) become,

$$
\mathbf{F} = \mathbf{Kd} \tag{5.51}
$$

In Figs. 5.25 and 5.26 the system of coordinates has been chosen in such a way that the x axis is parallel to the beam, its positive direction being the same as the direction 1–2. While this has the advantage of simplifying the expression of the various matrices involved, it is seldom possible to refer the beam to such a convenient coordinate system. Let us assume that the position of the beam, relative to a general coordinate system is as shown in Fig. 5.27. The force matrix F_1' referred to the new coordinates is

$$F_1' = \begin{bmatrix} X_1' \\ Y_1' \\ M_1' \end{bmatrix} = \begin{bmatrix} \cos\alpha & -\sin\alpha & 0 \\ \sin\alpha & \cos\alpha & 0 \\ 0 & 0 & 1 \end{bmatrix} \begin{bmatrix} X_1 \\ Y_1 \\ M_1 \end{bmatrix} = TF_1 \tag{5.52}$$

a similar equation holding for F_2' and F_2. The matrix T is called the transformation matrix. From Fig. 5.27(b) it can be shown that,

$$d_1 = \begin{bmatrix} \cos\alpha & \sin\alpha & 0 \\ -\sin\alpha & \cos\alpha & 0 \\ 0 & 0 & 1 \end{bmatrix} \begin{bmatrix} u_1' \\ v_1' \\ \theta_1' \end{bmatrix} = T^t d'$$

In general

$$F' = \begin{bmatrix} X_1' \\ Y_1' \\ M_1' \\ X_2' \\ Y_2' \\ M_2' \end{bmatrix} = \begin{bmatrix} \cos\alpha & -\sin\alpha & 0 & 0 & 0 & 0 \\ \sin\alpha & -\cos\alpha & 0 & 0 & 0 & 0 \\ 0 & 0 & 1 & 0 & 0 & 0 \\ 0 & 0 & 0 & \cos\alpha & -\sin\alpha & 0 \\ 0 & 0 & 0 & \sin\alpha & \cos\alpha & 0 \\ 0 & 0 & 0 & 0 & 0 & 1 \end{bmatrix} \begin{bmatrix} X_1 \\ Y_1 \\ M_1 \\ X_2 \\ Y_2 \\ M_2 \end{bmatrix} = \begin{bmatrix} T & O \\ O & T \end{bmatrix} F$$

where O is a 3×3 null matrix, with all its elements equal to zero

$$d = \begin{bmatrix} d_1 \\ d_2 \end{bmatrix} = \begin{bmatrix} T^t & O \\ O & T^t \end{bmatrix} d'$$

We can then write,

$$F' = \begin{bmatrix} T & O \\ O & T \end{bmatrix} F = \begin{bmatrix} T & O \\ O & T \end{bmatrix} Kd$$

$$= \begin{bmatrix} T & O \\ O & T \end{bmatrix} K \begin{bmatrix} T^t & O \\ O & T^t \end{bmatrix} d' = K'd' \tag{5.53}$$

Hence, the stiffness matrix for an elementary beam, referred to an arbitrary system of coordinates, is found to be,

$$K' = \begin{bmatrix} T & O \\ O & T \end{bmatrix} K \begin{bmatrix} T^t & O \\ O & T^t \end{bmatrix}$$

where T is given by equation (5.52).

200

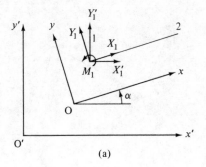

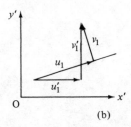

Fig. 5.27 Change of coordinate system

5.3.4 The stiffness matrix of a structure

Consider now the structure of Fig. 5.28(a) in which the elementary beam previously studied occupies the position AB, point A corresponding to end 1 and B to end 2. Other beams, CA, DA, FB, and EB are connected to AB and to anchors at S,D,E, and F. At points A and B external forces $\mathbf{F}'_A$ and $\mathbf{F}'_B$ are acting. In the diagram, the anchor points have all been labelled as end 1 for the corresponding beams.

The equilibrium conditions at A and B can be expressed in the form,

$$\mathbf{F}'_A = (\mathbf{F}'_1)_{AB} + (\mathbf{F}'_2)_{CA} + (\mathbf{F}'_2)_{DA}$$
$$\mathbf{F}'_B = (\mathbf{F}'_2)_{AB} + (\mathbf{F}'_2)_{EB} + (\mathbf{F}'_2)_{FB}$$

In these equations, the only significance of the primes is to denote that the reference system is $Ox'-Oy'$. From equations (5.49) and (5.50)

$$(\mathbf{F}'_1)_{AB} = (\mathbf{K}'_{11})_{AB}\,\mathbf{d}'_A + (\mathbf{K}'_{12})_{AB}\,\mathbf{d}'_B$$
$$(\mathbf{F}'_2)_{CA} = (\mathbf{K}'_{21})_{CA}\,\mathbf{d}'_C + (\mathbf{K}'_{22})_{CA}\,\mathbf{d}'_A$$

and the companion equations, where

$$(\mathbf{K}'_{ij})_{mn} = (\mathbf{T})_{mn}\,(\mathbf{K}_{ij})_{mn}\,(\mathbf{T}^t)_{mn}$$

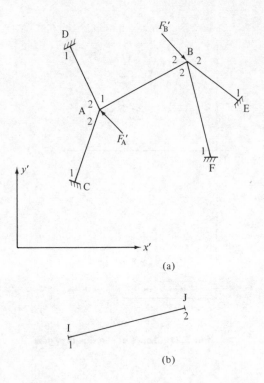

(a)

(b)

Fig. 5.28 (a) Typical open frame structure, (b) elementary beam in open-frame structure

and $i, j = 1$ or $2, m, n = $ AB, CA , . . . , FB. The nodal equations at A and B can therefore be written as a single matrix equation,

$$\begin{bmatrix} \mathbf{F}'_A \\ \mathbf{F}'_B \end{bmatrix} = \begin{bmatrix} \{(\mathbf{K}'_{11})_{AB} + (\mathbf{K}'_{22})_{CA} + (\mathbf{K}'_{22})_{DA}\}\{(\mathbf{K}'_{12})_{AB}\} \\ \{(\mathbf{K}'_{21})_{AB}\}\{(\mathbf{K}'_{22})_{AB} + (\mathbf{K}'_{22})_{EB} + (\mathbf{K}'_{22})_{FB}\} \end{bmatrix} \begin{bmatrix} \mathbf{d}'_A \\ \mathbf{d}'_B \end{bmatrix}$$

or

$$(\mathbf{F})_{\text{nodes}} = [\text{Stiffness Matrix}](\mathbf{d})_{\text{nodes}} \tag{5.54}$$

In structural analysis it is always possible to reduce the problem to the determination of the nodal displacements from the nodal forces or vice versa, although sometimes, the forces will be known at some nodes as well as the displacements of other nodes in which case the unknowns will consist of nodal forces and displacements. In any case, it is apparent that the solution of the problem is reduced to that of the simultaneous linear equations represented by equation (5.54) in which the stiffness matrix can be obtained by grouping together the

202

stiffnesses of the various constituents in the form that has been described. In a more general case, the nodal force matrix will be,

$$(\mathbf{F})_{\text{nodes}} = \begin{bmatrix} \mathbf{F}'_A \\ \mathbf{F}'_B \\ \vdots \\ \mathbf{F}'_I \\ \vdots \\ \mathbf{F}'_J \\ \vdots \\ \mathbf{F}'_N \end{bmatrix} \quad \text{and the nodal displacements,} \quad (\mathbf{d})_{\text{nodes}} = \begin{bmatrix} \mathbf{d}'_A \\ \mathbf{d}'_B \\ \vdots \\ \mathbf{d}'_I \\ \vdots \\ \mathbf{d}'_J \\ \vdots \\ \mathbf{d}'_N \end{bmatrix}$$

where I and J are two nodes connected by the single beam of Fig. 5.28(b), I being an end 1 and J an end 2 for the beam IJ. The contribution of this beam to the stiffness matrix consists in the addition of the term $(K'_{11})_{IJ}$ to the element situated on row I and column I, the terms $(K'_{12})_{IJ}, (K'_{21})_{IJ}, (K'_{22})_{IJ}$ being added to the elements on row I and column J, on row J and column I and on row J and column J respectively, i.e.,

$$[\text{Stiffness Matrix}] = \begin{bmatrix} & (I) & (J) \\ (I) - & (K'_{11})_{IJ} - & (K'_{12})_{IJ} - \\ (J) - & (K'_{21})_{IJ} - & (K'_{22})_{IJ} - \end{bmatrix} \tag{5.55}$$

with similar contributions from the remaining constituent beams.

5.3.5 Application of the method of analysis for open frames to continuous solids: General considerations   The method just described for the analysis of open frames can be divided into three clearly distinct stages:

(a) First the stiffness of each elementary beam is determined.

(b) The elements constituting the structure are assembled and the stiffness matrix of the structure as a whole is determined.

(c) The key equation—(5.54)—is solved, and the unknown nodal forces or displacements are found in terms of the known displacements or forces.

While in the problem treated all the external forces acted at the nodes, the method can easily be extended to cover more general cases, including distributed loads, etc. A detailed discussion of these problems belongs to the realm of structural engineering and is outside the scope of an introductory book on engineering elasticity. Some references of particular interest to those readers who wish to pursue the study of this subject are listed in the bibliography.

Intuitively, it would seem that the stresses and strains in a continuous solid can also be estimated by means of a similar method. However, while an open frame is a physical assembly of well defined beams and can therefore be sub-divided into its elementary components, the same is not strictly true in a con-tinuous solid in which the elementary components—grains, molecules—are of an infinitesimal size and can only be approximated by finite elements of an arbitrary size and shape. Obviously, the accuracy of the approximation will be improved by choosing the finite elements to be as small as possible.

The finite element technique consists in applying the matrix method of structural analysis to continuous solids. To this end the solid is subdivided into a number of small elements, not necessarily of uniform size, and usually of a triangular shape in plane stress or plane strain problems. In the general three-dimensional case, which will not be discussed here, small tetrahedrons or cubes may be used. It must be emphasized that the solid is not replaced by an open frame or netting but by an assembly of real two- or three-dimensional elements, triangular or otherwise, joined along their edges or faces. To match deformations and forces along contacting edges or faces in contiguous elements is clearly a task of extreme difficulty and for this reason a simplifying assumption must be made to reduce the problem to less formidable proportions. This assumption is that the elements deform in such a way that straight lines drawn on them remain straight, i.e., all straight boundaries remain straight and the continuity between two contiguous elements with two common corners is ensured when the common corners are displaced by the same amount. Furthermore, it is assumed that the boundary forces that have to be applied to each element in order to maintain equilibrium within the element while producing the necessary element deformation, is equivalent to a system of forces applied at the corners. These forces and displacements at the corners are similar to the nodal forces and displacements in the open frame structure, the corners common to several elements playing the same role as the nodes which connect several beams. The fundamental philosophy of the finite element technique is therefore to approximate the true deformation and edge loading of each element by means of corner forces and displacements, matching these forces and displacements in the same way as has been done for the open frame.

5.3.6 Stiffness of a triangular element

Consider the triangular element of Fig. 5.29. The displacement of a point P, of coordinates x and y must be pro-

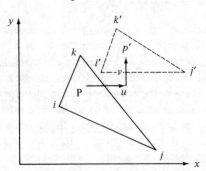

$i'\, j'\, k'$ = deformed triangular element

Fig. 5.29 Triangular element

portional to the coordinates of the point if the deformations of the element is such that straight lines are to remain straight. We can then write,

$$u = \alpha_1 + \alpha_2\, x + \alpha_3\, y$$

$$v = \alpha_4 + \alpha_5\, x + \alpha_6\, y$$

where $\alpha_1, \ldots, \alpha_6$ may be expressed in function of the displacements at the corners, u_i and v_i, u_j and v_j, u_k, and v_k. Applying the preceding equations to the corner points, we obtain,

$$
\begin{bmatrix}
1 & x_i & y_i & 0 & 0 & 0 \\
1 & x_j & y_j & 0 & 0 & 0 \\
1 & x_k & y_k & 0 & 0 & 0 \\
0 & 0 & 0 & 1 & x_i & y_i \\
0 & 0 & 0 & 1 & x_j & y_j \\
0 & 0 & 0 & 1 & x_k & y_k
\end{bmatrix}
\begin{bmatrix}
\alpha_1 \\ \alpha_2 \\ \alpha_3 \\ \alpha_4 \\ \alpha_5 \\ \alpha_6
\end{bmatrix}
=
\begin{bmatrix}
u_i \\ u_j \\ u_k \\ v_i \\ v_j \\ v_k
\end{bmatrix}
$$

Solving this system of equations we find,

$$\alpha_1 = [u_i(x_j y_k - x_k y_i) + u_j(x_k y_i - x_i y_k) + u_k(x_i y_j - x_j y_i)]/\Delta$$

$$\alpha_2 = [u_i(y_j - y_k) + u_j(y_k - y_i) + u_k(y_i - y_j)]/\Delta$$

$$\alpha_3 = [u_i(x_k - x_j) + u_j(x_i - x_k) + u_k(x_j - x_i)]/\Delta$$

$$\alpha_4 = [v_i(x_j y_k - x_k y_i) + v_j(x_k y_i - x_i y_k) + v_k(x_i y_j - x_j y_i)]/\Delta$$

$$\alpha_5 = [v_i(y_j - y_k) + v_j(y_k - y_i) + v_k(y_i - y_j)]/\Delta$$

$$\alpha_6 = [v_i(x_k - x_j) + v_j(x_i - x_k) + v_k(x_j - x_i)]/\Delta$$

$$\Delta = (x_j y_k + x_i y_j + x_k y_i) - (x_j y_i + x_i y_k + x_k y_j)$$

so that the general expression for the displacement of point P becomes,

$$u = \frac{u_i}{\Delta}[(x_j y_k - x_k y_i) + (y_i - y_k)x + (x_k - x_j)y]$$

$$+ \frac{u_j}{\Delta}[(x_k y_i - x_i y_k) + (y_k - y_i)x + (x_i - x_k)y]$$

$$+ \frac{u_k}{\Delta}[(x_i y_j - x_j y_i) + (y_i - y_j)x + (x_j - x_i)y] \tag{5.56}$$

a similar expression holding for the displacement along the y-direction, v, replacing u_i, u_j, u_k, by v_i, v_j, and v_k respectively. The strains at P e_{xx}, e_{yy}, and e_{xy} are,

$$e_{xx} = \frac{\partial u}{\partial x} = [u_i(y_j - y_k) + u_j(y_k - y_i) + u_k(y_i - y_j)]/\Delta$$

$$e_{yy} = \frac{\partial v}{\partial y} = [v_i(x_k - x_j) + v_j(x_i - x_k) + v_k(x_j - x_i)]/\Delta \qquad (5.57)$$

$$2e_{xy} = \frac{\partial u}{\partial y} + \frac{\partial v}{\partial x} = [u_i(x_k - x_j) + u_j(x_i - x_k) + u_k(x_j - x_i) + v_i(y_j - y_k)$$
$$+ v_j(y_k - y_i) + v_k(y_i - y_j)]/\Delta$$

These equations can be expressed in matrix form by defining a strain matrix **e**

$$\mathbf{e} = \begin{bmatrix} e_{xx} \\ e_{yy} \\ 2e_{xy} \end{bmatrix}$$

and a corner displacement matrix,

$$\mathbf{d} = \begin{bmatrix} u_i \\ v_i \\ u_j \\ v_j \\ u_k \\ v_k \end{bmatrix}$$

becoming,

$$\mathbf{e} = \mathbf{Bd} \qquad (5.58)$$

where **B** is the 3 x 6 matrix,

$$\mathbf{B} = \frac{1}{\Delta} \begin{bmatrix} (y_j - y_k) & 0 & (y_k - y_i) & 0 & (y_i - y_j) & 0 \\ 0 & (x_k - x_j) & 0 & (x_i - x_k) & 0 & (x_j - x_i) \\ (x_k - x_j) & (y_j - y_k) & (x_i - x_k) & (y_k - y_i) & (x_j - x_i) & (y_i - y_j) \end{bmatrix}$$

The stresses at P can be determined once the strains are known. Taking,

$$\mathbf{S} = \begin{bmatrix} \sigma_{xx} \\ \sigma_{yy} \\ \sigma_{xy} \end{bmatrix}$$

the proportionality between $\mathbf{S}$ and $\mathbf{e}$ implies that,

$$\mathbf{S} = \mathbf{De} \tag{5.59}$$

where

$$\mathbf{D} = \frac{E}{1 - v^2} \begin{bmatrix} 1 & v & 0 \\ v & 1 & 0 \\ 0 & 0 & \dfrac{1 - v}{2} \end{bmatrix}$$

in plane stress problems and

$$\mathbf{D} = \frac{E(1 - v)}{(1 + v)(1 - 2v)} \begin{bmatrix} 1 & v/(1 - v) & 0 \\ v/(1 - v) & 1 & 0 \\ 0 & 0 & (1 - 2v)/2(1 - v) \end{bmatrix}$$

in plane strain problems. On the boundary, the stresses must be balanced by edge forces which are assumed to be equivalent to a system of forces applied at the corners, X_i and Y_i, X_j and Y_j, X_k and Y_k. If we assume that there are no body forces, the corner forces constitute the only external load on the element. To find them, we write that,

Work of external forces = Work of internal stresses

for the displacement considered. It can be shown that in this way what we obtain is in fact an upper bound rather than the exact solution, but we need not go into details. The work of the external forces is,

$$2W_e = X_i u_i + Y_i v_i + X_j u_j + Y_j v_j + X_k u_k + Y_k v_k$$

$$= [u_i \, v_i \, u_j \, v_j \, u_k \, v_k] \begin{bmatrix} X_i \\ Y_i \\ X_j \\ Y_j \\ X_k \\ Y_k \end{bmatrix} = \mathbf{d}^t \, \mathbf{F}$$

while the work of the internal stresses per unit volume is,

$$2W_i = e_{xx} \sigma_{xx} + e_{yy} \sigma_{yy} + 2e_{xy} \sigma_{xy} = \mathbf{e}^t \mathbf{S} = (\mathbf{Bd})^t \, \mathbf{DBd}$$

from equations (5.58) and (5.59). Taking into account the rules given in section 5.3.2,

$$2W_i = \mathbf{d}^t \, \mathbf{B}^t \, \mathbf{DBd}$$

and equating W_e and W_i, we find,

$$\mathbf{F} = \int_{\text{volume}} (\mathbf{B}^t \mathbf{DBd})\, d(\text{volume}) = \mathbf{Kd} \qquad (5.60)$$

where

$$\mathbf{K} = \int_{\text{volume}} (\mathbf{B}^t \mathbf{DB})\, d(\text{volume})$$

since all the elements in this expression are independent of the coordinates depending solely on the geometry of the element and its elastic properties,

$$\mathbf{K} = (\mathbf{B}^t \mathbf{DB}) \times (\text{Area of triangle}) \times (\text{Thickness})$$

In these equations, $\mathbf{F}$ and $\mathbf{d}$ are column matrices with six elements and $\mathbf{K}$ is a 6×6 square matrix. Given the coordinates of i, j, and k, the value of Young's modulus and Poisson's ratio, the matrix $\mathbf{K}$ can be calculated quite easily, although it is quite obvious that the calculation is too tedious to be undertaken by any other means than by using a computer. Programming a computer to perform this calculation is a very simple problem indeed which can be simplified even further by using existing library programmes for matrix operations. Note that equation (5.60) is the same as (5.53). In both cases the stiffness matrix is referred to a general system of coordinates but, while $\mathbf{F}$ and $\mathbf{d}$ in the triangular element include only the corner forces and displacements of the three corners in the x-direction and the y-direction, $\mathbf{F}'$ includes forces and moments and $\mathbf{d}'$ displacements and rotations at the two ends of the beam.

5.3.7 Assembly of the complete stiffness matrix and solution of the problem

The assembly of the complete stiffness matrix follows the same lines as for the open frame. Consider for example Fig. 5.30 which shows some of the triangular elements into which a certain plate under plane stress or strain has been sub-

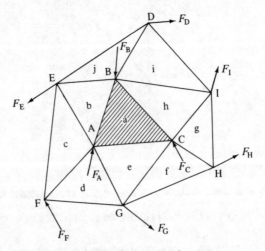

Fig. 5.30 Assembly of triangular elements

divided. It is also assumed that externally applied forces act in the manner shown. It is convenient to expand equation (5.60) in order to examine the contribution of each element to the overall stiffness, thus, for triangle ABC we write,

$$\begin{bmatrix} (F_A)_a \\ (F_B)_a \\ (F_C)_a \end{bmatrix} = \begin{bmatrix} (K_{AA})_a & (K_{AB})_a & (K_{AC})_a \\ (K_{BA})_a & (K_{BB})_a & (K_{BC})_a \\ (K_{CA})_a & (K_{CB})_a & (K_{CC})_a \end{bmatrix} \begin{bmatrix} d_A \\ d_B \\ d_C \end{bmatrix} \qquad (5.61)$$

where $(K_{AA})_a, \ldots, (K_{CC})_a$ are 2×2 matrices. Similar equations may be written for the remaining triangles. For equilibrium at A,

$$F_A = (F_A)_a + (F_A)_b + (F_A)_c + (F_A)_d + (F_A)_e$$

$$= [(K_{AA})_a + (K_{AA})_b + (K_{AA})_c + (K_{AA})_d + (K_{AA})_e]d_A$$

$$+ [(K_{AB})_a + (K_{AB})_b]d_B + [(K_{AC})_a + (K_{AC})_e]d_C$$

$$+ [(K_{AE})_b + (K_{AE})_c]d_C + [(K_{AF})_c + (K_{AF})_d]d + [(K_{AG})_d + (K_{AG})_e]d_G$$

similar equations holding for the remaining corners or nodes. Comparing this equation to the ones previously obtained in section 5.3.4 we conclude that the contribution to the stiffness matrix of a triangular element IJL consists in adding the term $(K_{IJ})_{IJL}$ to the element on row I and column J, etc., so that the stiffness matrix may be represented by,

$$\begin{bmatrix} & (I) & (J) & (L) & \\ & | & | & | & \\ (I) & - (K_{II})_{IJL} & - (K_{IJ})_{IJL} & - (K_{IL})_{IJL} & - \\ & | & | & | & \\ (J) & - (K_{JI})_{IJL} & - (K_{JJ})_{IJL} & - (K_{JL})_{IJL} & - \\ & | & | & | & \\ (L) & - (K_{LI})_{IJL} & - (K_{LJ})_{IJL} & - (K_{LL})_{IJL} & - \\ & | & | & | & \end{bmatrix}$$

all elements of which are 2×2 matrices. Thus, the elements in any one row, say row I, of the stiffness matrix can be expressed in the following general form,

$$(K_{IR})_{general} = \sum_N (K_{IR})_{IRN}$$

where the subindex R corresponds to a column in the general stiffness matrix and the summation is extended over all the triangles IRN, i.e., all the triangles having a common edge IR. When $I \neq R$, the summation is extended over all the triangles with a common corner I. It follows that the total number of non-zero elements in any given row will necessarily be very small, no matter what the size of the total matrix is, since only a few triangles will have a common edge IR or a common corner I. The total stiffness matrix is therefore very sparsely populated, most of its elements being zero and use is made of this property for the numerical solution of the problem, which, in general terms, is formulated,

$$\begin{bmatrix} \mathbf{F_A} \\ \mathbf{F_B} \\ \vdots \end{bmatrix} = [\text{Stiffness matrix}] \begin{bmatrix} \mathbf{d_A} \\ \mathbf{d_B} \\ \vdots \end{bmatrix} \tag{5.62}$$

As in the case of the open frame, the problem consists in finding the unknown forces or displacements from the known displacements or forces. For example, if the solid to be analysed is an infinitely large plate with a central hole, under uniaxial stress, it can be subdivided in the form shown in Fig. 5.31. The corner or nodal forces are then zero everywhere, except on the boundary Y-Y, where they are equal to the resultant of the applied stress. The displacements of every point are then obtained by solving equation (5.62) using one of the methods that will be briefly described in the next section in this chapter. Once the displacements are known, it is possible to calculate the strains from equation (5.58) and the stresses from (5.59).

The finite element thus described demands for its practical application the use of computers. While the actual programming does not present any conceptual difficulty, it still requires a considerable amount of tedious preparation, often more than in the case of the finite difference method. At present, a considerable effort is being put in the development of the method, whose potential and

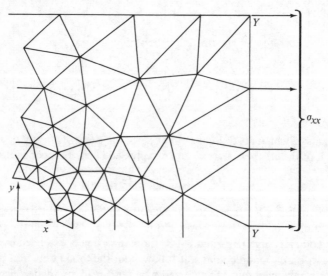

Fig. 5.31 Plate with a circular hole considered as an assembly of triangular elements

advantages over other numerical or experimental techniques remain to be assessed. It appears to be particularly useful when dealing with problems in non-linear elasticity and for the plane stress or strain analysis of heterogeneous or anisotropic solids.

5.4 Notes on the Solution of Simultaneous Linear Equations

Throughout this chapter we have seen that the problem of numerical stress analysis is reduced to the solution of a set of simultaneous linear equations, some typical examples being (5.5), (5.17), and (5.61). These equations may be represented in the following general form,

$$AX = C \tag{5.63}$$

where A is an $n \times n$ square matrix and X and C are column vectors with n elements. In this equation, the elements of X are unknown while those of A and C are known. The elements of X, x_i, may be the values of a stress function or a displacement at a number of points, A is the stiffness matrix or the matrix formed by the coefficients of the governing equation expressed in finite difference form, and C is a column of known constants, e.g., components of the nodal forces along the x- and y-direction or in general, values of a certain function of the coordinates at given points.

Both direct and successive approximations methods can be used for the solution of equation (5.63). Of the direct methods, the only one that has found some application is that of Choleski. The relaxation method, previously described, is only one of several methods based on the successive approximation to the exact solution, amongst which are included Jacobi's scheme, the Gauss–Seidel iteration method, and two particular types of the relaxation method, successive over relaxation (S.O.R.) and dynamic relaxation (D.R.). All these methods will now be described very briefly.

Choleski's method consists in transforming the original equation into

$$UX = E \tag{5.64}$$

where U is a unit upper triangular matrix, in which all the elements below the principal diagonal are zero and those on the diagonal are equal to unity. Once the transformation is achieved, the unknowns can easily be found, since equation (5.64) is the same as,

$$x_1 + \sum_{j=2}^{n} u_{1j} x_j = e_1$$

$$x_2 + \sum_{j=3}^{n} u_{2j} x_j = e_2$$

$$\cdots\cdots$$

$$x_n = e_n$$

i.e., x_n is first determined and by a process of back substitution $x_{n-1}, x_{n-2} \ldots x_j$ are all found in terms of the elements of U and E. The problem thus consists in finding these matrices. To this end, we assume that the transformation of equation (5.63) into (5.64) is defined by

$$L(UX - E) = AX - C \qquad (5.65)$$

or

$$LU = A \quad \text{and} \quad LE = C$$

in which we take L to be a lower triangular matrix, i.e., a square matrix with all its elements above the principal diagonal equal to zero. It is easy to see that equation (5.65) is equivalent to,

$$L[UE] = [AC]$$

or

$$\begin{bmatrix} l_{11} & 0 & \cdots & 0 \\ l_{21} & l_{22} & \cdots & 0 \\ \cdots & \cdots & \cdots & \cdots \\ l_{n1} & l_{n2} & \cdots & l_{nn} \end{bmatrix} \begin{bmatrix} 1 & u_{12} & \cdots & u_{1n} & e_1 \\ 0 & 1 & \cdots & u_{2n} & e_2 \\ \cdots & \cdots & \cdots & \cdots & \cdots \\ 0 & \cdots & \cdots & 1 & e_n \end{bmatrix} = \begin{bmatrix} a_{11} & \cdots & a_{1n} & c_1 \\ a_{21} & \cdots & a_{2n} & c_2 \\ \cdots & \cdots & \cdots & \cdots \\ a_{n1} & \cdots & a_{nn} & c_n \end{bmatrix}$$

$$l_{11} = a_{11}$$

$$l_{11} u_{12} = a_{12}, \quad u_{12} = \frac{a_{12}}{a_{11}}, \text{ in general, } u_{1j} = \frac{a_{1j}}{a_{11}}$$

$$l_{21} = a_{21}, \text{ in general, } l_{i1} = a_{i1}$$

$$l_{21} u_{12} + l_{22} = a_{22}, \quad l_{22} = a_{22} - a_{21} \frac{a_{12}}{a_{11}}$$

The following recurrence formulae summarise the computations required for the determination of the elements of $L, U,$ and C,

$$l_{ij} = a_{ij} - \sum_{r=1}^{j-1} l_{ir} u_{rj} \qquad l_{i1} = a_{i1}$$

$$u_{ij} = \frac{1}{l_{ii}} \left[a_{ij} - \sum_{r=1}^{i-1} l_{ir} u_{rj} \right] \qquad u_{1j} = \frac{a_{1j}}{a_{11}}$$

taking

$$a_{i, n+1} = c_i \qquad u_{i, n+1} = e_i$$

212

This method has general application and is suitable for machine computation although it requires a machine with a large storage capacity and is not particularly efficient since it does not make any use of an essential characteristic of the matrix $\mathbf{A}$, namely its sparseness, coupled with the fact that in each row there is one element, always situated on a different column, which is larger than the others. For example, when the governing equation is that of Laplace, row i corresponds to the coefficients of the finite difference approximation for node i in Fig. 5.32,

$$4x_i - x_j - x_k - x_l - x_m = 0$$

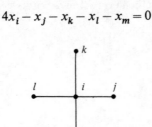

Fig. 5.32 Grid element

therefore, $a_{ii} = 4$, $a_{ij} = a_{ik} = a_{il} = a_{im} = -1$, all other elements being zero and $c_i = 0$. Similar, though slightly more complicated expressions are found in all other cases. This being so, it is advantageous to split $\mathbf{A}$ into two matrices $\mathbf{A}_1$ and $\mathbf{A}_2$ such that,

$$\mathbf{A}_1 = \begin{bmatrix} a_{11} & 0 & \dots & 0 \\ 0 & a_{22} & \dots & 0 \\ \dots & \dots & \dots & \dots \\ 0 & \dots & \dots & a_{nn} \end{bmatrix} \qquad \mathbf{A}_2 = \begin{bmatrix} 0 & a_{12} & \dots & a_{1n} \\ a_{21} & 0 & \dots & a_{2n} \\ \dots & \dots & \dots & \dots \\ a_{n1} & \dots & a_{n,n-1} & 0 \end{bmatrix}$$

in Jacobi's scheme and

$$\mathbf{A}_1 = \begin{bmatrix} a_{11} & 0 & \dots & 0 \\ a_{21} & a_{22} & \dots & 0 \\ \dots & \dots & \dots & \dots \\ a_{n1} & a_{n2} & \dots & a_{nn} \end{bmatrix} \qquad \mathbf{A}_2 = \begin{bmatrix} 0 & a_{12} & \dots & \dots & a_{1n} \\ 0 & 0 & a_{23} & \dots & a_{2n} \\ \dots & \dots & \dots & \dots & \dots \\ 0 & 0 & \dots & \dots & 0 \end{bmatrix}$$

in the Gauss-Seidel iteration method. In both cases, equation (5.63) is written in the form,

$$\mathbf{A}_1 \mathbf{X} = \mathbf{C} - \mathbf{A}_2 \mathbf{X}$$

and, given the way in which $\mathbf{A}_1$ and $\mathbf{A}_2$ have been defined, the elements of $\mathbf{X}$ that appear on the left-hand side do not appear on the right-hand side. Assigning

213

to the elements of X on the right-hand side arbitrary values, $X^{(r)}$, the corresponding values of the left-hand side are

$$A_1 X^{(r+1)} = C - A_2 X^r$$

The values $X^{(r+1)}$ thus obtained are then used on the right-hand side, and new values $X^{(r+2)}$ are derived. The iteration process is continued until the difference between the corresponding values of every unknown in two subsequent iterations is less than a specified error. The iteration formulae are,

$$a_{ii} x_i^{(r+1)} = c_i - \left[\sum_{j=1}^{i-1} a_{ij} x_j^{(r)} + \sum_{j=i+1}^{n} a_{ij} x_j^{(r)} \right] \tag{5.66}$$

according to Jacobi and

$$a_{ii} x_i^{(r+1)} = c_i - \left[\sum_{j=1}^{i-1} a_{ij} x_j^{(r+1)} + \sum_{j=i+1}^{n} a_{ij} x_j^{(r)} \right] \tag{5.67}$$

in the Gauss-Seidel method. While both methods converge ultimately into the exact solution provided that

$$a_{ii} \geqslant \sum_{j=1}^{i-1} a_{ij} + \sum_{j=i+1}^{n} a_{ij}$$

that of Gauss-Seidel converges more rapidly than Jacobi's.

The relaxation method, as described in section 5.1, is well suited for manual computation, e.g., by means of a desk calculator but it is not convenient for automatic machine computation since it relies to a very large extent on the judgement and experience of the computer if it is to converge rapidly. The methods of Jacobi and of Gauss-Seidel on the other hand rely on simple iteration formulae of a purely repetitive nature and are thus particularly suitable for automatic electronic computers. The convergence of the Gauss-Seidel iteration can be improved even further by using the successive over relaxation method. The S.O.R. method is based on an iteration formula similar to equation (5.67),

$$a_{ii} x_i^{(r+1)} = a_{ii} x_i^{(r)} (1 - \omega) + \omega \left\{ c_i - \left[\sum_{j=1}^{i-1} a_{ij} x_j^{(r+1)} + \sum_{j=i+1}^{n} a_{ij} x_j^{(r)} \right] \right\}$$

$$\tag{5.68}$$

where ω is an accelerating factor whose optimum value is given by,

$$\omega = \frac{2}{1 + \sqrt{(1 - \theta^2)}}$$

and θ is the largest latent root of,

$$\begin{bmatrix} 0 & (a_{12}/a_{11}) & \cdots & (a_{1n}/a_{11}) \\ (a_{21}/a_{22}) & 0 & \cdots & (a_{2n}/a_{22}) \\ \cdots & \cdots & \cdots & \cdots \\ (a_{n1}/a_{nn}) & \cdots & (a_{n,\,n-1}/a_{nn}) & 0 \end{bmatrix}$$

or the largest root of the equation defined by the following determinant,

$$\begin{vmatrix} \theta a_{11} & a_{12} & \cdots & a_{12} \\ a_{21} & \theta a_{22} & \cdots & a_{2n} \\ \cdots & \cdots & \cdots & \cdots \\ a_{n1} & \cdots & \cdots & \theta a_{nn} \end{vmatrix} = 0$$

in general $0 < \omega < 2$. Gauss -Seidel's iteration formula is a particular case of the S.O.R. formula for which $\theta = 0$, $\omega = 1$.

Dynamic relaxation is another iteration method recently developed which has, as yet, found a less wide application than either of the other methods previously described. Its convergence, when solving Laplace's equation, is faster than Jacobi's and Gauss-Seidel's, but less fast than that achieved by using the S.O.R. method with the optimum value of the accelerating factor. On the other hand, it has the advantage of being based on a simple iteration formula. If the governing equation is Laplace's, we may regard its solution as the steady state solution of the damped wave equation,

$$c^2 \left(\frac{\partial^2 \phi}{\partial x^2} + \frac{\partial^2 \phi}{\partial y^2} \right) = \frac{\partial^2 \phi}{\partial t^2} + \frac{K}{t} \frac{\partial \phi}{\partial t}$$

It can be shown that this equation is equivalent to,

$$\delta\phi_i^{(r+1)} = \{(1 - K/2)\, \delta\phi_i^{(r)} + \tfrac{1}{2}(\phi_k^{(r)} + \phi_m^{(r)} + \phi_j^{(r)} + \phi_l^{(r)} - 4\phi_i^{(r)})\}/(1 + K/2)$$

where

$$\phi_i^{(r+1)} = \phi_i^{(r)} + \delta\phi_i^{(r+1)}$$

using the notation of Fig. 5.32. The value of K should be chosen so as to optimize the convergence, and while no general rules can be given, it is better to choose a low value rather than a high one, i.e., it is better to under-damp the wave equation than to over-damp it. For simplicity, K may be taken to be equal to zero, in which case the D.R. iteration formula becomes,

$$\phi_k^{(r)} + \phi_m^{(r)} + \phi_j^{(r)} + \phi_l^{(r)} - 2(\phi_i^{(r+1)} + \phi_i^{(r-1)}) = 0$$

6. Discs, Cylinders, and Spheres

A significant number of engineering components exhibit rotational symmetry, both of shape and of loading. Amongst them can be included thick-walled cylinders and spheres under pressure, and also rotating discs, in which the loading may be a combination of shrinkage stresses, together with the effect of temperature gradients and of centrifugal forces.

In the stress analysis of these components no restriction is placed on the relative magnitude of the ruling dimensions. However, significant simplifications are achieved when one of these dimensions is considerably smaller than the rest. Thus, in thin discs, in which the thickness is much smaller than the radius, the existence of two free surfaces in close proximity justifies the assumption usually made of a plane stress system. Thin-walled cylinders or spheres, in which the wall thickness-to-radius ratio is less than 1:10 are widely used in pressure vessels and constitute the two most elementary examples of thin shells of revolution. Their study, also simplified by the assumption of plane stress, serves as an introduction to more complex shapes. In this chapter, discs, cylinders, and spheres will be discussed.

6.1 General Equations

Consider the element, shown in Fig. 6.1, under the effect of a centrifugal force $F = m\omega^2 r$, where m is the density, ω the rotation velocity (radians/second) and r the radius of rotation. For equilibrium, the resultants of all forces acting in the axial direction and in the radial direction must be zero, hence,

$$\left(\sigma_{zz} + \frac{\partial \sigma_{zz}}{\partial z}dz\right)rdrd\theta + \left(\sigma_{zr} + \frac{\partial \sigma_{zr}}{\partial r}dr\right)(r + dr)dzd\theta - \sigma_{zz}rdrd\theta$$

$$- \sigma_{zr}rd\theta dz = 0$$

$$\left(\sigma_{rr} + \frac{\partial \sigma_{rr}}{\partial r}dr\right)(r + dr)dzd\theta + \left(\sigma_{zr} + \frac{\partial \sigma_{zr}}{\partial z}dz\right)rdrd\theta + Frdrd\theta dz$$

$$- \sigma_{rr}rdzd\theta - \sigma_{zr}rdrd\theta - 2\sigma_{\theta\theta}drdz\frac{d\theta}{2} = 0$$

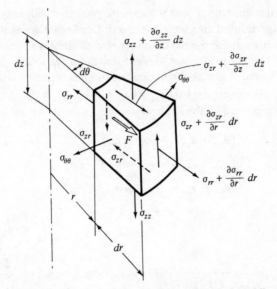

Fig. 6.1 Stresses in an element defined by cylindrical coordinates

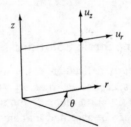

Fig. 6.2 Positive directions of displacements

These two equations reduce to the following equilibrium conditions,

$$\frac{\partial}{\partial r}(r\sigma_{zr}) + \frac{\partial}{\partial z}(r\sigma_{zz}) = 0$$

$$\frac{\partial}{\partial r}(r\sigma_{rr}) + \frac{\partial}{\partial z}(r\sigma_{zr}) - \sigma_{\theta\theta} + rF = 0$$

(6.1)

With the radial and axial components of displacement defined in Fig. 6.2, the strains are,

$$e_{rr} = \frac{\partial u_r}{\partial r} \qquad e_{\theta\theta} = \frac{u_r}{r} \qquad e_{zz} = \frac{\partial u_z}{\partial z}$$

$$e_{zr} = \tfrac{1}{2}\left(\frac{\partial u_r}{\partial z} + \frac{\partial u_z}{\partial r}\right)$$

(6.2)

The rotational symmetry of both loading and geometry will be noted. Having formulated the equilibrium conditions and the strain-displacement equations, we now proceed to establish the relationship between stresses and strains, assuming that the temperature distribution $T(r, z)$ throughout the solid is given by a certain function of the radial and axial coordinates. In a stress-free situation the strains, solely due to the temperature are $e_{zz} = e_{\theta\theta} = T$ with $e_{zr} = 0$. Adding to these strains those produced by the stresses,

$$Ee_{zz} = \sigma_{zz} - \nu(\sigma_{rr} + \sigma_{\theta\theta}) + E\alpha T$$

$$Ee_{rr} = \sigma_{rr} - \nu(\sigma_{zz} + \sigma_{\theta\theta}) + E\alpha T$$

$$Ee_{\theta\theta} = \sigma_{\theta\theta} - \nu(\sigma_{zz} + \sigma_{rr}) + E\alpha T$$

$$Ee_{zr} = (1 + \nu)\,\sigma_{zr}$$

$$(6.3)$$

equivalent to,

$$\sigma_{zz} = \frac{E}{(1 + \nu)(1 - 2\nu)}[(1 - \nu)\,e_{zz} + \nu e_{rr} + \nu e_{\theta\theta}] - \frac{E\alpha T}{1 - 2\nu} \qquad (6.4)$$

and the companion equations.

6.2 Plane Stress and Plane Strain Solutions

The general equations previously derived are easily solved when a state of plane stress or strain exists. The first case arises when σ_{zz} and σ_{zr} are very small compared to the other stresses, a situation that is found in a thin disc with stress-free faces. Furthermore, it may then be accepted that any variation through the thickness of stresses and strains is small and may be neglected. A long, thick-walled cylinder presents a somewhat different situation. Here, it may also be assumed that the stresses and strains remain unchanged along the axial direction, being thus solely a function of the radial coordinate. This implies that the shear stress σ_{zr} is zero, and, referring to the strain-displacement and stress-strain equations, that u_z remains constant through the thickness, u_r remaining independent of the axial coordinate. A plane strain condition is achieved if the ends of the cylinder are constrained between two rigid walls, in which case e_{zz} is zero. In general, the cylinder will be free to expand axially. As we shall see, the true condition of the cylinder may then be derived from a plane strain solution combined with an axial force.

6.2.1 **Thin disc**   In this case the equilibrium conditions are reduced to,

$$\frac{d}{dr}(r\sigma_{rr}) - \sigma_{\theta\theta} + rF = 0 \qquad (6.5)$$

and the strain-displacements become,

$$e_{rr} = \frac{du_r}{dr} \qquad e_{\theta\theta} = \frac{u_r}{r} \tag{6.6}$$

The axial strain, not necessarily equal to zero, is seldom required and it may be ignored. Its magnitude may be obtained from the stress-strain equations (6.4) setting $\sigma_{zz} = 0$,

$$Ee_{zz} = -\frac{E\nu}{1-\nu}(e_{rr} + e_{\theta\theta}) + \frac{1+\nu}{1-\nu}E\alpha T \tag{6.7}$$

The hoop and radial stresses are given by

$$\sigma_{\theta\theta} = \frac{E}{(1+\nu)(1-\nu)}(e_{\theta\theta} + \nu e_{rr}) - \frac{E\alpha T}{1-\nu}$$

$$\sigma_{rr} = \frac{E}{(1+\nu)(1-\nu)}(\nu e_{\theta\theta} + e_{rr}) - \frac{E\alpha T}{1-\nu}$$

Expressing in these equations the strains in terms of the radial displacement from (6.6) and using the expressions thus obtained in equation (6.5) we obtain,

$$\frac{d^2 u_r}{dr^2} + \frac{1}{r}\frac{du_r}{dr} - \frac{1}{r^2}u_r = (1+\nu)\alpha\frac{dT}{dr} - \frac{1-\nu^2}{E}m\omega^2 r \tag{6.8}$$

The general solution of this equation is,

$$u_r = Mr + \frac{N}{r} + (1+\nu)\frac{\alpha}{r}\int Tr dr - \frac{1-\nu^2}{8E}m\omega^2 r^3$$

where M and N are integration constants. From this equation, the stresses can be evaluated. To simplify the final expressions, the integration constants may be changed so that we obtain,

$$\sigma_{rr} = A - \frac{B}{r^2} - \frac{E\alpha}{r^2}\int Tr dr - \frac{3+\nu}{8}m\omega^2 r^2$$

$$\sigma_{\theta\theta} = A + \frac{B}{r^2} + \frac{E\alpha}{r^2}\int Tr dr - E\alpha T - \frac{1+3\nu}{8}m\omega^2 r^2 \tag{6.9}$$

$$Eu_r = (1-\nu)Ar + (1+\nu)\frac{B}{r} + (1+\nu)\frac{\alpha E}{r}\int Tr dr - \frac{1-\nu^2}{8}m\omega^2 r^3$$

The values of the new integration constants, A and B, will be chosen so as to satisfy the boundary conditions. In the most general case, take a turbine disc with an outside radius b, inside radius a, subjected to a pressure p_a on the inside face due, for example, to the disc being shrunk on to a shaft, while on the outside circumference uniformly distributed blades produce a centrifugal force p_b. It is clear that,

$$(\sigma_{rr})_{r=a} = -p_a \qquad (\sigma_{rr})_{r=b} = p_b$$

From these boundary conditions we find,

$$A = \frac{E\alpha}{b^2 - a^2} \int_a^b Tr\,dr + \frac{3+\nu}{8} m\omega^2(b^2 + a^2) + \frac{a^2}{b^2 - a^2} p_a + \frac{b^2}{b^2 - a^2} p_b$$

$$B = \frac{E\alpha a^2}{b^2 - a^2} \int_a^b Tr\,dr + \frac{3+\nu}{8} m\omega^2 a^2 b^2 + \frac{a^2 b^2}{b^2 - a^2}(p_a + p_b) \tag{6.10}$$

In equations (6.9), the integration takes place between a (lower limit) and r.

Equations (6.9) and (6.10) provide a complete solution for stresses and radial displacement in the most general case. From this solution, the determination of stresses or displacements in other cases follows immediately. Thus, for a solid disc $a = 0$ and,

$$A = \frac{E\alpha}{b^2} \int_0^b Tr\,dr + \frac{3+\nu}{8} m\omega^2 b^2 + p_b \tag{6.11}$$

$$B = 0$$

It is important to note that at the centre of the solid disc,

$$\sigma_{rr} = \sigma_{\theta\theta} + E\alpha T = \frac{E\alpha}{b^2} \int_0^b Tr\,dr + \frac{3+\nu}{8} m\omega^2 b^2 + p_b \tag{6.12}$$

On the other hand, on the inside surface of a disc with a small perforation in the centre, $(a \approx 0)$,

$$\sigma_{rr} = 0$$

$$\sigma_{\theta\theta} + E\alpha T = 2\left[\frac{E\alpha}{b^2} \int_0^b Tr\,dr + \frac{3+\nu}{8} m\omega^2 b^2 + p_b\right] \tag{6.13}$$

It is therefore apparent that the presence of the perforation increases the hoop stress while reducing the radial stress to zero.

6.2.2 **Thick-walled cylinder**   Consider a long cylinder of inside radius a, outside radius b, both small compared to the length. Intuitively, we can accept that plane sections, normal to the centre line, remain plane and normal, the axial strain being either zero or constant while the shear strain is zero. Furthermore, we also

220

assume that the stresses and strains are independent of the axial position. From equations (6.3) we then have,

$$\sigma_{zz} = \nu(\sigma_{rr} + \sigma_{\theta\theta}) - E\alpha T + Ee_{zz}$$

where e_{zz} is constant.

The equilibrium conditions and the strain-displacement equations are reduced to the same forms as those obtained for the plane stress case while the stress-strain equations may be written in the following form:

$$\sigma_{rr} = \frac{E}{(1+\nu)(1-2\nu)}[(1-\nu)e_{rr} + \nu e_{\theta\theta}] - \frac{E\alpha T}{1-2\nu} + \frac{\nu Ee_{zz}}{(1+\nu)(1-2\nu)}$$

$$\sigma_{\theta\theta} = \frac{E}{(1+\nu)(1-2\nu)}[\nu e_{\theta\theta} + (1-\nu)e_{rr}] - \frac{E\alpha T}{1-2\nu} + \frac{\nu Ee_{zz}}{(1+\nu)(1-2\nu)} \quad (6.14)$$

from which the following governing equation is derived following the same steps as in the previous section:

$$\frac{d^2 u_r}{dr^2} + \frac{1}{r}\frac{du_r}{dr} - \frac{1}{r^2}u_r = \frac{1+\nu}{1-\nu}\alpha\frac{dT}{dr} - \frac{(1+\nu)(1-2\nu)}{(1-\nu)E}m\omega^2 r \quad (6.15)$$

an equation that shows a strong similarity to the one previously derived. The stresses are found to be,

$$\sigma_{rr} = C - \frac{D}{r^2} - \frac{E\alpha}{(1-\nu)r^2}\int_a^r Trdr - \frac{3-2\nu}{8(1-\nu)}m\omega^2 r^2$$

$$\sigma_{\theta\theta} = C + \frac{D}{r^2} + \frac{E\alpha}{(1-\nu)r^2}\int_a^r Trdr - \frac{E\alpha T}{1-\nu} - \frac{1+2\nu}{8(1-\nu)}m\omega^2 r^2 \quad (6.16)$$

$$\sigma_{zz} = 2\nu C - \frac{E\alpha T}{1-\nu} - \frac{\nu}{2(1-\nu)}m\omega^2 r^2 + Ee_{zz}$$

Taking the internal pressure to be p_a and assuming an external pressure p_b, the boundary conditions are,

$$(\sigma_{rr})_{r=a} = -p_a \qquad (\sigma_{rr})_{r=b} = -p_b$$

From these boundary conditions we find,

$$C = \frac{E\alpha}{1-\nu}\frac{1}{b^2-a^2}\int_a^b Trdr + \frac{3-2\nu}{8(1-\nu)}m\omega^2(a^2+b^2) + \frac{a^2}{b^2-a^2}p_a - \frac{b^2}{b^2-a^2}p_b$$

$$D = \frac{E\alpha}{1-\nu}\frac{a^2}{b^2-a^2}\int_a^b Trdr + \frac{3-2\nu}{8(1-\nu)}m\omega^2 a^2 b^2 + \frac{a^2 b^2}{b^2-a^2}(p_a - p_b) \quad (6.17)$$

The axial stress and strain depend on the axial force applied to the cylinder. Thus in the case of an open ended hydraulic ram, in which the hydrostatic pressure is taken by the piston, the axial force is zero,

$$F = 0 = 2\pi \int_a^b \sigma_{zz} \, r dr \qquad (6.18)$$

while in a close-ended pressure vessel under internal and external pressure,

$$F = \pi a^2 p_a - \pi b^2 p_b = 2\pi \int_a^b \sigma_{zz} \, r dr \qquad (6.19)$$

In the absence of temperature gradients, the preceding equations lead to the following conditions,

$$Ee_{zz} = -2vC + \frac{v}{4(1-v)} m\omega^2 (b^2 + a^2)$$

$$\sigma_{zz} = \frac{v}{2(1-v)} m\omega^2 \left(\frac{b^2 + a^2}{2} - r^2 \right)$$

for the hydraulic cylinder while for the closed vessel,

$$Ee_{zz} = -2vC + \frac{v}{4(1-v)} m\omega^2 (b^2 + a^2) + \frac{1}{b^2 - a^2} (a^2 p_a - b^2 p_b)$$

$$\sigma_{zz} = \frac{v}{2(1-v)} m\omega^2 \left(\frac{b^2 + a^2}{2} - r^2 \right) + \frac{1}{b^2 - a^2} (a^2 p_a - b^2 p_b)$$

In either case, the radial displacement will be found from,

$$Eu_r = Ere_{\theta\theta} = r[\sigma_{\theta\theta} - v(\sigma_{rr} + \sigma_{zz}) + E\alpha T] \qquad (6.20)$$

6.2.3 Application to rotating discs and to cylindrical vessels

In order to illustrate the previous general treatment we now consider two practical examples. The first is a rotating disc, shrunk on to a shaft. The second is a vessel, consisting of two concentric cylinders shrunk together.

Example 1   A disc, of inside radius a and outside radius b is shrunk on to a shaft of radius $a + h$, the interference h being small in comparison with a. Both disc and shaft are at the same temperature and rotate with an angular velocity ω. The stresses in the disc are given by equations (6.9) and (6.10).

$$\sigma_{rr} = \frac{3+v}{8} m\omega^2 \left[a^2 \left(1 - \frac{b^2}{r^2} \right) + b^2 - r^2 \right] + \frac{a^2}{b^2 - a^2} p \left(1 - \frac{b^2}{r^2} \right)$$

$$\sigma_{\theta\theta} = \frac{3+v}{8} m\omega^2 \left[a^2 \left(1 + \frac{b^2}{r^2} \right) + b^2 - \frac{1+3v}{3+v} r^2 \right] + \frac{a^2}{b^2 - a^2} p \left(1 + \frac{b^2}{r^2} \right)$$

222

where p is the contact pressure between disc and shaft. The radial displacement of the disc, measured on the inside radius, is,

$$Eu_r = \frac{m\omega^2}{4}[(3+\nu)ab^2 + (1-\nu)a^3] + p\frac{1}{b^2-a^2}[(1+\nu)ab^2 + (1-\nu)a^3]$$

The shaft is a long cylinder without any axial restraint in which the stresses are given by equations (6.16), (6.17), and (6.18), suitably modified to take into account the fact that the inside radius is in fact equal to zero,

$$(\sigma_{rr})_{\text{shaft}} = \frac{3-2\nu}{8(1-\nu)}m\omega^2(a^2-r^2) - p$$

$$(\sigma_{\theta\theta})_{\text{shaft}} = \frac{3-2\nu}{8(1-\nu)}m\omega^2\left[a^2 - \frac{1+2\nu}{3-2\nu}r^2\right] - p$$

$$(\sigma_{zz})_{\text{shaft}} = \frac{\nu}{4(1-\nu)}m\omega^2(a^2-2r^2)$$

It can be checked that the axial force, required to balance the axial stress is zero. The radial displacement at $r = a$ is,

$$(Eu_r)_{\text{shaft}} = \frac{1-\nu}{4}m\omega^2 a^3 - (1-\nu)pa$$

Answers are now sought to the following questions:

(a) What are the stresses at standstill?
(b) At what velocity will the shrink fit loosen up and what will the stresses be?
(c) What are the stresses at some lower velocity?

The first question is readily answered by noting that at standstill,

$$Eh = Eu_r - (Eu_r)_{\text{shaft}} = p\frac{2ab^2}{b^2-a^2}$$

from which we obtain,

$$\sigma_{rr} = \frac{Eha}{2b^2}\left(1 - \frac{b^2}{r^2}\right) \qquad \sigma_{\theta\theta} = \frac{Eha}{2b^2}\left(1 + \frac{b^2}{r^2}\right)$$

$$(\sigma_{rr})_{\text{shaft}} = (\sigma_{\theta\theta})_{\text{shaft}} = -\frac{Eh}{2a}\left(1 - \frac{a^2}{b^2}\right)$$

The stress distribution is shown in Fig. 6.3 for the particular case when $b = 5a$, $h = a/1000$.

To answer the second question we take $p = 0$,

$$Eh = \frac{3+\nu}{4}ab^2 m\omega^2, \qquad \omega = \sqrt{\left\{\frac{4Eh}{(3+\nu)mab^2}\right\}}$$

223

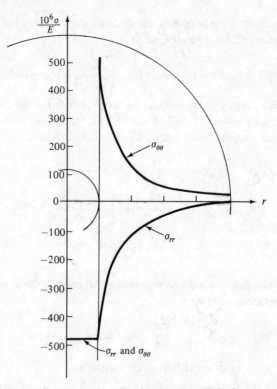

Fig. 6.3 Stress distribution in a disc shrunk onto a shaft. Static case

the stresses being,

$$(\sigma_{rr})_{\text{shaft}} = \frac{Eha}{2b^2} \frac{3 - 2\nu}{3 + \nu} \left(1 - \frac{r^2}{a^2}\right) \frac{1}{(1 - \nu)}$$

$$(\sigma_{\theta\theta})_{\text{shaft}} = \frac{Eha}{2b^2} \frac{3 - 2\nu}{3 + \nu} \left(1 - \frac{1 + 2\nu}{3 - 2\nu} \frac{r^2}{a^2}\right) \frac{1}{1 - \nu}$$

as shown in Fig. 6.4.

A simple procedure for the solution of the last part of this problem is based on the fact that the stresses and the radial displacement are proportional to the contact pressure and to the square of the angular velocity. At any given velocity,

$$Eh = \text{constant} = \omega^2 \left[\frac{m}{4}(3 + \nu) ab^2\right] + p \left[2a \frac{b^2}{b^2 - a^2}\right]$$

Since the contact pressure varies linearly with (ω^2) from a maximum at standstill to zero at the maximum angular velocity, the stresses induced by the contact

224

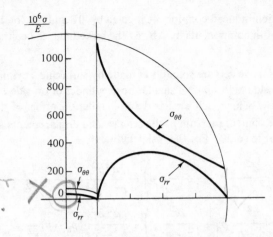

$\sigma_{\theta\theta} = \sigma_{rr}$
$t\tau = 0.$

Fig. 6.4 Stress distribution in a disc shrunk onto a shaft. Separation of disc from shaft

pressure alone also vary linearly with (ω^2) and the total stresses being the sum of the pressure induced component and of the rotation induced component are also proportional to (ω^2). For example, considering the hoop stress in the disc at the bore $(r = a)$, we obtain from the preceding equations and their graphical representations Figs. 6.3 and 6.4,

$(\sigma_{\theta\theta})_a = 0.52 \times 10^{-3} E$ at standstill, with $p = 0.48 \times 10^{-3} E$.
$(\sigma_{\theta\theta})_a = 10^{-3} E$ at separation, which occurs when $\omega^2 = 48.5 \times 10^{-6} E/ma^2$.

Representing the variation of stress against $(\omega a)^2 m/E$ as in Fig. 6.5, the pressure induced component for any given value of the angular velocity is defined by the

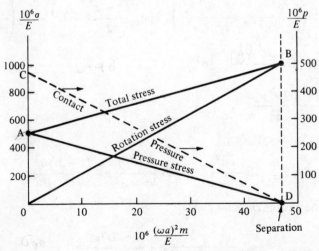

Fig. 6.5 Stress distribution in a disc shrunk onto a shaft. General solution

225

line AD. The rotation induced component is given by OB and the total stress, being the sum of both components by AB. At the same time, the contact pressure is defined by CD.

Example 2 Pressure vessels are sometimes made by shrinking a cylinder of outside radius c and inside radius b on a smaller bore cylinder of outside radius $b + h$ and inside radius a, h being much smaller than b. Under the effect of the internal pressure p_a and the contact pressure p, the stresses and displacements are given by equations (6.16) to (6.20). For the inside cylinder,

$$\sigma_{rr} = \left(\frac{a^2}{b^2 - a^2} p_a - \frac{b^2}{b^2 - a^2} p \right) - \frac{a^2 b^2}{b^2 - a^2} (p_a - p) \frac{1}{r^2}$$

$$\sigma_{\theta\theta} = \left(\frac{a^2}{b^2 - a^2} p_a - \frac{b^2}{b^2 - a^2} p \right) + \frac{a^2 b^2}{b^2 - a^2} (p_a - p) \frac{1}{r^2}$$

and for the outside cylinder,

$$\sigma_{rr} = \frac{b^2}{c^2 - b^2} p - \frac{b^2 c^2}{c^2 - b^2} p \frac{1}{r^2}$$

$$\sigma_{\theta\theta} = \frac{b^2}{c^2 - b^2} p + \frac{b^2 c^2}{c^2 - b^2} p \frac{1}{r^2}$$

The axial strain is the same in both cylinders, so that we can write,

$$2 \int_a^b \pi \sigma_{zz} r dr + 2 \int_b^c \pi \sigma_{zz} r dr = \pi a^2 p_a$$

$$\nu(b^2 - a^2) \left(\frac{a^2}{b^2 - a^2} p_a - \frac{b^2}{b^2 - a^2} p \right) + E e_{zz} \frac{b^2 - a^2}{2} + \nu(c^2 - b^2) \left(\frac{b^2}{c^2 - b^2} p \right)$$

$$+ E e_{zz} \frac{c^2 - b^2}{2} = \tfrac{1}{2} a^2 p_a$$

$$E e_{zz} = \frac{(1 - 2\nu) a^2 p_a}{c^2 - a^2}$$

and, for the inside cylinder,

$$\sigma_{zz} = 2\nu \frac{a^2}{b^2 - a^2} p_a - 2\nu \frac{b^2}{b^2 - a^2} p + \frac{(1 - 2\nu) a^2}{c^2 - a^2} p_a$$

while for the outside cylinder,

$$\sigma_{zz} = 2\nu \frac{b^2}{c^2 - b^2} p + \frac{(1 - 2\nu) a^2}{c^2 - a^2} p_a$$

226

The radial displacements at the interface are, for the inside cylinder,

$$\frac{1}{b}Eu_r = p_a\left(\frac{2(1-v^2)a^2}{b^2-a^2} - \frac{(1-2v)va^2}{c^2-a^2}\right) - p\frac{(1+v)[a^2+(1-2v)b^2]}{b^2-a^2}$$

and,

$$\frac{1}{b}Eu_r = p\frac{(1+v)[c^2+(1-2v)b^2]}{c^2-b^2} - \frac{v(1-2v)}{c^2-a^2}a^2 p_a$$

for the outside cylinder. In the same way as in the case of the shrunk-on disc,

$$\frac{1}{b}Eh = \frac{1}{b}(Eu_r)_{\text{outside}} - \frac{1}{b}(Eu_r)_{\text{inside}}$$

$$= \frac{2(1-v^2)}{(b^2-a^2)(c^2-b^2)}[pb^2(c^2-a^2) - p_a a^2(c^2-b^2)] \quad (6.21)$$

From this equation the contact pressure may be obtained. The whole process is, however, rather tedious and lengthy and can be simplified by noting that the stresses are proportional to $(1/r^2)$. Consider first the stresses caused by the contact pressure alone, obtained from the previous equations by taking $p_a = 0$. Plotting in Fig. 6.6(a) the values $1/a^2$, $1/b^2$, and $1/c^2$ on the horizontal axis, at either side of the vertical axis, point A is defined as shown, measuring contact pressure vertically downwards. The radial stresses are then represented by lines ABC and the hoop stresses by DE and FG. It is easy to see that the same construction can be applied to the general case, illustrated in Fig. 6.6(b), the only difference being that the contact pressure will first be found from equation (6.21).

6.3 Cylindrical Shells

In practice, pressure vessels and containment tanks consist of a main body, usually cylindrical, to which are welded the closure ends and a number of attachments for the support of the vessel itself and of any internal components. The equations previously derived can then be used to determine the stresses in the main cylindrical shell away from the attachments but they fail to give any indication of the stresses introduced by these, since for the exact solution of the problem it becomes necessary to take into account the possible variations of stresses and strains in the axial and hoop directions as well as in the radial direction.

A substantial simplification which makes practicable the solution of this problem is based on the fact that in the majority of the vessels the wall thickness is relatively small compared to the diameter, being usually less than one-tenth of the diameter. This means that terms such as $a^2/(b^2-a^2)$ can be replaced by

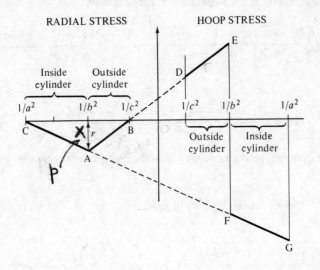

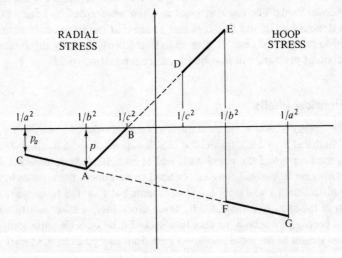

Fig. 6.6 Graphical determination of stresses in composite vessels

$R/2t$, where R is the mean radius and t the wall thickness. In the absence of temperature gradients or rotating forces, the approximate value of the hoop stress is,

$$\sigma_{\theta\theta} \approx \frac{pR}{t}$$

while the axial stress is,

$$\sigma_{zz} \approx \frac{pR}{2t}$$

where p is positive when the vessel is under internal pressure, negative when the pressure acts from the outside of the vessel. The maximum radial stress, equal to $(-p)$, can be neglected since it is less than one-tenth of the hoop stress.

Besides neglecting the radial stress, it is also assumed that the deformation due to shear stresses is small compared to the extensional strains. This implies that plane sections, normal to the mid-shell surface, remain plane and normal to the deformed surface. The variation of σ_{zz} and $\sigma_{\theta\theta}$ through the thickness is linear.

Under rotationally symmetrical loading the stresses acting on a small element are illustrated in Fig. 6.7. In general stresses vary with r and z.

In the analysis of thin shells it is found convenient to work with the stress resultants rather than with the actual stresses. These resultants are defined by equations of the type,

$$N_{zz} = \int_{R-t/2}^{R+t/2} \sigma_{zz}\, dr \tag{6.22a}$$

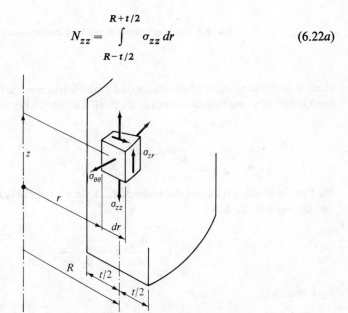

Fig. 6.7 Defining coordinates for a cylindrical shell element

229

for the forces and,

$$M_{zz} = \int\limits_{R-t/2}^{R+t/2} \sigma_{zz}\, r\, dr \qquad (6.22b)$$

for the moments. Both force and moment resultants are expressed per unit length. Positive force resultants correspond to tensile stresses and positive moments to tension on the outside face, as shown in Fig. 6.8. The equilibrium conditions, in terms of the resultant forces and moments, may be obtained

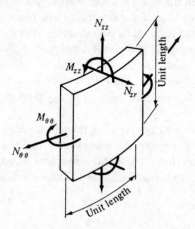

Fig. 6.8 Stress resultants acting on a shell element

directly or by integration of equations (6.1), which represent the conditions of equilibrium of a small element within the shell. Taking the first equation,

$$\int\limits_{R-t/2}^{R+t/2} \frac{\partial}{\partial r}(r\sigma_{zr})\, dr + \int\limits_{R-t/2}^{R+t/2} r\frac{\partial \sigma_{zz}}{\partial z}\, dr = 0$$

the first integral is zero since the value of σ_{zr} for $r = R \pm (t/2)$ is zero. The second integral can be approximated by,

$$R \int\limits_{R-t/2}^{R+t/2} \frac{\partial \sigma_{zz}}{\partial z}\, dr = R\frac{\partial N_{zz}}{\partial z} = 0$$

equivalent to,

$$RN_{zz} = V \qquad (6.23)$$

230

where V is an integration constant whose physical meaning is the externally applied axial force per radian. In a closed cylinder under internal pressure the total axial force is $(\pi R^2 p)$,

$$V = \frac{pR^2}{2}$$

In the absence of a centrifugal force, the integration of the second of equations (6.1) gives another equilibrium condition in terms of forces,

$$R\frac{dN_{zr}}{dz} - N_\theta = -pR \qquad (6.24)$$

noting that σ_{rr} equals $(-p)$ on the inside wall surface and is zero on the outside.

Moment equilibrium is obtained by multiplying the first equation (6.1) by $(r - R)$ prior to integrating. Since,

$$\int_{R-t/2}^{R+t/2} (r - R)\frac{\partial}{\partial r}(r\sigma_{zr})\,dr = \int_{R-t/2}^{R+t/2} \frac{\partial}{\partial r}[(r - R)\,r\sigma_{zr}]\,dr - \int_{R-t/2}^{R+t/2} r\sigma_{zr}\,dr$$

and the first integral on the right is zero on account of σ_{zr} being zero at both limits, the remaining terms are approximately equal to,

$$-R\int_{R-t/2}^{R+t/2} \sigma_{zr}\,dr + R\int_{R-t/2}^{R+t/2} (r - R)\frac{\partial\sigma_{zz}}{\partial z}\,dr = 0$$

providing the third equilibrium condition,

$$-N_{zr} + \frac{dM_{zz}}{dz} = 0 \qquad (6.25)$$

The next step is to establish stress-strain equations. Since σ_{rr} may be neglected, these are,

$$Ee_{zz} = \sigma_{zz} - v\sigma_{\theta\theta} + E\alpha T$$

$$Ee_{\theta\theta} = \sigma_{\theta\theta} - v\sigma_{zz} + E\alpha T$$

or

$$\sigma_{zz} = \frac{E}{(1 - v^2)}(e_{zz} + ve_{\theta\theta}) - \frac{E\alpha T}{1 - v} \qquad (6.26)$$

$$\sigma_{\theta\theta} = \frac{E}{(1 - v^2)}(e_{\theta\theta} + ve_{zz}) - \frac{E\alpha T}{1 - v}$$

together with,

$$Ee_{zr} = (1 + v)\,\sigma_{zr}$$

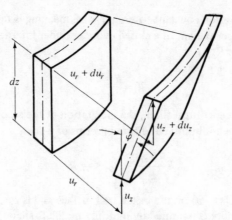

Fig. 6.9 Deformation of a shell element

In the same way as the stresses on an element are replaced by the resultant forces and moments, it is convenient to define the displacements of a point by means of those of the mid-shell surface and the rotation experienced by the shell—see Fig. 6.9. It is clear from the figure that the displacements $(u_z)_r$ and $(u_r)_r$ at point r are,

$$(u_z)_r = u_z - (r - R)\varphi \qquad (u_r)_r = u_r$$

where u_z, u_r, and φ are the axial and radial displacements and the rotation of the mid-shell surface respectively. Noting that $\varphi = du_r/dz$,

$$e_{zz} = \frac{du_z}{dz} - (r - R)\frac{d^2 u_r}{dz^2} \qquad e_{\theta\theta} = \frac{u_r}{r}$$

From the stress-strain equations (6.26), together with those defining the stress resultants (6.22a and b), we obtain,

$$N_{zz} = \frac{E}{1 - v^2}\left[t\frac{du_z}{dz} + vu_r\ln\frac{R + t/2}{R - t/2}\right] - N$$

$$\approx \frac{Et}{1 - v^2}\left[\frac{du_z}{dz} + v\frac{u_r}{R}\right] - N$$

$$N_{\theta\theta} = \frac{Et}{1 - v^2}\left[v\frac{du_z}{dz} + \frac{u_r}{R}\right] - N$$

$$M_{zz} = -\frac{Et^3}{12(1 - v^2)}\frac{d^2 u_r}{dz^2} - M$$

$$M_{\theta\theta} = -v\frac{Et^3}{12(1 - v^2)}\frac{d^2 u_r}{dz^2} - M \tag{6.27}$$

232

where

$$N = \int_{R-t/2}^{R+t/2} \frac{\alpha ET}{1-\nu} \, dr$$

$$M = \int_{R-t/2}^{R+t/2} \frac{\alpha ET}{1-\nu} (r-R) \, dr$$

are temperature induced terms.

In the analysis of thin shells, the stress resultants are determined first. The stresses are then derived by means of the following equations,

$$\sigma_{zz} = \frac{N_{zz}}{t} + \frac{12 M_{zz}(r-R)}{t^3} + \sigma$$

$$\sigma_{\theta\theta} = \frac{N_{\theta\theta}}{t} + \frac{12 M_{\theta\theta}(r-R)}{t^3} + \sigma$$

where

$$\sigma = \frac{N}{t} + \frac{12 M(r-R)}{t^3} - \frac{\alpha ET}{1-\nu}$$

It will be noted that this temperature induced term is zero when the temperature remains constant through the shell thickness.

The above expressions for the stresses are a direct result of the assumption made concerning the linear variation through the thickness. The shear stress is seldom, if ever, required. It may be found from N_{zr}, assuming a parabolic distribution similar to the shear-stress variation in a beam of rectangular cross-section.

6.3.1 Governing equation The two equilibrium equations (6.24) and (6.25) can be combined into a single equation, eliminating the derivative of the shear force between (6.24) and the derivative of (6.25),

$$R \frac{d^2 M_{zz}}{dz^2} - N_{\theta\theta} = -pR$$

and expressing in this equation the stress resultants in terms of the mid-shell displacements (equation (6.27)). We find,

$$\frac{d^4 u_r}{dz^4} + \frac{12}{R^2 t^2} u_r = \frac{1}{DR} \left[pR - R \frac{d^2 M}{dz^2} + N \right] - \frac{12\nu}{t^2 R} \frac{du_z}{dz}$$

where D is the *shell stiffness,* defined by the expression,

$$D = \frac{Et^3}{12(1-\nu^2)}$$

16

From the first equation (6.27),

$$\frac{du_z}{dz} = \frac{1-v^2}{Et}[N_{zz} + N] - v\frac{u_r}{R} = \frac{1-v^2}{Et}\left[\frac{V}{R} + N\right] - v\frac{u_r}{R}$$

and taking the *shell characteristic* β to be given by

$$\beta^4 = \frac{3(1-v^2)}{t^2}R^2$$

the preceding equation becomes,

$$\frac{d^4 u_r}{dz^4} + \frac{4\beta^4}{R^4}u_r = \frac{1}{DR}\left[pR - v\frac{V}{R} - R\frac{d^2 M}{dz^2} + (1-v)N\right] \qquad (6.28)$$

Equation (6.28) governs the overall behaviour of the cylindrical shell. From its solution, once the radial displacement is known, it is possible to determine (du_z/dz) and the stress resultants from equations (6.27). The actual stresses then follow immediately. The direct solution of this governing equation may be rather difficult, depending on the form of its right-hand side, i.e., on the pressure and temperature distributions. For this reason it is invariably found easier to use the indirect method that will be described. This is based on the solution of the governing equation in two particular cases (a) in the absence of any moment resultant with the temperature term $M = 0$ and a linear variation of the pressure p and the temperature term N along the axis, (b) when the only external loading consists of forces and moments applied at the ends. The complete solution is obtained by combining both particular solutions. In case (a) the shell is said to behave as a membrane, being subjected solely to stresses that are uniform through thickness, with zero moment resultants. Case (b) corresponds to the edge loading solution.

6.3.2 Membrane solution Consider the cylindrical shell of Fig. 6.10 under an internal pressure that varies linearly between p_1 and p_2 at ends 1 and 2 respectively. The axial force is taken to be $2\pi V$ and it is assumed that the temperature, uniform through the shell thickness, varies linearly between T_1 and T_2. From the definition of N and M it is seen that, any given point,

$$M = 0 \qquad N = \frac{\alpha ETt}{1-v}$$

A solution of the governing equation is,

$$u_r = \frac{R}{Et}\left[pR + \alpha ETt - v\frac{V}{R}\right]$$

Since p and T vary linearly with the axial coordinate z, $(d^2 u_r/dz^2)$ is zero. It follows, from equations (6.27) that all bending moments are zero. The stresses

234

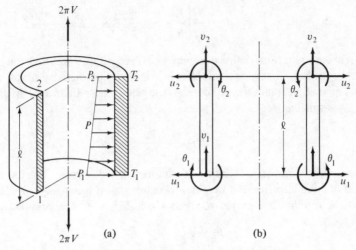

Fig. 6.10 Cylindrical shell under distributed load

are then uniform through the shell thickness and given by,

$$\sigma_{zz} = \frac{N_{zz}}{t} = \frac{V}{Rt}$$

$$\sigma_{\theta\theta} = \frac{N_{\theta\theta}}{t} = \frac{pR}{t}$$

(6.29)

It will be noted that, for a closed cylinder under uniform internal pressure,

$$\sigma_{zz} = \frac{pR}{2t}$$

The radial displacements u at ends 1 and 2 are,

$$u_1 = \frac{R}{Et}\left[p_1 R + \alpha Et T_1 - v\frac{V}{R}\right]$$

$$u_2 = \frac{R}{Et}\left[p_2 R + \alpha Et T_2 - v\frac{V}{R}\right]$$

(6.30)

The sign convention is given in Fig. 6.10(b).

The end rotations are,

$$\theta_1 = \theta_2 = -\frac{du_r}{dz} = \frac{R}{Etl}[(p_1 - p_2)R + \alpha Et(T_1 - T_2)]$$

(6.31)

Apart from a rigid body movement, the axial displacement of the ends is defined by $(w_2 - w_1)$,

235

$$w_2 - w_1 = \int_1^2 e_{zz}\, dz = \frac{l}{Et}\left[\frac{V}{R} - \nu\frac{p_1 + p_2}{2}R + E\alpha t\frac{T_1 + T_2}{2}\right] \qquad (6.32)$$

The preceding equations define a membrane behaviour of the cylindrical shell, in which the loading—mechanical and thermal—induce a particularly simple state of stress. The shell becomes slightly conical, the generators remain straight and rotate by a small angle,

$$\theta = \frac{u_1 - u_2}{l}$$

As an example, consider a cylindrical container, closed at end 1, open at end 2 and full of a liquid of specific weight m. Assume that it is hanging from its upper rim, end 2, and that the temperature varies as in Fig. 6.10. The pressure is,

$$p_2 = 0 \qquad p_1 = lm$$

The vertical force is equal to the weight of the fluid plus that of the container itself, say W,

$$2\pi V = W + \pi R^2\, lm$$

and from equations (6.29) to (6.32),

$$\sigma_{zz} = \frac{W}{2\pi Rt} + \frac{lmR}{2t}$$

$\sigma_{\theta\theta} = pR/t$, zero at end (2), attains linearly a maximum at end (1) equal to lmR/t.

$$u_1 = \frac{R}{Et}\left[\left(1 - \frac{\nu}{2}\right)lmR + \alpha EtT_1 - \frac{\nu W}{2\pi R}\right]$$

$$u_2 = \frac{R}{Et}\left[\alpha EtT_1 - \frac{\nu}{2}lmR - \frac{\nu W}{2\pi R}\right]$$

$$\theta_1 = \theta_2 = \frac{R}{Etl}[lmR + \alpha Et(T_1 - T_2)]$$

$$w_2 - w_1 = \frac{l}{Et}\left[\frac{W}{2\pi R} + \tfrac{1}{2}(1 - \nu)lmR + E\alpha t\frac{T_1 + T_2}{2}\right]$$

This represents the complete and exact solution of the problem provided that no restraints are imposed at end (1) by the closure head and at end (2) by the supports. If this is not the case, the true end displacements and rotations may no longer be equal to those calculated—they could in fact be equal to zero if both ends were clamped to perfectly stiff components. In general, there will always be some interaction between the cylindrical shell on the one part and the closure

236

head or the support on the other. The membrane solution that has been obtained provides an indication of the free or unrestrained end displacements but it fails to assess the importance of these restraints. How to do this will be described next.

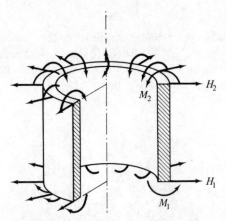

Fig. 6.11 Edge loading of cylindrical shell

6.3.3 Edge loading solution The restraining effect of closure heads, rings, flanges, or any other components attached to the ends of the cylindrical shell is equivalent to the imposition of a radial force and a moment, distributed along the edge as shown in Fig. 6.11. This edge loading, expressed in units of force or moment *per radian,* is positive when acting as shown. When the shell is loaded solely at the edges, the right-hand side of the governing equation (6.28) is zero and the equation is simplified to,

$$R^4 \frac{d^4 u_r}{dz^4} + 4\beta^4 u_r = 0$$

whose solution is,

$$u_r = C_1 f_1 + C_2 f_2 + C_3 f_3 + C_4 f_4 \tag{6.33}$$

where $C_1, \ldots, C_4$ are integration constants and

$$f_1 = e^{\beta z/R} \cos\frac{\beta z}{R} \qquad f_3 = e^{-\beta z/R} \cos\frac{\beta z}{R}$$

$$f_2 = e^{\beta z/R} \sin\frac{\beta z}{R} \qquad f_4 = e^{-\beta z/R} \sin\frac{\beta z}{R}$$

The integration constants are fixed by the end conditions. When these are stated in terms of forces,

$$\left.\begin{array}{l} RN_{zr} = -H_1 \\ RM_{zz} = -M_1 \end{array}\right\} \text{end 1} \qquad \left.\begin{array}{l} RN_{zr} = H_2 \\ RM_{zz} = M_2 \end{array}\right\} \text{end 2}$$

while in terms of displacements and rotations,

$$u_r = u_1 \atop \dfrac{du_r}{dr} = -\theta_1 \Bigg\} \ \text{end 1} \qquad u_r = u_2 \atop \dfrac{du_r}{dr} = -\theta_2 \Bigg\} \ \text{end 2}$$

the axial forces and displacement are included in the membrane solution. Using the matrix notation that was introduced in the previous chapter, and bearing in mind the expressions for $f_1, \ldots, f_4$, it has been shown by Bickell and Ruiz, *Pressure Vessel Analysis and Design,* that,

$$
\begin{bmatrix} C_1 \\ C_2 \\ C_3 \\ C_4 \end{bmatrix} = \frac{1}{4}
\begin{bmatrix}
(2\lambda_1 - \mu_1 - 1) & 2(\mu_1 - \nu_1) & (2\lambda_2 + \mu_2) & 2(\mu_2 - \nu_2) \\
(1 - \mu_1) & 2(1 - \nu_1) & \mu_2 & -2\nu_2 \\
(2\lambda_1 + \mu_1 + 1) & 2(\mu_1 + \nu_1) & (2\lambda_2 - \mu_2) & 2(\mu_2 + \nu_2) \\
(1 - \mu_1) & -2(1 + \nu_1) & \mu_2 & -2\nu_2
\end{bmatrix}
\begin{bmatrix} \dfrac{R^2 H_1}{2\beta^3 D} \\[2ex] \dfrac{R M_1}{2\beta^2 D} \\[2ex] \dfrac{R^2 H_2}{2\beta^3 D} \\[2ex] \dfrac{R M_2}{2\beta^2 D} \end{bmatrix}
$$

$$(6.34)$$

where $\lambda_1, \mu_1, \nu_1, \lambda_2, \mu_2, \nu_2$ are tabulated in Table 6.1. The derivation of these functions is a straightforward algebraic problem which does not require any explanation. It is important to note that in long cylindrical shells, λ_1, μ_1, and ν_1 are equal to unity while the remaining functions tend towards zero. Equations (6.34) then simplify to:

$$C_1 = C_2 = 0$$

$$C_3 = \frac{R H_1}{2\beta^3 D} + \frac{R M_1}{2\beta^2 D}$$

$$C_4 = -\frac{R M_1}{2\beta^2 D}$$

valid for $\beta l/R$ larger than about 3. The physical meaning of these expressions is that the behaviour of end 1 is not influenced by whatever happens at the remote end 2.

Once the integration constants have been determined, it becomes possible to find the radial displacement at any point from equation (6.33), while the strains and the stress resultants are given, as functions of the radial displacement, by the corresponding equations (6.26) and (6.27).

At the ends, the radial displacements and rotations are obtained by taking $z = 0$ (end 1) and $z = 1$ (end 2) in the general equations. It is then found,

238

Table 6.1

$\beta\dfrac{l}{R}$	λ_1	μ_1	ν_1	λ_2	μ_2	ν_2
0·35	5·715	24·50	70·10	− 2·857	24·48	69·93
0·37	5·406	21·93	59·36	− 2·702	21·91	59·18
0·39	5·129	19·74	50·72	− 2·563	19·71	50·52
0·41	4·879	17·86	43·68	− 2·438	17·84	43·48
0·43	4·653	16·24	37·89	− 2·324	16·21	37·68
0·45	4·446	14·84	33·09	− 2·221	14·80	32·86
0·47	4·257	13·60	29·07	− 2·126	13·57	28·83
0·49	4·084	12·52	25·68	− 2·039	12·48	25·44
0·51	3·924	11·56	22·81	− 1·959	11·52	22·55
0·53	3·776	10·71	20·35	− 1·885	10·66	20·08
0·55	3·640	9·949	18·24	− 1·816	9·899	17·96
0·57	3·512	9·268	16·41	− 1·752	9·214	16·13
0·59	3·394	8·655	14·83	− 1·692	8·597	14·53
0·61	3·283	8·101	13·44	− 1·636	8·039	13·14
0·63	3·179	7·600	12·23	− 1·584	7·534	11·92
0·65	3·082	7·145	11·17	− 1·535	7·074	10·84
0·67	2·991	6·730	10·22	− 1·488	6·655	9·889
0·69	2·905	6·351	9·388	− 1·445	6·272	9·044
0·71	2·824	6·004	8·645	− 1·403	5·920	8·291
0·73	2·747	5·685	7·983	− 1·364	5·597	7·618
0·75	2·675	5·392	7·389	− 1·327	5·299	7·015
0·80	2·510	4·754	6·156	− 1·243	4·648	5·757
0·85	2·365	4·230	5·205	− 1·169	4·101	4·783
0·90	2·236	3·788	4·449	− 1·101	3·654	4·000
0·95	2·121	3·421	3·856	− 1·042	3·274	3·331
1·00	2·019	3·104	3·370	− 0·986	2·939	2·873
1·05	1·927	2·836	2·980	− 0·936	2·654	2·458
1·10	1·843	2·605	2·660	− 0·890	2·405	2·115
1·15	1·768	2·406	2·397	− 0·848	2·188	1·827
1·20	1·699	2·232	2·178	− 0·809	1·996	1·585
1·25	1·637	2·081	1·996	− 0·773	1·825	1·379
1·50	1·395	1·562	1·435	− 0·621	1·200	0·706
1·75	1·239	1·284	1·187	− 0·501	0·805	0·355
2·00	1·138	1·134	1·076	− 0·400	0·535	0·155
2·25	1·074	1·057	1·029	− 0·312	0·341	0·037
2·50	1·037	1·020	1·010	− 0·235	0·200	− 0·032
2·75	1·016	1·005	1·005	− 0·168	0·098	− 0·069
3·00	1·007	1·000	1·004	− 0·113	0·028	− 0·085
3·50	1·001	1·001	1·003	− 0·035	− 0·042	− 0·078
4·00	1·001	1·002	1·002	0·004	− 0·056	− 0·052
4·50	1·001	1·001	1·001	0·017	− 0·043	− 0·027
5·00	1·000	1·000	1·000	0·017	− 0·026	− 0·009
5·50	1·000	1·000	1·000	0·011	− 0·012	0·001
6·00	1·000	1·000	1·000	0·006	− 0·003	0·003

$$
\begin{bmatrix} u_1 \\ \theta_1 \\ u_2 \\ \theta_2 \end{bmatrix} =
\begin{bmatrix}
\dfrac{R^2 \lambda_1}{2\beta^3 D} & \dfrac{R\mu_1}{2\beta^2 D} & \dfrac{R^2 \lambda_2}{2\beta^3 D} & \dfrac{R\mu_2}{2\beta^2 D} \\[2ex]
\dfrac{R\mu_1}{2\beta^2 D} & \dfrac{\nu_1}{\beta D} & -\dfrac{R\mu_2}{2\beta^2 D} & \dfrac{\nu_2}{\beta D} \\[2ex]
\dfrac{R^2 \lambda_2}{2\beta^3 D} & -\dfrac{R\mu_2}{2\beta^2 D} & \dfrac{R^2 \lambda_1}{2\beta^3 D} & -\dfrac{R\mu_1}{2\beta^2 D} \\[2ex]
\dfrac{R\mu_2}{2\beta^2 D} & \dfrac{\nu_2}{\beta D} & -\dfrac{R\mu_1}{2\beta^2 D} & \dfrac{\nu_1}{\beta D}
\end{bmatrix}
\begin{bmatrix} H_1 \\ M_1 \\ H_2 \\ M_2 \end{bmatrix} = \mathbf{AF}
$$

$$(6.35)$$

where $\mathbf{A}$ is the flexibility matrix. In a long cylinder,

$$
\mathbf{A} =
\begin{bmatrix}
\dfrac{R^2}{2\beta^3 D} & \dfrac{R}{2\beta^2 D} & 0 & 0 \\[2ex]
\dfrac{R}{2\beta^2 D} & \dfrac{1}{\beta D} & 0 & 0 \\[2ex]
0 & 0 & \dfrac{R^2}{2\beta^3 D} & -\dfrac{R}{2\beta^2 D} \\[2ex]
0 & 0 & -\dfrac{R}{2\beta^2 D} & \dfrac{1}{\beta D}
\end{bmatrix}
$$

As an example, we take a vessel, closed at both ends by rigid closure heads, hanging from the top, with the following dimensions:

$$R = 2 \text{ m} \qquad l = 10 \text{ m} \qquad t = 0\cdot5 \text{ cm}$$

take $E = 2 \times 10^5$ MN/m^2, $\nu = 0\cdot3$. From the defining equations, $\beta = 25\cdot708$, $D = 2290$. From Table 6.1, $\lambda_1 = \mu_1 = \nu_1 = 1$. It is assumed that the vessel is filled with a fluid of density 1500 kg/m^3, unpressurized.

The total weight of vessel and contents hanging from the top (end 2) is 2 MN.

If the vessel behaves in a membrane manner, from equations (6.29) to (6.32) we find,

$\left. \begin{array}{l} (\sigma_{zz})_{\text{mem}} = 31\cdot8 \text{ MN/m}^2 \\ (\sigma_{\theta\theta})_{\text{mem}} = 58\cdot8 \text{ MN/m}^2 \end{array} \right\}$ end 1 $\qquad \left. \begin{array}{l} (\sigma_{zz})_{\text{mem}} = 31\cdot8 \text{ MN/m}^2 \\ (\sigma_{\theta\theta})_{\text{mem}} = 0 \end{array} \right\}$ end 2

$(u_1)_{\text{mem}} = 0\cdot492$ mm $\qquad\qquad\qquad (u_2)_{\text{mem}} = -0\cdot096$ mm

$(\theta_1)_{\text{mem}} = (\theta_2)_{\text{mem}} = 0\cdot588 \times 10^{-4}$ rad $\qquad w_2 - w_1 = 1\cdot15$ mm

From these results, it is apparent that while satisfying the conditions of equilibrium we have violated the end conditions, since both radial displacements and

240

rotations must be zero at the junction with the rigid closure heads. This can only be achieved if end reactions exerted by the closure heads are added to the membrane solution. The radial displacements and rotations produced by this end loading must be such that,

$$(u_1)_{\text{total}} = (u_1)_{\text{membrane}} + (u_1)_{\text{end loading}} = 0$$

and the equivalent conditions for u_1, θ_1, and θ_2. Equations (6.35) provide a means for the calculation of the end forces and moments,

$$
\begin{bmatrix}
-0{\cdot}492 \times 10^{-3} \\
-0{\cdot}588 \times 10^{-4} \\
0{\cdot}096 \times 10^{-3} \\
-0{\cdot}588 \times 10^{-4}
\end{bmatrix}
$$

$$
=
\begin{bmatrix}
0{\cdot}514 \times 10^{-7} & 0{\cdot}66 \times 10^{-6} & 0 & 0 \\
0{\cdot}66 \times 10^{-6} & 1{\cdot}7 \times 10^{-5} & 0 & 0 \\
0 & 0 & 0{\cdot}514 \times 10^{-7} & -0{\cdot}66 \times 10^{-5} \\
0 & 0 & -0{\cdot}66 \times 10^{-5} & 1{\cdot}7 \times 10^{-5}
\end{bmatrix}
\begin{bmatrix}
H_1 \\
M_1 \\
H_2 \\
M_2
\end{bmatrix}
$$

from which we find

$$H_1 = -19\,000 \text{ N/radian} \qquad M_1 = 735 \text{ Nm/radian}$$
$$H_2 = 3640 \text{ N/radian} \qquad M_2 = 138 \text{ Nm/radian}$$

The integration constants $C_1, \ldots, C_4$ are determined from equation (6.34) and, once known, the radial displacement due to these loads and hence the stress resultants can be found. Usually it is sufficient to determine the stresses at the ends, since it is there that they will reach a maximum. From equations (6.27), in general,

$$N_{zz} = \frac{V}{R} + N \qquad N_{\theta\theta} = v\,\frac{V}{R} - (1-v)N + \frac{Et}{R}u_r$$

$$M_{zz} = -D\,\frac{d^2 u_r}{dz^2} - M \qquad M_{\theta\theta} = -vD\frac{d^2 u_r}{dz^2} - M$$

where u_r and $(d^2 u_r / dz^2)$ are obtained from equations (6.33) and (6.34). In long cylinders,

$$N_{zz} = 0 \qquad N_{\theta\theta} = \frac{Et}{R}\left[\frac{R^2 H_1}{2\beta^3 D} + \frac{R M_1}{2\beta^2 D}\right]$$

$$M_{zz} = -\frac{M_1}{R} - M \qquad M_{\theta\theta} = -v\frac{M_1}{R} - M$$

at end (1), and,

$$N_{zz} = 0 \qquad N_{\theta\theta} = \frac{Et}{R} \left[\frac{R^2 H_2}{2\beta^3 D} - \frac{RM_1}{2\beta^2 D} \right]$$

$$M_{zz} = \frac{M_2}{R} - M \qquad M_{\theta\theta} = \nu \frac{M_2}{R} - M$$

at end (2).

In this case, $\qquad \sigma_{zz} = \mp 88 \cdot 2 \, \text{MN/m}^2$

$$\sigma_{\theta\theta} = -98 \cdot 4 \, \text{MN/m}^2 \mp 26 \cdot 46 \, \text{MN/m}^2$$

at end (1). At end (2),

$$\sigma_{zz} = \pm 16 \cdot 56 \, \text{MN/m}^2$$

$$\sigma_{\theta\theta} = 19 \cdot 2 \, \text{MN/m}^2 \pm 4 \cdot 97 \, \text{MN/m}^2$$

The top sign corresponds to the stress on the outside face. The total stress is the combination of the membrane and the edge loading stresses. In this case, the maximum value of the total stress is $\sigma_{zz} = 120 \, \text{MN/m}^2$, on the inside face of end 1.

The rapidity with which the stresses caused by edge loading die out can be assessed by noting that at a distance z approximately equal to $2R/\beta$ their value drops down to one-tenth of their peak value. In the example considered, $2R/\beta = 0 \cdot 156$ m. The membrane solution is thus valid over most of the vessel.

6.3.4 Temperature variation across thickness

The membrane solution, treated in section 6.3.2, has been shown to provide an acceptable representation of the behaviour of a cylindrical shell under linearly varying pressure and temperature along the axis, provided that the ends are free to deform. We have seen that, having obtained this unrestrained state of the shell, the total stresses are found by adding to the membrane stresses those induced by the end loads required to satisfy the end restraints. It will be noted that the axial temperature variation is only used for the determination of the unrestrained displacements—and therefore affects the restraining end displacements. No temperature terms are however included in the edge loading expressions.

Consider now a shell in which the temperature varies linearly across the thickness as well as along the axis. At any given point, by definition,

$$N = \frac{\alpha E}{1 - \nu} \frac{T_0 + T_i}{2} t \qquad M = \frac{\alpha E}{1 - \nu} \frac{T_0 - T_i}{12} t^2$$

where T_0 is the temperature on the outside face and T_i the temperature inside. We also find $\sigma = 0$ (p. 233). The radial displacement is given by,

$$u_r = \frac{R}{Et} \left[\alpha E \frac{T_0 + T_i}{2} t \right] = \alpha R \frac{T_0 + T_i}{2}$$

$$u_z = \int_0^z \alpha \frac{T_0 + T_i}{2} dz, \quad \text{hence} \quad \frac{du_z}{dz} = \alpha \frac{T_0 + T_i}{2}$$

242

and from equations (6.27),

$$N_{zz} = N_{\theta\theta} = 0$$

given the linear variation of the temperature, $(d^2 u_r/dz^2)$ is zero, hence

$$M_{zz} = M_{\theta\theta} = -\frac{\alpha E}{1-\nu}\frac{T_0 - T_i}{12}t^2 = -M$$

and the stresses are,

$$\sigma_{zz} = \sigma_{\theta\theta} = -\frac{\alpha E}{1-\nu}(T_0 - T_i)\frac{r-R}{t}$$

being maximum at the inside face, where

$$(\sigma_{zz})_{in} = (\sigma_{\theta\theta})_{in} = \frac{\alpha E}{1-\nu}\frac{T_0 - T_i}{2}$$

while

$$(\sigma_{zz})_{out} = (\sigma_{\theta\theta})_{out} = -\frac{\alpha E}{1-\nu}\frac{T_0 - T_i}{2}$$

the value of $(T_0 - T_i)$ being the one corresponding to the particular position chosen.

If the cylindrical shell is completely unrestrained, at the free ends we must obviously have $(\sigma_{zz})_{end} = 0$, a situation that is easily achieved by adding moments $M_1 = -M_2 = -MR$ at the free ends. The end stresses are then,

$$(\sigma_{zz})_{end} = 0$$

$$(\sigma_{\theta\theta})_{max} = -\frac{\alpha E}{1-\nu}\frac{T_0 - T_i}{2}\left(1 - \nu + \frac{\sqrt{(1-\nu^2)}}{\sqrt{3}}\right)$$

and the end displacements are, from equations (6.35)

$$u_1 = \alpha R \left(\frac{T_0 + T_i}{2}\right)_1 - \frac{R^2}{2\beta^2 D}M$$

$$\theta_1 = \frac{\alpha R}{l}\left[\left(\frac{T_0 + T_i}{2}\right)_1 - \left(\frac{T_0 + T_i}{2}\right)_2\right] - \frac{R}{\beta D}M$$

6.3.5 Cylinders joined in series

Cylindrical shells, of various thicknesses and mean radii, are usually joined together to form flanges or to provide supporting ledges in pressure vessels. Also, tapering hubs, of the type shown in Fig. 6.12, can be visualized as a series of cylinders joined in series. Similar arrangements are found in all shell-like components, e.g., winding drums, pistons in reciprocating engines, etc.

Figure 6.13 shows a junction between two cylinders, n and $(n-1)$, of a series. In the figure, end 1 of cylinder n is joined to end 2 of $(n-1)$ and external loads

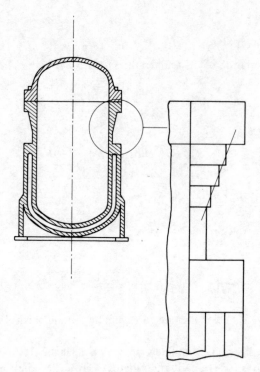

Fig. 6.12 Typical jacketed vessel

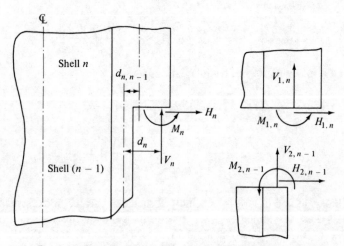

Fig. 6.13 Junction of two cylindrical shells of different thickness

are applied at the junction. In order to satisfy continuity of displacements, end forces $H_{1,n}, V_{1,n}, H_{2,n-1}$, and $V_{2,n-1}$ and moments $M_{1,n}, M_{2,n-1}$ must be included. The end forces act along the mid surfaces of the corresponding shells and, for equilibrium,

$$H_{2,n-1} + H_{1,n} = H_n$$
$$V_{2,n-1} + V_{1,n} = V_n \tag{6.36}$$
$$M_{2,n-1} + M_{1,n} + V_{1,n} d_{n,n-1} = M_n + V_n d_n$$

In general,

$$V_{2,n-1} = -V_{1,n-1} \quad \text{and} \quad V_{1,n} = -V_{2,n}$$

The condition of continuity of displacement implies the satisfaction of the following equations,

$$u_{1,n} = u_{2,n-1} + \int_0^{d_{n,n-1}} \alpha T dr = u_n + \int_{d_n}^{d_{n,n-1}} \alpha T dr$$

$$\theta_{1,n} = \theta_{2,n-1} = \theta_n \tag{6.37}$$

$$w_{1,n} = w_{2,n-1} + \theta_n d_{n,n-1} = w_n - \theta_n(d_n - d_{n,n-1})$$

At the end of the series, on end 1 of the first shell,

$$H_{1,1} = H_1$$
$$M_{1,1} = M_1 + V_1 d_1$$
$$u_{1,1} = u_1 - \int_0^{d_1} \alpha T dr$$
$$\theta_{1,1} = \theta_1$$
$$w_{1,1} = w_1 - d_1 \theta_1$$

where H_1, M_1, V_1 are the external loads, u_1 and θ_1 the radial displacement and rotation of the end. Similar conditions hold for end 2 of the last shell of the series, N. In general, the external forces are given. In addition, each component shell may be subjected to distributed loading, varying linearly along the axis. The problem then consists in the determination of the displacements and edge loads, from which the complete stress distribution may be obtained. To do this, equations (6.36) and (6.37) are written for every junction and solved by eliminating first the displacements $u_{1,n}, u_{2,n-1}$, etc. When the number of shells in the series is large, it is found convenient to introduce a shorthand notation utilizing matrices. In the first place, displacement matrices $\mathbf{D}_{1,n}, \mathbf{D}_{2,n}$ are defined,

$$\begin{bmatrix} u_{1,n} \\ \theta_{1,n} \end{bmatrix} = \mathbf{D}_{1,n} \qquad \begin{bmatrix} u_{2,n} \\ \theta_{2,n} \end{bmatrix} = \mathbf{D}_{2,n}$$

and, since
$$u_{1,n} = (u_{1,n})_{\text{unrestrained}} + (u_{1,n})_{\text{edge load}}$$

we also define displacement matrices $(\mathbf{D}_{1,n})_{\text{unrestrained}}$, $(\mathbf{D}_{1,n})_{\text{edge load}}$, etc., such that,

$$\mathbf{D}_{1,n} = (\mathbf{D}_{1,n})_{\text{unrestrained}} + (\mathbf{D}_{1,n})_{\text{edge load}}$$

for ends 1 and 2 of each shell. In all cases, the unrestrained end displacements are easily determined, since they depend solely of the distributed load over each shell, considered separately from the rest. The axial displacement may also be included in the unrestrained solution. The edge loads produce displacements given by equation (6.35) which can be expressed as follows,

$$(\mathbf{D}_{1,n})_{\text{edge load}} = \mathbf{A}_{1,1}^{(n)} \mathbf{F}_{1,n} + \mathbf{A}_{1,2}^{(n)} \mathbf{F}_{2,n}$$
$$(\mathbf{D}_{2,n})_{\text{edge load}} = \mathbf{A}_{2,1}^{(n)} \mathbf{F}_{1,n} + \mathbf{A}_{2,2}^{(n)} \mathbf{F}_{2,n}$$

where

$$\mathbf{F}_{1,n} = \begin{bmatrix} H_{1,n} \\ M_{1,n} \end{bmatrix}, \qquad \mathbf{F}_{2,n} = \begin{bmatrix} H_{2,n} \\ M_{2,n} \end{bmatrix}$$

and,

$$\mathbf{A}_{1,1}^{(n)} = \begin{bmatrix} \dfrac{R^2 \lambda_1}{2\beta^3 D} & \dfrac{R\mu_1}{2\beta^2 D} \\[2ex] \dfrac{R\mu_1}{2\beta^2 D} & \dfrac{\nu_1}{\beta D} \end{bmatrix} \qquad \mathbf{A}_{1,2}^{(n)} = \begin{bmatrix} \dfrac{R^2 \lambda_2}{2\beta^3 D} & \dfrac{R\mu_2}{2\beta^2 D} \\[2ex] -\dfrac{R\mu_2}{2\beta^2 D} & \dfrac{\nu_2}{\beta D} \end{bmatrix}$$

$$\mathbf{A}_{2,1}^{(n)} = \begin{bmatrix} \dfrac{R^2 \lambda_2}{2\beta^3 D} & -\dfrac{R\mu_2}{2\beta^2 D} \\[2ex] \dfrac{R\mu_2}{2\beta^2 D} & \dfrac{\nu_2}{\beta D} \end{bmatrix} \qquad \mathbf{A}_{2,2}^{(n)} = \begin{bmatrix} \dfrac{R^2 \lambda_1}{2\beta^3 D} & -\dfrac{R\mu_1}{2\beta^2 D} \\[2ex] -\dfrac{R\mu_1}{2\beta^2 D} & \dfrac{\nu_1}{\beta D} \end{bmatrix}$$

calculated for shell (n).

It can be shown that the elimination of the unrestrained end displacements result in the following equations:

$$\begin{bmatrix} \{\mathbf{A}_{2,2}^{(1)} + \mathbf{A}_{1,1}^{(2)}\} & -\mathbf{A}_{1,2}^{(2)} & \mathbf{O} & \dots & & & & & \\ \dots & \dots & \dots & & & & & & \\ \mathbf{O} & \mathbf{O} & \mathbf{O} & -\mathbf{A}_{2,1}^{(n-1)} & \{\mathbf{A}_{2,2}^{(n-1)} + \mathbf{A}_{1,1}^{(n)}\} & -\mathbf{A}_{1,2}^{(n)} & \mathbf{O} & \dots & \\ \dots & & & & & & & & \\ \mathbf{O} & & & & \mathbf{O} & -\mathbf{A}_{2,1}^{(N)} & \{\mathbf{A}_{2,2}^{(N)} + \mathbf{A}_1^{(}{}} \end{bmatrix}$$

$$\times \begin{bmatrix} \mathbf{F}_{1,2} \\ \vdots \\ \mathbf{F}_{1,n} \\ \vdots \\ \mathbf{F}_{1,N} \end{bmatrix} = \begin{bmatrix} \mathbf{X}_2 + \mathbf{A}_{2,1}^{(1)}\mathbf{F}_{1,1} \\ \vdots \\ \mathbf{X}_n \\ \vdots \\ \mathbf{X}_N \end{bmatrix}$$

(6.38)

where

$$\mathbf{X}_n = (\mathbf{D}_{2,n-1})_{\text{unrestrained}} - (\mathbf{D}_{1,n})_{\text{unrestrained}} + \begin{bmatrix} \displaystyle\int_0^{d_{n,n-1}} \alpha T dr \\ \mathbf{0} \end{bmatrix}$$

$$+ \mathbf{A}_{2,2}^{(n-1)}\left\{ \mathbf{F}_n + \begin{bmatrix} 0 \\ d_n \end{bmatrix} V_n - \begin{bmatrix} 0 \\ d_{n,n-1} \end{bmatrix} V_{1,n} \right\}$$

$$- \mathbf{A}_{1,2}^{(n)}\left\{ \mathbf{F}_{n+1} + \begin{bmatrix} 0 \\ d_{n+1} \end{bmatrix} V_{n+1} - \begin{bmatrix} 0 \\ d_{n+1,n} \end{bmatrix} V_{1,n+1} \right\}$$

Besides its obvious application to vessels of the type illustrated in Fig. 6.12, splitting a single cylindrical shell into small components makes possible the stress analysis when the axial variation of pressure or temperature is non-linear. All that is required is to approximate the true distribution by a number of straight lines, so that the variation along each part of the whole shell is linear. The set of simultaneous linear equations (6.38) is simplified when all the components of the series have the same dimensions and also when they are sufficiently long so that no interaction between their ends take place. In either case, the method of solution is the same as detailed in the previous chapter, the only unknowns being the end forces $\mathbf{F}_1, \ldots, \mathbf{F}_N$.

It must be pointed out that the method just described does not take into consideration the stress concentration introduced by the sharp transition between the shells.

6.4 Spherical Shells

Thin-walled spherical shells are used as closure heads in pressure vessels, domes and bulkheads, etc. The same assumptions that were made concerning the deformation of the cylindrical shells are valid with spheres. Stress resultants, defined in exactly the same form, are also used for the analysis in preference to the stresses themselves.

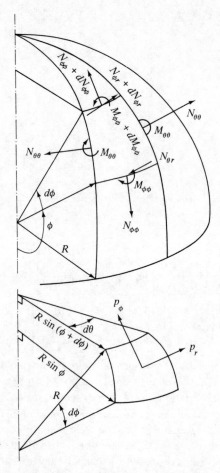

Fig. 6.14 Stress resultants and distributed loads in spherical shells

The equilibrium conditions, in terms of the stress resultants, can be derived by referring to Fig. 6.14, which represents an element, defined by the meridional and circumferential coordinates ϕ and θ, of area $(R^2 \sin \phi \, d\phi \, d\theta)$. Loading is assumed to be rotationally symmetrical with respect to axis $\phi = 0$. A distributed load, of radial and meridional components p_r and p_ϕ acts as shown in the figure. For equilibrium in the meridional direction,

$$N_{\phi\phi} R \sin \phi \, d\theta - [N_{\phi\phi} + dN_{\phi\phi}] R \sin(\phi + d\phi) \, d\theta - N_{\phi r} R \sin \phi \, d\theta \frac{d\phi}{2}$$

$$- [N_{\phi r} + d N_{\phi r}] R \sin(\phi + d\phi) \, d\theta \frac{d\phi}{2} + N_{\theta\theta} \, d\theta \cos \phi R \, d\phi$$

$$= p_\phi R^2 \sin \phi \, d\phi \, d\theta$$

248

For equilibrium in the radial direction,

$$-N_{\phi r} R \sin\phi\, d\theta + [N_{\phi r} + dN_{\phi r}]\, R\sin(\phi + d\phi)\, d\theta - N_{\phi\phi} R \sin\phi\, d\theta \frac{d\phi}{2}$$

$$-[N_{\phi\phi} + dN_{\phi\phi}]\, R\sin(\phi + d\phi)\, d\theta \frac{d\phi}{2} - N_{\theta\theta} \sin\phi R\, d\phi\, d\theta$$

$$= -p_r R^2 \sin\phi\, d\phi\, d\theta$$

Moment equilibrium (hoop direction only),

$$M_{\phi\phi} R \sin\phi\, d\theta - [M_{\phi\phi} + dM_{\phi\phi}]\, R\sin(\phi + d\phi)\, d\theta$$
$$+ N_{\phi r} R^2 \sin\phi\, d\theta\, d\phi + M_{\theta\theta} R \cos\phi\, d\phi\, d\theta = 0$$

The derivation of the last equation is quite simple by referring to Fig. 6.14. The only term that may appear a little obscure is the one containing $M_{\theta\theta}$ in the last equation. Using double arrows for moments in Fig. 6.15(a), $M_{\theta\theta}$ is the sum of $-M_{\theta\theta}\cos\phi$ and $M_{\theta\theta}\sin\phi$. Only the first component need be considered when setting out equilibrium conditions as in Fig. 6.15(b).

Simplifying the preceding equations,

$$\frac{d}{d\phi}(N_{\phi\phi}\sin\phi) - N_{\theta\theta}\cos\phi + N_{\phi r}\sin\phi = -p_\phi R \sin\phi$$

$$\frac{d}{d\phi}(N_{\phi r}\sin\phi) - N_{\theta\theta}\sin\phi - N_{\phi\phi}\sin\phi = -p_r R\sin\phi \qquad (6.39)$$

$$\frac{d}{d\phi}(M_{\phi\phi}\sin\phi) - M_{\theta\theta}\cos\phi - RN_{\phi r}\sin\phi = 0$$

In spherical shells, as in cylinders, the displacement of point r can be expressed in terms of mid-shell displacements and rotation. If these are u_ϕ, u_r, and ω at point A (Fig. 6.16) and ω is positive when clockwise,

$$\omega = -\frac{u_\phi}{R} + \frac{1}{R}\frac{du_r}{d\phi}$$

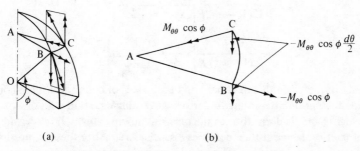

| (a) | (b) |

Fig. 6.15 Effect of $M_{\theta\theta}$ on moment equilibrium

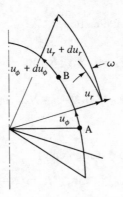

Fig. 6.16 Displacements of two points on the same meridian in a spherical shell

The meridional and hoop strains $e_{\phi\phi}$ and $e_{\theta\theta}$ are,

$$e_{\phi\phi} = \frac{(u_r)_r}{R} + \frac{(du_\phi)_r}{R\,d\phi} = \frac{u_r}{R} + \frac{1}{R}\frac{d}{d\phi}[u_\phi - (r-R)\omega]$$

$$= \frac{u_r}{R} + \frac{r}{R^2}\frac{du_\phi}{d\phi} - \frac{r-R}{R^2}\frac{d^2 u_r}{d\phi^2}$$

$$e_{\theta\theta} = \frac{(u_r)_r}{R} + \frac{(u_\phi)_r}{R\tan\phi} = \frac{u_r}{R} + \frac{ru_\phi}{R^2\tan\phi} - \frac{r-R}{R^2\tan\phi}\frac{du_r}{d\phi}$$

and following the same reasoning as in cylindrical shells, it is found that the stress resultants can be expressed in terms of the mid-shell displacements by means of equations (6.40),

$$N_{\phi\phi} = \frac{Et}{(1-\nu^2)R}\left[u_r(1+\nu) + \frac{du_\phi}{d\phi} + \frac{\nu u_\phi}{\tan\phi}\right]$$

$$N_{\theta\theta} = \frac{Et}{(1-\nu^2)R}\left[u_r(1+\nu) + \nu\frac{du_\phi}{d\phi} + \frac{u_\phi}{\tan\phi}\right]$$

$$M_{\phi\phi} = \frac{-Et^3}{12(1-\nu^2)R}\left[\frac{d\omega}{d\phi} + \frac{\nu}{\tan\phi}\omega\right]$$

$$M_{\theta\theta} = \frac{-Et^3}{12(1-\nu^2)R}\left[\nu\frac{d\omega}{d\phi} + \frac{1}{\tan\phi}\omega\right]$$

(6.40)

The similarity between this treatment and that used for cylindrical shells could be carried further by attempting to derive a governing equation with one of the displacements and its derivatives as the only unknowns. This is, however, found to be impracticable and it then becomes essential to simplify these expressions before lumping them together into a single governing equation. The general solution is then undertaken by combining the membrane solution and the effect of edge loading, incorporating all the simplifications that result from either one or the other types of behaviour into equations (6.39) and (6.40).

6.4.1 Membrane solution

When the shell behaves as a membrane the bending moments are zero and the equilibrium equations reduce to

$$N_{\phi r} = 0$$

$$\frac{d}{d\phi}(N_{\phi\phi} \sin\phi) - N_{\theta\theta} \cos\phi = -p_\phi \, R \sin\phi$$

$$N_{\theta\theta} + N_{\phi\phi} = p_r R$$

whose solution is given by,

$$N_{\phi\phi} = \frac{1}{\sin^2\phi}\left[C + R \int_{\phi_0}^{\phi} (p_r\cos\phi - p_\phi \sin\phi) \sin\phi \, d\phi\right]$$

$$N_{\theta\theta} = p_r R - N_{\phi\phi}$$

(6.41)

where C is an integration constant. The mid-shell displacements are found from the first two equations (6.40). Eliminating u_r between them,

$$\frac{du_\phi}{d\phi} - \frac{u_\phi}{\tan\phi} = \sin\phi \frac{d}{d\phi}\left(\frac{u_\phi}{\sin\phi}\right) = \frac{1+v}{E}\frac{R}{t}(N_{\phi\phi} - N_{\theta\theta})$$

and eliminating $(du_\phi/d\phi)$,

$$u_r = \frac{1}{E}\frac{R}{t}(N_{\theta\theta} - v N_{\phi\phi}) - \frac{u_\phi}{\tan\phi}$$

the rotation of the mid-shell is then found to be,

$$\omega = \frac{R}{Et}\left[(1 + v)\,p_\phi + \frac{dp_r}{d\phi}\right]$$

For the membrane solution to be valid, $M_{\phi\phi}$ and $M_{\theta\theta}$ must be zero, a condition that implies that

$$\omega = 0 \qquad (1 + v)\,p_\phi + \frac{dp_r}{d\phi} = 0$$

(6.42)

This imposes a serious limitation to the validity of the simple membrane solution and it is often found that the shell seldom behaves as a true membrane. The use of this solution is then only justified when the bending stresses, whilst not zero, are sufficiently small when compared to the calculated stress values. Let us

251

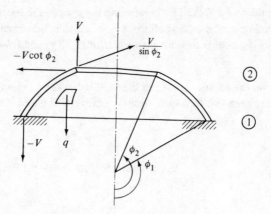

Fig. 6.17 Spherical dome under distributed vertical load

consider, for example, the open dome of Fig. 6.17 under its own weight. Taking the weight per unit shell area to be q,

$$p_r = q \cos \phi \qquad p_\phi = -q \sin \phi$$

and from equation (6.41),

$$N_{\phi\phi} = \frac{C + qR(\cos\phi_1 - \cos\phi)}{\sin^2\phi}$$

since there is a free surface at $\phi = \phi_2$,

$$(N_{\phi\phi})_{\phi_2} = 0 \qquad C = -qR(\cos\phi_1 - \cos\phi_2)$$

The membrane solution is therefore

$$N_{\phi\phi} = qR\frac{\cos\phi_2 - \cos\phi}{\sin^2\phi}$$

$$N_{\theta\theta} = qR\left(\cos\phi - \frac{\cos\phi_2 - \cos\phi}{\sin^2\phi}\right)$$

The displacements may be defined by horizontal and vertical components (normal and parallel to the axis of rotation). Thus,

$$\delta h = u_\phi \cos\phi + u_r \sin\phi = \frac{qR^2}{Et}\left[\sin\phi\cos\phi + \frac{1+\nu}{\sin\phi}(\cos\phi - \cos\phi_2)\right]$$

$$\delta v = u_\phi \sin\phi - u_r \cos\phi$$

$$= \frac{qR^2}{Et}\left[(1+\nu)\left(\ln\frac{\sin\phi}{\sin\phi_1} + \frac{\cos\phi_2}{2}\ln\frac{1+\cos\phi_1}{1-\cos\phi_1}\frac{1-\cos\phi}{1+\cos\phi}\right) + \cos^2\phi_1 - \cos^2\phi\right]$$

and the rotation of the mid-shell,

$$\omega = -\frac{qR}{Et}(2+\nu)\sin\phi$$

not being zero shows the approximate nature of the membrane solution. In this example, the meridional bending stress is found to be $\pm q(2+\nu)\cos\phi/2(1-\nu)$ which is small compared with the maximum direct stress induced by $N_{\phi\phi}$ and may be neglected.

When a vertical force V (per radian) is applied at edge 2, in the manner shown in Fig. 6.17, and $-V$ is applied at 1 in order to maintain equilibrium, at edge 2 noting that $N_{\phi\phi}$ is force per unit length,

$$(N_{\phi\phi})_{\phi_2} = \frac{V}{R\sin^2\phi_2} = \frac{C}{\sin^2\phi_2}$$

hence $C = V/R$

$$N_{\phi\phi} = \frac{V}{R\sin^2\phi} \qquad N_{\theta\theta} = -N_{\phi\phi}$$

The displacements are then,

$$\delta h = -\frac{1+\nu}{Et}\frac{V}{\sin\phi}$$

$$\delta v = \frac{1+\nu}{Et}V\left[\frac{1}{2}\ln\frac{1+\cos\phi_1}{1-\cos\phi_1}\frac{1-\cos\phi}{1+\cos\phi}\right]$$

The stress resultants thus determined balance the forces $V/\sin\phi_2$ and the equivalent at edge 1. The horizontal components $-V\cot\phi_2$, $-V\cot\phi_1$ are self-balancing and are ignored when membrane behaviour is assumed. However, it is clear that they will produce an edge bending effect of the same type as was considered in the cylindrical shell under edge loading. This effect will die out quite rapidly but must be considered when examining in detail the unrestrained stresses and deformations of the dome. It is also important to note that the membrane solution, which usually gives a good estimate for the cylindrical shells with free edges, fails to do so in the case of the sphere. Even in the rather elementary example we have treated, the unrestrained condition requires the addition of edge loads to the membrane solution.

6.4.2 Edge loading solution Consider now a spherical shell, similar to the open dome of Fig. 6.17, in which $p_r = p_\phi = 0$ and the only loading consists of forces and moments distributed along the edges. From the first two equilibrium equations (6.39), it is easily seen that,

253

$$N_{\phi\phi} = N_{\phi r} \cot \phi \quad \text{and} \quad N_{\theta\theta} = \frac{dN_{\phi r}}{d\phi} \tag{6.43}$$

It can also be shown (Bickell and Ruiz, loc. cit.) that,

$$\omega = \frac{1}{Et}\left[\frac{d^2 N_{\phi r}}{d\phi^2} + \cot\phi \frac{dN_{\phi r}}{d\phi} - N_{\phi r} \cot^2\phi + \nu N_{\phi r}\right] \tag{6.44}$$

and substituting $M_{\phi\phi}$ and $M_{\theta\theta}$ in the third equilibrium equation (6.39) by their expressions from (6.40),

$$N_{\phi r} = \frac{-Et^3}{12(1-\nu^2)R^2}\left[\frac{d^2\omega}{d\phi^2} + \cot\phi\frac{d\omega}{d\phi} - \omega\cot^2\phi - \nu\omega\right]$$

Defining the operator $L^2 = (d^2/d\phi^2) + \cot\phi\,(d/d\phi) - \cot^2\phi$ the two preceding equations reduce to,

$$(L^2 + \nu)\,N_{\phi r} = Et\omega \qquad (L^2 - \nu)\,\omega = -\frac{12(1-\nu^2)}{Et^3}R^2\,N_{\phi r}$$

equivalent to

$$(L^4 - \nu^2)\,N_{\phi r} = -4\beta^4\,N_{\phi r}$$

where β is the shell characteristic previously defined. Neglecting ν^2 with respect to $4\beta^4$, the following governing equation is obtained,

$$(L^4 + 4\beta^4)\,N_{\phi r} = 0 \tag{6.45}$$

Once $N_{\phi r}$ has been found, $N_{\phi\phi}$ and $N_{\theta\theta}$ are derived from equation (6.43), ω from (6.44), and $M_{\phi\phi}$ and $M_{\theta\theta}$ from (6.40). As in the case of cylinders, the general solution of the governing equation has the form,

$$-N_{\phi r} = C_1 f_1 + C_2 f_2 + C_3 f_3 + C_4 f_4$$

where $C_1, \ldots, C_4$ are integration constants and $f_1, \ldots, f_4$ are functions of ϕ. The values of the integration constants are chosen so as to satisfy the boundary conditions. Referring to Fig. 6.18, the edge force (per radian) H_1 is balanced by a shear stress resultant $-N_{\phi r} R \sin\phi_1$ and a meridional stress resultant, included in the membrane solution, $N_{\phi\phi} R \sin\phi_1$ such that, at edge 1, $N_{\phi r}R = -H_1$ and $N_{\phi\phi} R \tan\phi_1 = -H_1$, while at edge 2, with the sign convention of Fig. 6.14, $N_{\phi r} R = H_2$, and $N_{\phi\phi} R \tan\phi_2 = H_2$.

The meridional stress resultants are included in the membrane solution and the deformations and stresses they produce are calculated following the method previously described. Only the shear resultants need be considered in the edge loading solution.

In addition to the forces H_1 and H_2, bending moments M_1 and M_2 (per radian) are balanced by moment stress resultants at edge 1, $M_{\phi\phi} R \sin\phi_1 = -M_1$ and at edge 2, $M_{\phi\phi} R \sin\phi_2 = M_2$ with the sign conventions of Figs. 6.14 and 6.18.

254

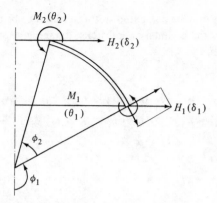

Fig. 6.18 Edge loads, rotations, and displacements in spherical shell

The edge displacements are,

$$\delta h = \delta_1 \quad \text{and} \quad \omega = -\theta_1 \text{ at edge 1,} \quad \text{and} \quad \delta h = \delta_2$$

and $\omega = -\theta_2$ at edge 2.

The exact solution of the governing equation involves the use of slowly convergent series and for this reason several approximations are essential. The most commonly used approach was first used by Hetenyi. Functions $f_1, \ldots, f_4$ are the real and imaginary parts of the solutions of

$$(L^2 - 2i\beta^2) f = 0$$

i.e., $(f_1 + if_2), (f_3 - if_4)$.

This equation can be simplified by a change of variable. Writing $j = f\sqrt{\sin\phi}$ leads to,

$$j'' - \left[2i - \frac{1}{2\beta^2} + \frac{3}{4}\left(\frac{\cot\phi}{\beta}\right)^2 \right] j = 0$$

In all thin-walled spheres, $(\frac{1}{2}\beta^2)$ is small and assuming that $(\cot\phi)/\beta$ is also small and neglecting both terms, we obtain,

$$f_1 = -\frac{1}{\sqrt{\sin\phi}} e^{\beta(\phi-\phi_1)} [\cos\beta(\phi-\phi_1) + \sin\beta(\phi-\phi_1)]$$

$$f_2 = \frac{1}{\sqrt{\sin\phi}} e^{\beta(\phi-\phi_1)} [\cos\beta(\phi-\phi_1) - \sin\beta(\phi-\phi_1)]$$

$$f_3 = \frac{1}{\sqrt{\sin\phi}} e^{-\beta(\phi-\phi_1)} [\cos\beta(\phi-\phi_1) - \sin\beta(\phi-\phi_1)]$$

$$f_4 = \frac{1}{\sqrt{\sin\phi}} e^{-\beta(\phi-\phi_1)} [\cos\beta(\phi-\phi_1) + \sin\beta(\phi-\phi_1)]$$

These expressions may be compared to those found to be applicable to cylinders. In order to satisfy the boundary conditions, the integration constants $C_1, \ldots, C_4$ should be given by,

$$
\begin{bmatrix} C_1 \\ C_2 \\ C_3 \\ C_4 \end{bmatrix} = \tfrac{1}{4}
\begin{bmatrix}
(2\lambda_1 - \mu_1 - 1) & (\mu_1 - \nu_1) & (2\lambda_2 + \mu_2) & (\mu_2 - \nu_2) \\
(1 - \mu_1) & (1 - \nu_1) & \mu_2 & -\nu_2 \\
(2\lambda_1 + \mu_1 + 1) & (\mu_1 + \nu_1) & (2\lambda_2 - \mu_2) & (\mu_2 + \nu_2) \\
(1 - \mu_1) & -(1 + \nu_1) & \mu_2 & -\nu_2
\end{bmatrix}
\begin{bmatrix}
\dfrac{H_1 \sqrt{\sin\phi_1}}{R} \\[2ex]
\dfrac{2\beta M_1}{R^2 \sqrt{\sin\phi_1}} \\[2ex]
\dfrac{H_2 \sqrt{\sin\phi_2}}{R} \\[2ex]
\dfrac{2\beta M_2}{R^2 \sqrt{\sin\phi_2}}
\end{bmatrix}
\tag{6.46}
$$

In this expression, similar to equation (6.34) for cylinders, the terms $\lambda_1, \mu_1, \ldots$ are equal to those tabulated in Table 6.1, changing the variable $(\beta l/R)$ into $\beta(\phi_2 - \phi_1)$. When this is large, equation (6.46) becomes,

$$C_1 = C_2 = 0$$

$$C_3 = \frac{H_1 \sqrt{\sin\phi_1}}{R} + \frac{\beta M_1}{R^2 \sqrt{\sin\phi_1}}$$

$$C_4 = -\frac{\beta M_1}{R^2 \sqrt{\sin\phi_1}}$$

The end displacements and rotations are expressed by,

$$
\begin{bmatrix} \delta_1 \\ \theta_1 \\ \delta_2 \\ \theta_2 \end{bmatrix} = \mathbf{A}
\begin{bmatrix} H_1 \\ M_1 \\ H_2 \\ M_2 \end{bmatrix}
\tag{6.47}
$$

where the flexibility matrix $\mathbf{A}$ is as shown in Table 6.2. When $\beta(\phi_2 - \phi_1)$ is larger than about 3,

$$
\mathbf{A} =
\begin{bmatrix}
\dfrac{R^2 \sin\phi_1}{2\beta^3 D} & \dfrac{R}{2\beta^2 D} & 0 & 0 \\[2.5ex]
\dfrac{R}{2\beta^2 D} & \dfrac{1}{\beta D \sin\phi_1} & 0 & 0 \\[2.5ex]
0 & 0 & \dfrac{R^2 \sin\phi_2}{2\beta^3 D} & \dfrac{-R}{2\beta^2 D} \\[2.5ex]
0 & 0 & \dfrac{-R}{2\beta^2 D} & \dfrac{1}{\beta D \sin\phi_2}
\end{bmatrix}
$$

Table 6.2

	H_1	M_1	H_2	M_2
δ_1	$\lambda_1 \dfrac{R^2 \sin\phi_1}{2\beta^3 D}$	$\mu_1 \dfrac{R}{2\beta^2 D}$	$\lambda_2 \dfrac{R^2\sqrt{(\sin\phi_1 \sin\phi_2)}}{2\beta^3 D}$	$\mu_2 \dfrac{R}{2\beta^2 D}\sqrt{\dfrac{\sin\phi_1}{\sin\phi_2}}$
θ_1	*	$\dfrac{\nu_1}{\beta D \sin\phi_1}$	$-\mu_2 \dfrac{R}{2\beta^2 D}\sqrt{\dfrac{\sin\phi_2}{\sin\phi_1}}$	$\dfrac{\nu_2}{\beta D\sqrt{(\sin\phi_1 \sin\phi_2)}}$
δ_2	*	*	$\lambda_1 \dfrac{R^2\sin\phi_2}{2\beta^3 D}$	$-\mu_1 \dfrac{R}{2\beta^2 D}$
θ_2	*	*	*	$\dfrac{\nu_1}{\beta D \sin\phi_2}$

* Denotes symmetrical elements.

it then becomes a simple task to find the stress resultants,

$$N_{\phi r} = -\frac{\sqrt{\sin\phi_1}}{R} f_3 H_1 + \frac{\beta}{R_2\sqrt{\sin\phi_1}}(f_4 - f_3) M_1$$

$$M_{\phi\phi} = -\frac{\sqrt{\sin\phi_1}}{2\beta}(f_4 - f_3) H_1 - \frac{1}{R\sqrt{\sin\phi_1}} f_4 M_1$$

$$N_{\phi\phi} = N_{\phi r}\cot\phi \tag{6.48}$$

$$N_{\theta\theta} = \frac{Eh\delta}{R\sin\phi}$$

$$M_{\theta\theta} = \nu M_{\phi\phi}$$

$$\delta = \frac{R^2 \sin\phi\sqrt{\sin\phi_1}}{2\beta^3 D}\frac{f_3 + f_4}{2} H_1 + \frac{R}{2\beta^2 D}\frac{\sin\phi}{\sqrt{\sin\phi_1}} f_3 M_1$$

The solution thus obtained is only valid for small values of $(\cot\phi)/\beta$, i.e., for thin, deep shells. It is commonly accepted that satisfactory results are obtained if $\beta\phi\sqrt{2} \gg 6$. This implies that, for a relatively thick shell for which $R/h = 20$, ϕ must be included between 45 and 135°. For a thin shell, $R/h = 100$, ϕ must be between 20 and 160°. Since the effect of edge loads dies off away from the edge, the solution is also applicable for spherical shells with only one edge provided that its angle, ϕ_1 or ϕ_2, lies between those limits.

In shallow shells, where ϕ or $(\pi - \phi)$ are very small, it is possible to take $\phi \approx \tan\phi \approx \sin\phi$. It is then found that the governing equation accepts the

257

following solutions,

$$f_1 = (\ker x)' \qquad f_2 = (\kei x)'$$
$$f_3 = (\ber x)' \qquad f_4 = -(\bei x)'$$

where $x = \beta\phi\sqrt{2}$ and $\ker x$, $\kei x$, $\ber x$, $\bei x$ are Kelvin functions. (See, for example, H. H. Lowell, N.A.S.A. *Technical Report* No. R.32.)

The shallow shell approximation is of particular interest for the analysis of a spherical shell with a small circular hole. The boundary conditions require,

$$\delta h = \delta_1 \qquad \omega = -\theta_1 \qquad RN_{\phi r} = -H_1 \qquad M_{\phi\phi}\,R\sin\phi = -M_1$$

at the edge of the hole (edge 1). Since f_1 and f_2 increase with ϕ, $C_1 = C_2 = 0$. It is then found,

$$C_3 = \frac{H_1}{R}\left(\frac{g_3}{f_3\,g_3 + f_4\,g_4}\right) + \frac{2\beta}{R^2\sin\phi_1}\,M_1\left(\frac{f_4}{f_3\,g_3 + f_4\,g_4}\right)$$

$$C_4 = \frac{H_1}{R}\left(\frac{g_4}{f_3\,g_3 + f_4\,g_4}\right) + \frac{2\beta}{R^2\sin\phi_1}\,M_1\left(\frac{-f_3}{f_3\,g_3 + f_4\,g_4}\right)$$

where

$$g_3 = f_3' + \frac{\nu}{\beta\phi_1}f_3$$

$$g_4 = f_4' + \frac{\nu}{\beta\phi_1}f_4$$

and the terms inside the brackets are all evaluated for $\phi = \phi_1$.

6.4.3 Junction of spherical and cylindrical shells

As a simple example, consider the vessel of Fig. 6.19 in which a spherical cap is joined to a long cylindrical barrel, the whole being under hydrostatic pressure only. It is first assumed that the shells are separate from one another, with entirely unrestrained edges and freely supported by the axial forces shown in Fig. 6.20(a). The value of these forces, just sufficient to balance the hydrostatic pull, is $pR/2$ per radian.

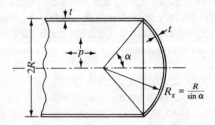

Fig. 6.19 Junction of spherical and cylindrical shells

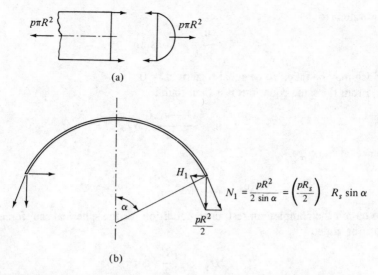

(a)

(b)

Fig. 6.20 Forces at the junction between both shells required for equilibrium

The unrestrained behaviour of the cylindrical shell is given by the membrane solution, equations (6.29) to (6.31),

$$(\sigma_{zz})_{\text{free cyl.}} = \frac{pR}{2t} \qquad V = \frac{pR^2}{2}$$

$$(\sigma_{\theta\theta})_{\text{free cyl.}} = \frac{pR}{t}$$

$$(u_2)_{\text{free cyl.}} = \frac{pR^2}{Et}\left(1 - \frac{\nu}{2}\right)$$

In the spherical shell, the supporting force $(pR^2/2)$ is equivalent to the tangential force N_1 together with the normal force H_1. Under the action of N_1 and the internal pressure, the shell behaves in a membrane-like manner, the equations of section 6.4.1 being applicable taking $p_r = p$, $p_\phi = 0$.

From equations (6.41),

$$N_{\phi\phi} = \frac{pR_s}{2} = N_{\theta\theta} \qquad (\sigma_{\phi\phi})_{\text{mem. sph.}} = (\sigma_{\theta\theta})_{\text{mem. sph.}} = \frac{pR_s}{2t}$$

and the rotation ω is zero. From the first two equations (6.40),

$$N_{\phi\phi} - N_{\theta\theta} = \frac{Eh}{(4\nu)R_s}\left[\frac{du_\phi}{d\phi} - \frac{u_\phi}{\tan\phi}\right] = 0$$

259

equivalent to

$$\frac{u_\phi}{\sin\phi} = \text{constant}$$

which may be taken to be equal to zero, $u_\phi = 0$.

From the same equations it is then found,

$$u_r = \frac{(1-\nu)R_s}{Et} N_{\phi\phi}$$

hence, at the edge,

$$(\delta_1)_{\text{mem. sph.}} = \frac{pR_s^2}{2tE} \sin\alpha(1-\nu)$$

To obtain the complete unrestrained condition for the spherical cap, forces H_1 must be added,

$$H_1 = -\frac{pR^2}{2} \cot\alpha$$

In order to simplify the notation, the parameter β for the sphere must be expressed in function of the characteristics of the cylinder,

$$(\beta)_{\text{sphere}} = \beta \frac{1}{\sqrt{\sin\alpha}}$$

also,

$$(D)_{\text{sphere}} = D$$

Assuming that $(\beta\alpha)_{\text{sphere}}$ is large, equations (6.46) et seq. are applicable. In particular,

$$(\delta_1)_{H_1} = -\frac{pR^4}{4\beta^3 D} \cot\alpha \sqrt{\sin\alpha} \quad (\theta_1)_{H_1} = -\frac{pR^3}{2\beta^2 D} \cot\alpha$$

and the total unrestrained displacement and rotation at edge 1 are,

$$(\delta_1)_{\text{free sph.}} = (\delta_1)_{\text{mem. sph.}} + (\delta_1)_{H_1} = \frac{pR^2}{2tE}\frac{1-\nu}{\sin\alpha} - \frac{pR^4}{4\beta^3 D} \cot\alpha \sqrt{\sin\alpha}$$

$$(\theta_1)_{\text{free sph.}} = 0 + (\theta_1)_{H_1} = -\frac{pR^3}{2\beta^2 D} \cot\alpha$$

the unrestrained stresses are,

$$(\sigma_{\phi\phi})_{\text{free sph.}} = (\sigma_{\phi\phi})_{\text{mem. sph.}} + (\sigma_{\phi\phi})_{H_1}$$

$$(\sigma_{\theta\theta})_{\text{free sph.}} = (\sigma_{\theta\theta})_{\text{mem. sph.}} + (\sigma_{\theta\theta})_{H_1}$$

where the stresses caused by H_1 are obtained from equations (6.48). (Note that in equations (6.48) H_1 was taken positive when outwards.)

260

In order to maintain the continuity of displacement and rotation between both shells, edge loads must be added to the free edges as shown in Fig. 6.21. For equilibrium,

$$H_{1,2} = -H_{2,1} = H$$
$$M_{1,2} = -M_{2,1} = M$$

and for continuity of displacement and rotation,

$$(u_2)_{\text{free cyl.}} + (u_2)_{H,\,M\,\text{cyl.}} = (\delta_1)_{\text{free sph.}} + (\delta_1)_{H,\,M\,\text{sph.}}$$
$$(\theta_2)_{H,\,M\,\text{cyl.}} = (\theta_1)_{\text{free sph.}} + (\theta_1)_{H,\,M\,\text{sph.}}$$

Using the flexibility matrices of equations (6.35) and (6.47) for cylinder and sphere respectively,

$$H = \frac{pR^2}{4\beta}\left[2 - \nu - \frac{1-\nu}{\sin\alpha} + 2\beta \cot\alpha\sqrt{\sin\alpha}\right]\frac{1}{1 + \sqrt{\sin\alpha}}$$

$$M = \frac{pR^3}{4\beta}\frac{\sqrt{\sin\alpha}}{1 + \sqrt{\sin\alpha}}\cot\alpha$$

The complete stress system is the combination of the unrestrained state and the edge loading. A typical situation, for $\alpha = 50°$ is illustrated in Fig. 6.22.

6.4.4 Thermal stresses in spheres　Thermal stresses are set up when the temperature is not uniform through the whole shell. It is quite unnecessary to retrace here all the steps that led to the solution of this problem in the case of cylindrical shells, since the method is virtually the same. The only changes that need be introduced in equations (6.40) consist in the addition of terms $(-N)$ or $(-M)$, as in equation (6.27) and defined in the same way. The equilibrium conditions remain unchanged. Considering only the membrane solution, equations (6.41) still apply, while the displacement

$$h = u_0 \cos\phi + u_r \sin\phi$$

is increased by

$$\frac{1-\nu}{E}\frac{R}{h}N\sin\phi$$

and the rotation is increased by

$$\frac{1-\nu}{Eh}\frac{dN}{d\phi}$$

it is clear that an additional restriction to the validity of the membrane solution has been introduced. One way in which this difficulty can be avoided is to split

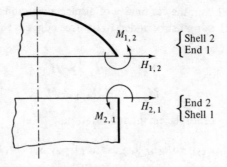

Fig. 6.21 Edge loads at the junction

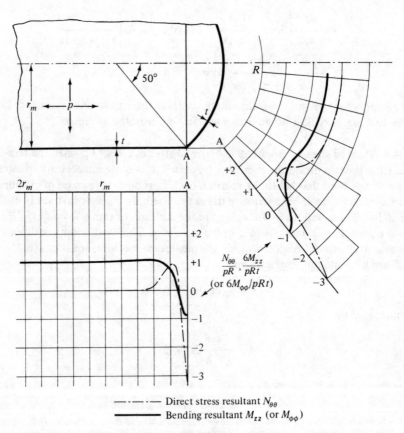

--- · --- Direct stress resultant $N_{\theta\theta}$
―――― Bending resultant M_{zz} (or $M_{\phi\phi}$)

Fig. 6.22 Stress distribution in a cylindrical shell joined to a spherical closure head

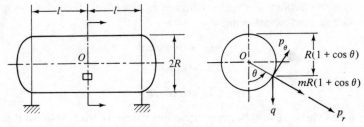

Fig. 6.23 Horizontal tank

the shell into small bands, each one being under uniform temperature, matching end displacements, and rotations as was done for the cylinders joined in series.

6.5 Lateral Loading of Cylindrical and Spherical Shells

The effect of lateral loading, e.g., wind load on domes or vertical tanks, self-weight, and weight of contents in horizontal tanks, will be considered next. As shown in Fig. 6.23, on an element of a cylindrical shell of weight q per unit area and completely filled with a liquid of specific weight m, the hoop components of the distributed load are,

$$p_r = (q + mR) \cos \theta + mR$$

$$p_\theta = -q \sin \theta$$

When a vertical cylinder is subjected to wind loading, as in Fig. 6.24, the compression produced on the windward side together with the suction on the leeward side can be approximated to a distributed load q',

$$p_r = q' \cos \theta$$

$$p_\theta = -q' \sin \theta$$

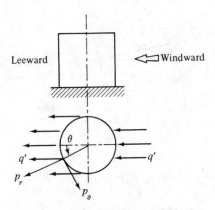

Fig. 6.24 Wind loading on vertical tank

263

Wind loading on a spherical dome produces a similar effect, the radial, hoop, and meridional component being,

$$p_r = (q' \sin \phi) \cos \theta$$
$$p_\theta = (-q') \sin \theta$$
$$p_\phi = (q' \cos \phi) \cos \theta$$

see Fig. 6.25. The similarity between these three types of loads leads to the definition of general lateral loading as,

$$p_r = \bar{p}_r \cos \theta \quad p_\theta = \bar{p}_\theta \sin \theta \quad p_\phi = \bar{p}_\phi \cos \theta \quad \text{(zero in cylindrical shells)}.$$

These expressions can be used to represent any of the examples described by giving the appropriate values to the amplitudes $\bar{p}_r, \bar{p}_\theta, \bar{p}_\phi$. It will be noted that the term (mR) in the radial component of load in the horizontal tank is equivalent to a rotationally symmetrical load and may be treated as such.

In the following, only the membrane behaviour will be considered. The treatment, valid for regions sufficiently far away from the edges of the shell or any other discontinuities will be restricted to the type of loading that has first been described. For more accurate treatment, including the bending of the shell and the effect of edge loading, the reader is referred to specialized books on the subject such as those by Flügge and by Novozhilov. The application of the general principles to the shells used in pressure vessels is covered in *Pressure Vessel Design and Analysis* by Bickell and Ruiz.

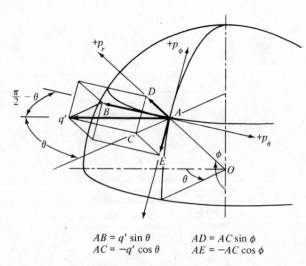

$$AB = q' \sin \theta \qquad AD = AC \sin \phi$$
$$AC = -q' \cos \theta \qquad AE = -AC \cos \phi$$

Fig. 6.25 Lateral loading on spherical shell

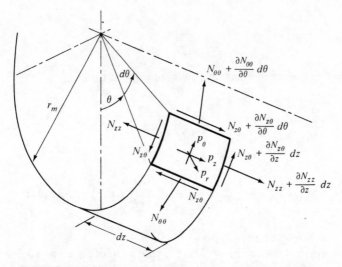

Fig. 6.26 Stress resultants in cylindrical shell under lateral loading

6.5.1 Cylindrical shell

The absence of moment resultants simplifies the setting up of equilibrium conditions. Referring to Fig. 6.26,

$$R\frac{\partial N_{zz}}{\partial z} + \frac{\partial N_{z\theta}}{\partial \theta} = -Rp_z$$

$$\frac{\partial N_{\theta\theta}}{\partial \theta} + R\frac{\partial N_{z\theta}}{\partial z} = -Rp_\theta \qquad (6.48)$$

$$N_{\theta\theta} = Rp_r$$

in these equations, the term $p_z = \bar{p}_z \cos\theta$ has been included. If the load is as previously described, $p_z = 0$. From the third equation, it is apparent that we can take

$$N_{\theta\theta} = \bar{N}_{\theta\theta}\cos\theta = R\bar{p}_r\cos\theta$$

where $\bar{N}_{\theta\theta}$ is a function of z but not of θ. The second equilibrium equation can then be written in the form,

$$-\bar{N}_{\theta\theta} + R\frac{d\bar{N}_{z\theta}}{dz} = -R\bar{p}_\theta$$

taking

$$N_{z\theta} = \bar{N}_{z\theta}\sin\theta$$

($\bar{N}_{z\theta}$ independent of θ). Finally, the first equation becomes,

$$R\frac{d\bar{N}_{zz}}{dz} + \bar{N}_{z\theta} = -R\bar{p}_z$$

with

$$N_{zz} = \overline{N}_{zz} \cos\theta$$

By integration, the equilibrium equations are reduced to,

$$N_{\theta\theta} = R\bar{p}_r$$

$$\overline{N}_{z\theta} = C_1 + \int_0^z (\bar{p}_r - \bar{p}_\theta)\,dz$$

(6.49)

$$-\overline{N}_{zz} = -C_2 + \frac{1}{R}C_1 z + \int_0^z \bar{p}_z\,dz + \frac{1}{R}\int_0^z dz \int_0^z (\bar{p}_r - \bar{p}_\theta)\,dz$$

C_1 and C_2 are integration constants.

In the horizontal tank of Fig. 6.23, the application of equations (6.49) gives the following results,

$$\overline{N}_{\theta\theta} = R(q + mr) \qquad N_{\theta\theta} = t\sigma_{\theta\theta} = R(q + mR)\cos\theta$$

to which a term (mR) has to be added to include the rotationally symmetrical component

$$N_{z\theta} = C_1 + (2q + mR)z$$

The total vertical shear resultant

$$S = \int_0^{2\pi} RN_{z\theta} \sin\theta\,d\theta = \pi R[C_1 + (2q + mR)z]$$

must be zero for $z = 0$ due to symmetry, therefore $C_1 = 0$ and

$$N_{z\theta} = (2q + mR)z \sin\theta$$

at the ends; it is easily checked that the total shear resultant equals the weight of vessel and contents.

$$\overline{N}_{zz} = C_2 - \frac{(2q + mR)z^2}{2R}$$

At the ends,

$$\int_0^{2\pi} N_{zz} R\cos\theta\,d\theta = 0$$

hence

$$C_2 = \frac{(2q + mR)\,l^2}{2R}$$

$$N_{zz} = t\sigma_{zz} = \frac{2q + mR}{2R}(l^2 - z^2)\cos\theta$$

The solution thus found corresponds to the simple beam theory followed in elementary strength of materials. Apart from the limitation imposed by the fact that bending has been ignored and the effect of a restriction in the free displacement of the ends likely to be imposed by the closure heads and supports, the circular cross-section tends to become slightly oval. This results in bending in the hoop direction coupled with hoop forces that may be quite high.

6.5.2 Spherical shell   The equilibrium equations—see Fig. 6.27—are,

$$N_{\phi\phi}\cos\phi + \frac{\partial N_{\phi\phi}}{\partial\phi}\sin\phi + \frac{\partial N_{\phi\theta}}{\partial\theta} - N_{\theta\theta}\cos\phi = -R\sin\phi\,p_\phi$$

$$\frac{\partial N_{\theta\theta}}{\partial\theta} + 2N_{\phi\theta}\cos\phi + \frac{\partial N_{\phi\theta}}{\partial\phi}\sin\phi = -R\sin\phi\,p_\theta$$

$$N_{\theta\theta} + N_{\phi\phi} = Rp_r$$

Taking, as in the case of the cylinder,

$$N_{\theta\theta} = \bar{N}_{\theta\theta}\cos\theta \qquad N_{\phi\phi} = \bar{N}_{\phi\phi}\cos\theta$$

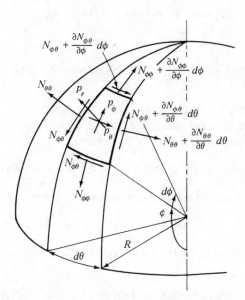

Fig. 6.27 Stress resultants in spherical shell under lateral loading

the last equilibrium equation becomes,

$$\overline{N}_{\theta\theta} + \overline{N}_{\phi\phi} = R\bar{p}_r$$

taking $N_{\phi\theta} = \overline{N}_{\phi\theta} \sin\theta$, the second equation becomes,

$$\frac{d}{d\phi}(\overline{N}_{\phi\theta}\sin\phi) - \overline{N}_{\theta\theta} + \overline{N}_{\phi\theta}\cos\phi = -R\sin\phi\,\bar{p}_\theta$$

and the first equation is finally,

$$\frac{d}{d\phi}(\overline{N}_{\phi\phi}\sin\phi) + \overline{N}_{\phi\theta} - \overline{N}_{\theta\theta}\cos\phi = -R\sin\phi\,\bar{p}_\phi$$

These equations reduce to,

$$\overline{N}_{\theta\theta} = R\bar{p}_r - \overline{N}_{\phi\phi}$$

$$\overline{N}_{\phi\phi} = \frac{A}{\sin^3\phi} + \frac{B\cos\phi}{\sin^3\phi}$$

$$\overline{N}_{\phi\theta} = \frac{A\cos\phi}{\sin^3\phi} + \frac{B}{\sin^3\phi}$$

(6.50)

where

$$A = C_1 + R\int_{\phi_0}^{\phi}(\bar{p}_\theta\cos\phi - \bar{p}_\phi)\sin\phi\,d\phi$$

$$B = C_2 + R\int_{\phi_0}^{\phi}(\bar{p}_r\sin\phi + \bar{p}_\phi\cos\phi - \bar{p}_\theta)\sin\phi\,d\phi$$

and C_1, C_2 are integration constants.

In a hemispherical dome under wind loading, $\phi_1 = \pi/2$, $\phi_2 = \pi$ and the preceding equations lead to the following results,

$$A = C_1 + q'R(1 - \sin^2\phi) \qquad B = C_2 - 2q'R\cos\phi$$

For the stress resultants to remain finite at the apex, $\phi = \pi$

$$A - B = 0 \qquad C_1 = C_2 + q'R$$

The total shear at the support must balance the total wind load, $(2\pi q'R^2)$,

$$\int_0^{2\pi} RC_2\sin^2\theta\,d\theta = -2\pi q'R^2 \qquad C_2 = -2q'R$$

Finally,

$$N_{\phi\phi} = t\sigma_{\phi\phi} = -q' R \cos\theta \frac{(1 + \cos\phi)^2}{\sin^3\phi}$$

$$N_{\theta\theta} = t\sigma_{\theta\theta} = q' R \cos\theta \left[\frac{(1 + \cos\phi)^2}{\sin^3\phi} + \sin\phi\right]$$

are the meridional and hoop stresses, assuming that the sphere behaves as a membrane.

7. Experimental Stress Determination

In the past, the serious limitations of analytical solutions have been overcome by means of experimental techniques. While these have lost some of their importance due to the widespread use of electronic computers for the numerical solution of the basic elastic equations, they continue to be particularly useful on many occasions. They also have some advantages over numerical or analytical methods which have resulted in their continued improvement and a revival of interest in them. One advantage of experimental techniques is their obvious appeal to designers due to the fact that the information is presented in a direct way and is easily interpreted. The testing of small, inexpensive models is generally accepted as providing good insight into the behaviour of complex engineering components. In addition, the accuracy of the result of a numerical or analytical solution depends on how close to the true behaviour of the material and structure is the mathematical model on which the analysis is based. Very often this accuracy can only be assessed by comparing the theoretical predictions with experimental measurements. Limits for the experimental errors, on the other hand, can easily be estimated. It is therefore not surprising to find that the experimental determination of stresses in critical components is considered to be essential.

In this chapter, the various experimental techniques for the determination of stresses will be briefly reviewed. The use of resistance strain gauges and of photoelasticity will be considered in some detail.

7.1 Experimental Techniques Currently Used

The experimental techniques may be divided into two broad groups. In the *whole field methods*, the strains or displacements are measured over a certain area, while in *point-by-point methods* they are measured at selected positions. The whole field methods currently used will be described first.

7.1.1 Whole field methods   (a) *Grid method.* A grid is inscribed on the unloaded component or model and the displacements of the nodal points upon loading are measured, usually from photographs taken first in the unloaded and then in the loaded positions. Only large deformations, as occur in the plastic field or when the material creeps can be determined.

(b) *Brittle coatings.* The component is painted with a lacquer which forms a brittle coating when it dries. Upon straining, the coating cracks along lines approximately perpendicular to the direction of the maximum tensile strain. The value of the strain required to produce a crack depends on the composition of the lacquer and on the conditions under which it has been applied. Lacquers designed to crack at strains of about 50×10^{-6} (50 microstrain) are commercially available. The method is best suited for qualitative analysis since its accuracy is affected by small variations in the temperature, humidity, and other variables beyond the control of the experimentalist.

(c) *Moiré fringes.* A grating consisting of equidistant parallel bars and slits of pitch p (lines/unit length) is fixed to the surface of a component and an identical reference grating is either placed or projected on the surface. If the reference and model gratings are parallel, and the bars in both coincide, dark bands appear when the model grating is strained in the direction perpendicular to that of the bars. These dark bands correspond to those positions where the slits of the reference grating are obscured by bars in the strained model grating. Figure 7.1 illustrates the mechanism.

If the pitch of the strained model grating is p', the strain is

$$e_{xx} = \frac{1/p' - 1/p}{1/p}$$

If the distance between two consecutive fringes is l_x, the number of lines between fringes is $l_x p = l_x p' + 1$ and we obtain for the strain,

$$e_{xx} = \frac{1}{pl_x - 1} \approx \frac{1}{pl_x}$$

since p is usually a large number (2 to 80 lines/mm). It is apparent that the strain in any direction, e_{ii}, may be determined by means of a grating oriented at right angles to the direction i. When a shear strain exists in addition to the extension strain, the model grating rotates with respect to the fixed reference and fringes as shown in Fig. 7.2 are formed. Measuring the distance l_y between consecutive fringes along the y-direction, we have rotation

$$\alpha_1 = \frac{1}{pl_y}$$

From a second set of gratings, at right angles to the first, the strain

$$e_{yy} = \frac{1}{pl_y'}$$

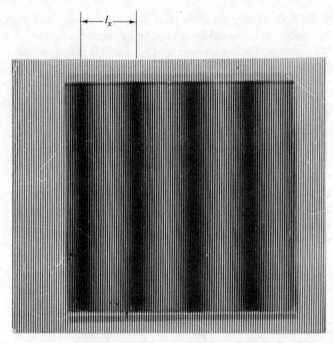

Fig. 7.1 Moiré fringes formed by two parallel gratings of slightly different pitch

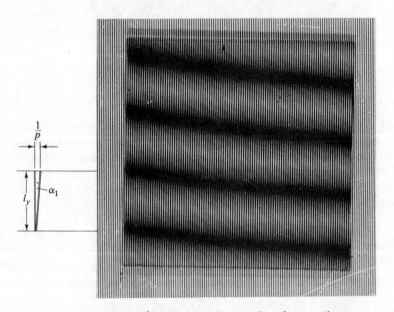

Fig. 7.2 Moiré fringes formed by rotation of two gratings

and the rotation

$$\alpha_2 = \frac{1}{pl'_x}$$

are obtained, measuring l'_x and l'_y along the original x- and y-directions between consecutive fringes, as formed by the second set of gratings. The shear strain is then,

$$e_{xy} = \tfrac{1}{2}(\alpha_1 + \alpha_2) = \frac{1}{2p}\left(\frac{1}{l'_y} + \frac{1}{l'_x}\right)$$

The main limitation of the Moiré method is its relatively low resolution. Using gratings of 1000 lines per inch,

$$e = \frac{10^{-3}}{l} = \frac{10^3}{l} \text{ microstrain}$$

is the strain required to produce two fringes l in. apart. The method is thus suitable for the direct determination of large strains. By indirect techniques, it is possible to improve the resolution of this method which then becomes applicable to the determination of small elastic strains.

(d) *Photoelasticity*. The use of photoelasticity may be considered as essential to the development of new designs and the dimensioning of complex components. The principles on which it is based are discussed in this chapter.

(e) *Interferometers and laser holograms*. Optical interferometers are used for the measurement of the deflection of flat, polished plates and of the changes in thickness in transparent photoelastic models. Recently, conventional light sources have been replaced by lasers, and laser interferometers have several advantages over other types. A pattern, called a hologram, is reproduced on a photographic plate, and represents the interference between light reflected by the deformed body and a reference beam from the source itself. While it is not possible to visualize the object by simply looking at the hologram, a true three-dimensional image is obtained when the hologram is viewed with the same light source that was used for its production, or with any other monochromatic source. The hologram constitutes a permanent true record of the deformation of the model under load.

7.1.2 Point-by-point methods   The extensional strain around a given point may be determined by means of extensometers or strain gauges. These instruments measure the relative displacement between two points, the distance between the points being the so-called gauge length.

In extensometers, the input is in all cases the displacement of one point relative to another. The output will be the displacement of a dial pointer, a recorder pen, an oscilloscope trace, or a spot of light projected on a scale. The *sensitivity* is defined as the ratio (output/input). When both output and input have the same dimensions, the ratio is usually called the *gain* or *magnification*.

The relationship between output and input is often required to be linear. This *linearity* may be defined in two ways (a) by specifying a maximum departure between idealized (linear) output and real output equal to a certain per cent of the full-scale deflection or (b) by specifying this maximum departure as a certain per cent of the idealized output. The *resolution* is defined as the smallest possible input required to produce an observable variation in output. Most instruments exhibit a *hysteresis* loop when they are subjected to cyclic variations of input. The hysteresis may be caused by Coulomb friction in which case the output for a given input is always less when the input is increasing than when it is decreasing. When the hysteresis is caused by viscous friction, the friction force depends on the rate of change of the input resulting in a time lag in the output.

(a) *Mechanical extensometers.* In dial gauges, the gain is achieved by means of gears. In the typical construction indicated in Fig. 7.3(a), the plunger A is toothed in the manner of a rack, and engages with gear B. Motion is transmitted through gears C and D, the last being attached to the pointer. Springs pre-load the gear train to eliminate backlash. The normal gain is about 1000, with a resolution of the order of 25×10^{-4} mm. In the Huggenberger instruments the gain is achieved by means of two levers as shown in Fig. 7.3(b), the fixed pivots being marked P. Their gain and resolution are similar to those of dial gauges. The most sensitive mechanical extensometer is the Johansson, in which a gain of 10 000 and resolution 25×10^{-5} mm are achieved. The construction is outlined in Fig. 7.3(c), and includes a pointer attached to a twisted strip. As the direction of twist is different on each side of the pointer, stretching the strip tends to rotate the pointer. Moving parts are reduced to a minimum and Coulomb friction forces are considerably smaller than in other mechanical extensometers. Viscous friction is used to damp the pointer movement.

(b) *Optical extensometers.* In the most elementary type of optical extensometer a light beam is used to replace the lever multiplication of the conventional mechanical extensometers. Other optical extensometers based on the interference of light have resolution of about 25×10^{-6} mm.

(c) *Electrical extensometers.* These are based on the measurement of the variation of resistance, inductance or capacitance in an electric displacement gauge or transducer.

In variable resistance transducer (Fig. 7.4), a closely coiled resistor has the two end terminals connected to a constant voltage source. A stem, measuring the displacement, moves a wiper connected to a third terminal and in contact with the coil. The resolution is of the order of 25×10^{-3} mm with sensitivities of 400 mV/mm.

The simplest variable inductance transducer consists of two coils in a Wheatstone bridge, with two inductors. The relative impedance of the two coils is

274

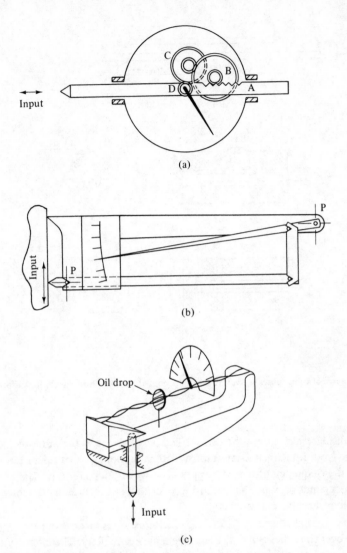

(a)

(b)

Oil drop

Input

(c)

Fig. 7.3 Typical mechanical extensometers (a) dial gauge, (b) Huggenberger extensometer, (c) Johansson extensometer

measured by means of a movable core, the change in inductance being proportional to the core displacement.

A more complex instrument in very wide use consists of a central coil coupled by means of a movable core to two secondary coils (Fig. 7.4). The instrument is known as a differential transformer. The secondary coils are wired in such a way that the induced voltages cancel each other when the core is in its central position. When the core is displaced, one secondary voltage increases while the other decreases, in an essentially linear dependence with the core movement

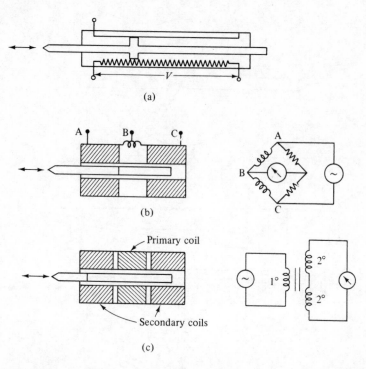

Fig. 7.4 Electrical transducers (a) variable resistance, (b) variable inductance, (c) differential transformer

for small displacements from the centre. The sensitivity is of the order of 2-5 volt output/volt input/inch displacement. The input voltage is of the order of 10 V and the frequency of the AC supply is between 50-10 000 Hz. The resolution depends only on the measuring, indicating or recording instrument to which the transducer is connected.

(d) *Variable capacitance transducer.* Capacitance transducers offer the advantage over the other types of their mechanical simplicity. The circuitry they require is on the other hand considerably more complex due mainly to their high impedance, considerably higher than that of most commercial indicating or recording instruments. This high impedance also results in difficulties caused by motion of leads or the approach of personnel.

The sensitivity of these transducers is very high and the resolution of commercial equipment is of the order of 25×10^{-8} mm.

7.2 Variable Resistance Strain Gauges

The electrical resistance of a wire increases when the wire is stretched. The ratio between the percent change in resistance and the corresponding percent increase

276

in length is called the *gauge factor k*,

$$k = \frac{\% \text{ change in resistance}}{\% \text{ change in length}} = \frac{\Delta R/R}{e}$$

For the metals in common use, k varies between 0·5 and 5. Semiconductor gauges of silicon or germanium have gauge factors as high as 150.

A resistance strain gauge consists of a conductor, bonded to a carrier which is, in turn, fixed on to the structure or machine (base). The carrier may be in the form of a foil, a sheath or a frame and it may be cemented or welded to

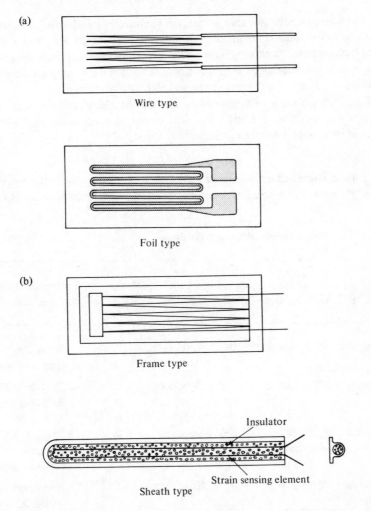

(a)

Wire type

Foil type

(b)

Frame type

Insulator

Strain sensing element

Sheath type

Fig. 7.5 (a) Resistance strain gauges of the bonded type, (b) resistance strain gauges of the unbonded type

the base (Fig. 7.5). The sheath and frame types are usually referred to as unbonded gauges.

When the temperature of gauge and base change by ΔT, the gauge grid stretches by $\alpha \Delta T$ over 1 in. while the carrier and base stretch by $\beta \Delta T$ α and β being respectively the thermal expansion coefficients of the grid and of the base metal. At the same time, the change in temperature results in a change in resistivity of $\gamma \Delta T$ (ohm/ohm). Assuming that the base is far more rigid than the gauge itself, the strain to which the gauge grid is subjected is $(\beta - \alpha)\Delta T$ and the total change in resistance is,

$$\Delta R/R = [(\beta - \alpha)k + \gamma]\,\Delta T$$

In some commercially available gauges, the parameters α and γ are chosen in such a way that the term in brackets in the previous expression is zero, for a given base material. Such gauges are called self-compensated. Often, no self-compensation is attempted, in which case, the temperature compensation is achieved by means of a dummy gauge, as will be shown later.

Table 7.1 lists some of the materials used for the construction of grids and their properties. Table 7.2 lists the carriers most commonly used and Table 7.3 some of the cements for fixing gauges to the base.

7.2.1 Parameters influencing gauge behaviour   *Moisture* affects the behaviour of gauges since it may cause changes in insulation resistance, which should be

Table 7.1   Strain Gauge Materials: Grids

Material	Composition (%)	k	Linearity	Max. strain (%)	Temperature range applications
Constantan	45Ni; 55Cu	2	Excellent	15	−100 to 300°C (self-compensated)
Nichrome	80Ni; 20Cr	2·2	Very good	5	−270 to 850°C (self-compensated 20 to 300°C)
Isoelastic	36Ni; 8Cr; 0·5Mo; 55·5Fe	3·6	Good	2	−50 to 300°C. Dynamic measurement
Platinum alloy	92Pt; 8W	4	Good	5	High temperature (600-800°C)

Table 7.2   Strain Gauge Materials: Carriers

Material	Electrical insulation	Temperature range	Max strain (%)	Application
1. Paper	Fair	−50 to 50%	1	General
2. Resin-impregnated paper, cellulose fibres	Good	As above	2	As (1)
3. Epoxy resin foils	Excellent	−270 to 120°C	2	As (1)
			15	Post-yield gauges
4. Epoxy/fibreglass	As (3)	−270 to 300°C	As (3)	As (3)
5. Ceramic	As (3)	High temperature gauges		

Table 7.3   Strain Gauge Materials: Cements

Type	Maximum temperature	Carrier	Drying time	Application
Cellulose acetate or nitrate	50°C	Paper	24-48 hr	Where accuracy is not essential
Epoxy	200°C	Any	24-48 hr	General, strong bond
Acrylic	70°C	Any (except paper)	15 min	General
Ceramic	−	Ceramic		High temperature gauges

at least 10 000 megohms. At low temperatures, waterproofing is achieved by covering the gauge with strain gauge cement (epoxy) or silicone grease. High temperature epoxy cements, with fibreglass cloth may be used up to their temperature limit (300°C).

When a gauge is fixed to a base and strained to, say, 2000 microstrain, a complex mechanism of creep of the cement, cold working of the strain sensing material, and hysteresis results in a *zero drift* of up to 50 microstrain upon removal of the strain. This drift decreases upon subsequent strain cycles.

The fatigue life of a gauge is usually limited by failure at the connections between grid and terminals. Warning of incipient failure is given by sharp increase in resistance, i.e., an apparent increase in gauge factor. For maximum long-term stability connections may be made by jumper wires to gauge terminals covering all with epoxy cement and curing for at least a week at 80°C. Fibreglass reinforcement increases the long term stability.

Hydrostatic pressure may also affect the gauge readings.

Leads form a dead ballast and their *lead resistance* may have to be taken into account. The correction term is given by $(1 + R_1/R)$, where R_1 is the lead resistance and R is the gauge resistance.

$$R_l = 16 \times 10^{-3} \, \frac{l(\text{m})}{A(\text{mm}^2)} \text{ ohm for copper leads}$$

More important than the lead resistance are the contact resistances between gauge and leads, leads and measuring instrument, and within the instrument itself. All connections should be soldered if possible. Switches must be placed in zero arms or in series with high resistances. High resistance gauges are advantageous.

Electromagnetic fields cause spurious transient effects (noise), and to avoid this, cables should be screened and earthed. Loops in cables should be avoided, and the gauges themselves should be screened, as they form inductive loops.

The junction between the gauge and the leads constitutes a *thermocouple*, giving a signal of 43×10^{-3} mV/°C (Cu – Cu/Ni) and of 22×10^{-3} mV/°C (Cu-Nichrome).

The *gauge length* has the obvious effect of averaging the strain in static measurements. This effect is more pronounced in dynamic measurements. The strain gauge factor is usually slightly different from the gauge factor that would be obtained for a single straight wire of the same strain sensing material. This is because of the *transverse sensitivity* of the gauge grid. When a gauge is under biaxial strain (e_{xx}, e_{yy}), with its axis along the x-direction,

$$\frac{\Delta R}{R} = k_x e_{xx} + k_y e_{yy} = k_{\text{true}} e_{xx}$$

the transverse sensitivity factor is $t = (k_y/k_x)$ and varies between 0·005 and 0·04 for most gauges. Gauges are calibrated under uniaxial stress conditions

$(e_{yy} = -\nu e_{xx})$. It can be shown that,

$$k_{\text{true}} = k_{\text{calibrated}} (1 + te_{yy}/e_{xx})/(1 - \nu t)$$

Stress gauges are strain gauges made with an intentionally high transverse sensitivity, so that $t = \nu$. Independently of the state of strain, such gauges give a stress value

$$\sigma_x = Ek_{\text{calibrated}} (\Delta R/R)$$

7.2.2 Strain rosettes   When principal strains e_1 and e_2 have been found, principal stresses σ_1 and σ_2 follow from the equations for a state of plane stress,

$$\sigma_1 = \frac{E}{1 - \nu^2}(e_1 + \nu e_2) \qquad \sigma_2 = \frac{E}{1 - \nu^2}(e_2 + \nu e_1)$$

The directions of principal axes of stress are, of course, the same as those for strain, and if they should happen to be known, the strains can be determined from two gauges positioned along these two mutually perpendicular directions. When the principal axes are not known, it is necessary to use a rosette of three gauges. It is usual to adopt either a rectangular or delta arrangement.

(a) *Rectangular rosette*. Gauges are inclined at 45° to each other, as shown in Fig. 7.6(a), and give strain values e_a, e_b, and e_c. Suppose that the (a) direction is inclined at angle α to the principal axis (1). Three transformation equations of the type (1.6) can be written down,

$$e_a = e - e' \cos 2\alpha$$
$$e_b = e - e' \cos 2(\alpha + 45°)$$
$$e_c = e - e' \cos 2(\alpha + 90°)$$

where

$$e = \tfrac{1}{2}(e_1 + e_2) \qquad e' = \tfrac{1}{2}(e_2 - e_1)$$

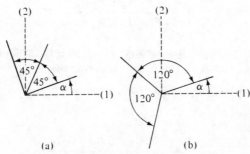

Fig. 7.6 Strain gauge rosettes (a) rectangular, (b) delta

These equations may be solved to obtain

$$e = \tfrac{1}{2}(e_a + e_c)$$

$$e'^2 = \tfrac{1}{2}(e_a^{\,2} + 2e_b^{\,2} + e_c^{\,2}) - e_b(e_a + e_c)$$

$$\tan 2\alpha = (e_a - 2e_b + e_c)/(e_a - e_c)$$

The angle α thus found refers ambiguously to either the axis of greater or lesser principal strain, and the angular relationship must be clarified by sketching the strain circle diagram, which has a mean strain e and radius e'. Numerically, the principal strains e_1 and e_2 are given by $e + e'$ and $e - e'$.

(b) *Delta rosette.* A similar procedure may be followed for this arrangement, which is shown in Fig. 7.6(b), by using the transformation equations

$$e_a = e - e' \cos 2\alpha$$

$$e_b = e - e' \cos 2(\alpha + 120°)$$

$$e_c = e - e' \cos 2(\alpha + 240°)$$

The results are

$$e = \tfrac{1}{3}(e_a + e_b + e_c)$$

$$e'^2 = \tfrac{4}{9}(e_a^{\,2} + e_b^{\,2} + e_c^{\,2} - e_a e_b - e_b e_c - e_c e_a)$$

$$\tan 2\alpha = \sqrt{3}(e_b - e_c)/(2e_a - e_b - e_c)$$

7.2.3 Strain gauge circuits   Two basic circuits are used: the *Wheatstone* bridge and the *potentiometer*. The former is used for static measurements (null balance method in general) and dynamic measurements (deflection method). The latter is only used for dynamic measurements (with temperature compensated gauges usually).

In the bridge shown in Fig. 7.7, the terminals are connected to input source (V), BD to output (E). Arms' resistance AB = R_1, BC = R_2, CD = R_3, DA = R_4. Output is given by,

$$E = \frac{R_1 R_3 - R_2 R_4}{(R_1 + R_2)(R_3 + R_4)} V$$

When the bridge is initially balanced, $R_1 R_3 = R_2 R_4$. In the null balance method the bridge is continuously balanced in such a way that,

$$(R_1 + \Delta R_1)(R_3 + \Delta R_3) = (R_2 + \Delta R_2)(R_4 + \Delta R_4)$$

For small changes,

$$\frac{\Delta R_1}{R_1} + \frac{\Delta R_3}{R_3} = \frac{\Delta R_2}{R_2} + \frac{\Delta R_4}{R_4}$$

282

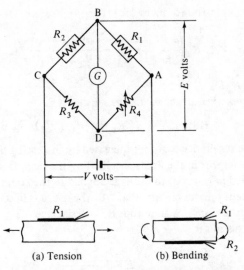

(a) Tension (b) Bending

Fig. 7.7 Wheatstone bridge circuit for the measurement of resistance changes in strain gauges

If R_4 is the measuring resistor, such as a decade box, potentiometer, or slide wire with ballast resistor, and R_1 is the active gauge, the sensitivity is

$$\Delta R_4/e = kR_4$$

If $R_1 = R_2 = R$ are active gauges fixed to the top and the bottom surfaces of a beam of symmetrical cross-section under bending,

$$\Delta R_1 = -\Delta R_2 = \Delta R$$

and
$$\Delta R_4/e = 2KR_4$$

If $R_1 = R_2 = R$ and only R_1 is active, R_2 being dummy, a change in temperature does not result in any unbalance, provided both gauges are identical and fixed to identical material. The previous circuit is therefore self-compensating.

In the deflection method, the output E is measured by means of a voltmeter of resistance R_L. The voltmeter reading will be

$$E' = R_L I_L = V \frac{R_1 R_3 - R_2 R_4}{(R_1 + R_2)(R_3 + R_4) + [R_1 R_3(R_2 + R_4) + R_2 R_4(R_1 + R_3)]/R_L}$$

$$E' \simeq E \text{ if } R_L \gg \frac{R_1 R_3(R_2 + R_4) + R_2 R_4(R_1 + R_3)}{(R_1 + R_2)(R_3 + R_4)}$$

this condition is satisfied in general. Valve voltmeters have R_L measured in megohms. If the bridge is initially balanced, the signal received by the voltmeter when changes in resistance take place is approximately,

$$\text{Output} = VR_1 R_2 \left(\frac{\Delta R_1}{R_1} + \frac{\Delta R_3}{R_3} - \frac{\Delta R_2}{R_2} - \frac{\Delta R_4}{R_4}\right) \Big/ (R_1 + R_2)(R_3 R_4)$$

with $R_1 R_3 = R_2 R_4$.

If R_1 is the active gauge, the bridge sensitivity is

$$\text{Output}/e = KVR_1 R_2/(R_1 + R_2)(R_3 + R_4)$$

The bridge sensitivity is always increased by increasing the current through the strain gauge, within the limits specified by the manufacturer. The sensitivity may be defined in terms of the voltage applied to the strain gauge.

In the potentiometer circuit—Fig. 7.8—the gauge is in series with a ballast resistor. Input terminals A,C, output B,C. Resistance AB = R_2, BC = R_1 (gauge). Open circuit output is

$$E = V\frac{R_1}{R_1 + R_2}$$

The circuit cannot be balanced, so that the initial reading of the measuring instrument (large impedance) is E. Upon application of strain,

$$E + \Delta E = V\frac{R_1 + \Delta R_1}{R_1 + R_2 + \Delta R_1} \simeq V\frac{R_1 + \Delta R_1}{R_1 + R_2}$$

hence

$$\Delta E = V\frac{\Delta R_1}{R_1 + R_2}$$

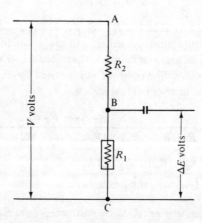

Fig. 7.8 Potentiometer circuit for the measurement of dynamic strains

Since V is of the order of 10 V, ΔE of the order of a few mV, the measuring instrument must have high resolution plus extended full-scale range. For this reason, only dynamic measurements are possible with DC current and filter to block the steady state output, letting only ΔE through (ΔE cyclic).

7.3 Photoelasticity

The first theory to explain the behaviour and the nature of light was the corpuscular theory, devised by Newton. It was assumed that light consisted of small corpuscules, travelling along straight lines. This theory explains the formation of shadow and the phenomenon of reflection. It also provides a partial explanation to the phenomenon of refraction but fails to account for the fact that the velocity of propagation of light in different materials is slower the denser the material. To explain this, as well as the diffraction and polarization of light, Huygens assumed the existence of a fluid, the *aether* that would fill all voids and would have the same characteristics as an elastic material. Light would propagate in this fluid in the form of *transverse waves*, aether particles oscillating irregularly in a plane normal to the direction of propagation of the light or ray. The existence of such an improbable fluid as the aether has been proved to be unnecessary to the acceptance of the wave nature of light. Light may be considered to have the same *electromagnetic* wave nature as radio waves or X-rays, the only basic difference being the wavelength.

White light is a superposition of the colours of the rainbow, each colour having a certain wavelength λ. The wave band covered by the monochromatic constituents of the white light ranges from $4\cdot34 \times 10^{-7}$ m (violet) to $6\cdot563 \times 10^{-7}$ m (red) in air (velocity of light 3×10^{8} m/s). The wavelength of monochromatic yellow light given by sodium is $5\cdot893 \times 10^{-7}$ m and the green light of a mercury lamp has a wavelength of $5\cdot461 \times 10^{-7}$ m, both in air.

When the light is such that all oscillations take place on one plane, it is said to be plane polarized—Fig. 7.9(b). It can then be described as a standing wave similar to the ripples emanating from a source of disturbance in a pond. The oscillation of a point is then given by the equation,

$$y = a \sin(2\pi/\lambda)vt = a \sin \omega t$$

where y = instantaneous deflection of the point considered, a = amplitude, v = wave velocity, λ = wavelength, t = time measured from an arbitrary moment t_0, $\omega = (2\pi v/\lambda)$ is the angular frequency.

The period is defined by $T = \lambda/v$ and the frequency, $N = 1/T = v/\lambda$. Monochromatic light is defined by its frequency (cycles per second or Hz).

When the instantaneous deflection of all points remains constant and the oscillation is such that all points are constrained to move along circles perpendicular to the direction of the ray, the light is said to be circularly polarized (Fig. 7.9(c)).

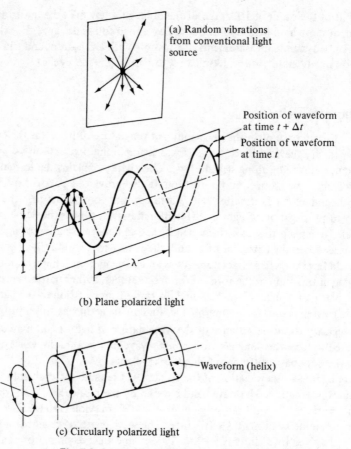

(a) Random vibrations from conventional light source

Position of waveform at time $t + \Delta t$

Position of waveform at time t

λ

(b) Plane polarized light

Waveform (helix)

(c) Circularly polarized light

Fig. 7.9 Conventional light source and polarized light

Certain crystals such as fluor-spar transmit two different rays, polarized in mutually perpendicular planes propagating through the crystal with different velocities. Other materials such as glass and some plastics show this double refraction or *birefringence*, when they are stressed. This is the property used in the *photoelastic* method to measure the stresses in a model.

The photoelastic bench consists of an instrument for the production and detection of polarized light called the *polariscope* and a *loading rig*. There are two basic types of polariscope illustrated in Fig. 7.10, (a) the diffused light type and (b) the transmission type.

The polarizers in most general use consist of thin sheets of polyvinyl alcohol heated and stretched before bonding to a supporting sheet of cellulose acetate. The polyvinyl face is then coloured by a liquid containing iodine. These filters split the incident light rays into two polarized rays, one of which is absorbed. The analyser is similar to the polarizer. When these two elements are positioned

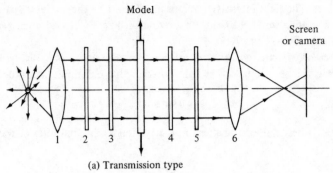

(a) Transmission type

1 - Collimating lens; 2 - Polarizer; 3; 4 - Quarter-wave plates
5 - Analyser; 6 - Condenser lens

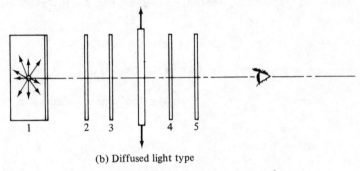

(b) Diffused light type

1 - Lamphouse source with diffusing glass. Other components as above.

Fig. 7.10 General arrangement of two types of photoelastic benches

in such a way that the plane of polarization of the light emerging from the polarizer is perpendicular to the polarization plane of the analyser, no light will be transmitted through the analyser. Polarizer and analyser are then said to be crossed. When the polarization planes coincide, the light is transmitted through both analyser and polarizers which are then said to be parallel.

A mica plate or a thin sheet of oriented polyvinyl alcohol laminated between two sheets of cellulose acetate is often placed in between the polarizer and analyser. This plate is birefringent and is dimensioned in such a way that the relative retardation between the two emerging rays is $\lambda/4$. The two emerging rays are polarized on mutually perpendicular planes at $45°$ to the polarization plane of the polarizer. A second *quarter-wave plate*, similar to the one described, is placed after the analyser with its planes of polarization crossed with respect to the first plate.

7.3.1 The path of light through a polariscope   (a) *Plane polariscope.* When the polariscope does not include quarter-wave plates, the light is plane polarized by the polarizer. The light intensity remains unaffected by the analyser (parallel polarizer and analyser—light field) or is reduced to zero (crossed polarizer and analyser—dark field) (Fig. 7.11(a)).

(b) *Circular polariscope.* When quarter-wave plates are added, the light becomes circularly polarized. The light emerging from the polarizer is defined by

$$y_p = a \sin \omega t$$

and its components along the axes of polarization of the first quarter-wave plate are

$$y_p' = a \frac{\sqrt{2}}{2} \sin \omega t$$

Of these components, one emerges from the quarter-wave plate with a relative retardation of $(\lambda/4)$ with respect to the other (see Fig. 7.11(b)). The retardation in time, is,

$$\Delta t = (\lambda/4v) = (\pi/2\omega)$$

and the two emerging oscillations may be represented by,

$$y_{1F} = a \frac{\sqrt{2}}{2} \sin \omega t \qquad \text{on the 'fast' axis of polarization}$$

$$y_{1S} = a \frac{\sqrt{2}}{2} \sin \left(\omega t - \frac{\pi}{2} \right) \qquad \text{on the 'slow' axis of polarization.}$$

These two oscillations have a resultant given by a vector of magnitude

$$A_1 = a \frac{\sqrt{2}}{2}$$

and inclined by

$$\gamma = (1/\omega t)$$

the light vector has therefore a fixed magnitude and its tip describes a helix when travelling along the x-direction. The light emerging from the polarizer + quarter-wave plate system is then circularly polarized.

The effect of the second wave plate, crossed with respect to the first, is to reproduce plane polarized light, since this second plate causes a retardation on y_{1F} of $\pi/2$ relative to y_{1S},

$$y_{2F} = a \frac{\sqrt{2}}{2} \sin \left(\omega t - \frac{\pi}{2} \right) \qquad \text{in the same plane as } y_{1S},$$

$$y_{2S} = a \frac{\sqrt{2}}{2} \sin \left(\omega t - \frac{\pi}{2} \right) \qquad \text{in the same plane as } y_{1F}$$

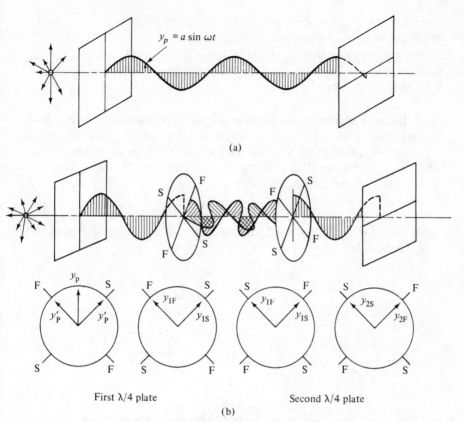

(a)

First λ/4 plate   Second λ/4 plate

(b)

Fig. 7.11 Path of light through polariscope

and the resultant is,

$$y_2 = a \sin\left(\omega t - \frac{\pi}{2}\right)$$

on the plane of polarization of the polarizer.

The following table summarises the previous results.

Plane polariscope

polarizer/analyser	crossed	dark field
„ „	parallel	light field

Circular polariscope

polarizer/analyser	crossed	
(λ/4) plates	crossed	dark field
polarizer/analyser	parallel	
(λ/4) plates	crossed	light field

The quarter-wave plates are always used with their optical axes at $45°$ to those of the polarizer and analyser and mutually crossed.

The quarter-wave plates are usually designed for green light ($5·461 \times 10^{-7}$ m wavelength in air). Using them crossed provides total compensation, independently of the wavelength.

7.3.2 Insertion of birefringent model in the plane polariscope

A birefringent model stressed under a biaxial stress field defined by the principal stresses σ_1 and σ_2, splits the polarized ray, into two components oscillating in the directions of the principal stresses. If β is the angle between the plane of polarization and the σ_1-direction the rays entering the model are

$$y_p' = a \cos\beta \sin\omega t \qquad y_p'' = a \sin\beta \sin\omega t$$

as shown in Fig. 7.12. The time lag between the two rays emerging from the model is Δt and the rays are

$$y_M' = a \cos\beta \sin\omega t$$

$$y_M'' = a \sin\beta \sin\omega(t - \Delta t)$$

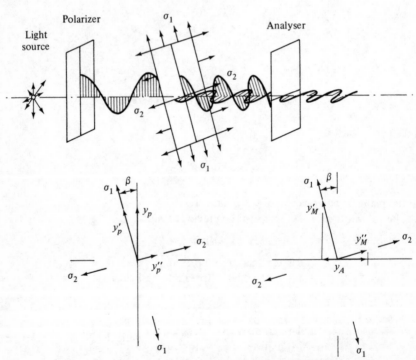

Fig. 7.12 Path of light through photoelastic bench, for a plane, crossed polariscope

The analyser only transmits components along its plane of polarization. If analyser and polarizer are crossed, the light ray emerging from the analyser will be,

$$y_A = a \sin\beta \cos\beta [\sin\omega t - \sin(\omega t - \omega \Delta t)]$$
$$= a \frac{\sin 2\beta}{2} [\sin\omega t - \sin\omega(t - \Delta t)]$$

while if they are parallel,

$$y_A = a \cos^2\beta \sin\omega t + a \sin^2\beta \sin\omega(t - \Delta t)$$

The condition for extinction of the emerging light is, in the case of crossed polarizer-analyser, either $\beta = 0$ or $\Delta t = (2N\pi/\omega)$, where $N = 0, 1, \ldots$.

Only the crossed polarizer-analyser position is generally used.

The conditions $\beta = 0$ and $\Delta t = 0$ are independent of the wavelength. The condition $\Delta t = 2\pi/\omega$, $4\pi/\omega$, etc., depends on the wavelength. When monochromatic light is used, the dark lines produced when this condition is satisfied are called *fringes*. With white light, the condition is only satisfied by one of the constituent colours, so that the remaining unsuppressed colours will give a coloured line called a *monochromatic*. The dark lines produced when the condition $\beta = 0$ is satisfied are called the isoclinics. They are the loci of all points where the principal stress directions coincide with the axes of the polarizer and the analyser. We shall see that the condition $\Delta t = 0$ is satisfied in neutral axes, in regions of zero stress, or when $\sigma_1 = \sigma_2$. Points where this is true are called *isotropic points*.

It can be shown that the conditions for extinction of light in the circular polariscope are,

$\Delta t = 2N\pi/\omega$ for the crossed polarizer-analyser set-up, and
$\Delta t = (2N + 1)\pi/\omega$ for the parallel polarizer-analyser system.

Only fringes (or monochromatics) and isotropic points are then detected.

7.3.3 Temporary birefringence of plastics under load   It has already been mentioned that when a plastic is under a biaxial stress system (σ_1, σ_2) it becomes birefringent. Maxwell found that the difference between the refraction indices of the two refracted rays is proportional to the difference between the principal stresses. It is therefore possible to write,

$$\frac{c}{v_1} - \frac{c}{v_2} = C(\sigma_1 - \sigma_2)$$

where c = velocity of propagation in vacuum; v_1, v_2 = velocities of rays (1) and (2) and C = stress optical coefficient of the material, sometimes expressed in brewsters (1 brewster = 10^{-13} cm^2/dyne = 69×10^{-10} in^2/lb = 10^{-12} m^2/N).

If the thickness of the model is h, the time spent by ray (1) to cross it is (h/v_1) while ray (2) takes (h/v_2). The relative retardation is then

$$\omega\Delta T = \omega\left(\frac{h}{v_1} - \frac{h}{v_2}\right) = \frac{2\pi c}{\lambda} h\left(\frac{1}{v_1} - \frac{1}{v_2}\right) = \frac{2\pi h}{\lambda} C(\sigma_1 - \sigma_2)$$

Fringes or isochromatics appear when

$$\sigma_1 - \sigma_2 = 2N\pi\left/\frac{2\pi\lambda}{\lambda} C\right. = \frac{N}{h}\frac{\lambda}{C}$$

in the crossed polariscopes, and

$$\sigma_1 - \sigma_2 = (2N + 1)\pi\left/\frac{2\pi h}{\lambda} C\right. = \frac{2N + 1}{2h}\frac{\lambda}{C}$$

in the parallel polariscope with quarter-wave plates.

The term (λ/C) is usually denoted by the letter f and is called the *material fringe constant*. It is expressed in fringes/lb/in^2/in thickness (lb/in) or fringes/MN/m^2/mm (N/mm).

Table 7.4 lists the properties of some plastic materials used in photoelasticity. These properties are only approximate and it is always essential to determine the fringe constant of the material used, with the same light as will be used in the photoelastic bench.

Table 7.4   Properties of some photoelastic materials

Type	f (lb/in)	E (lb/in^2 $\times 10^6$)	E/f (1/in)	Y (lb/in^2)	Y/f (1/in)	Machin.	Creep	Time edge effects
Cr 39	90–120	0·25–0·30	2800–3300	3000	30 0·42	P	P	F
Polyester	40	0·34	8500	6000	150 0·38	G	G	E
Epoxy	60	0·45	7500	9000	150 0·36	G	G	G

Note: E = Young's modulus; Y = limit of proportionality; E = Excellent; G = good; F = fair; P = poor. 1 lb/in = $1·742 \times 10^2$ N/m, 1 lb/in^2 = 6·895 kN/m^2.

7.3.4 Calibration methods   The material fringe constant is determined by loading a specimen in which the stress distribution is known. The specimens most commonly used are shown in Fig. 7.13.

(a) Tensile specimen: $\sigma_1 = F/bh$, $\sigma_2 = 0$, $b = a$, $4a/3\cdot5$, ...
(b) Bending specimen: $\sigma_1 = 12My/bh^3$, $\sigma_2 = 0$
(c) Disc under compression. At the centre: $\sigma_1 = 2F/\pi hD$, $\sigma_2 = -6F/\pi hD$

Condition for dark fringe or monochromatic,

$$\sigma_1 - \sigma_2 = \frac{Nf}{h} \qquad\qquad \text{crossed polariscope}$$

$$\sigma_1 - \sigma_2 = \frac{(2N+1)f}{h} \qquad\qquad \text{parallel polariscope}$$

(a) Stepped tension specimen

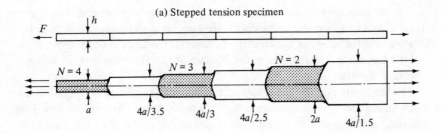

(b) Beam under pure bending

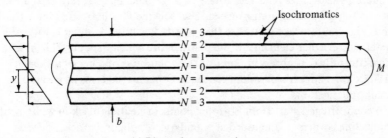

(c) Disc under compression

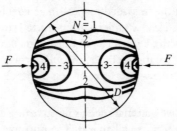

Fig. 7.13 Typical specimens for the calibration of photoelastic materials

Using white light it is possible to obtain a table for measurement of the stress difference by colour comparison, the isochromatics corresponding to the fringes in dark field systems are characterized by the *tint of passage*, sharp lines between red and indigo or green, for low fringe orders.

7.3.5 Interpretation of the stress pattern   (a) *The fringes (or monochromatics).* Zero order fringes ($\sigma_1 = \sigma_2$) are isotropic points or neutral axes. They are always black and can thus be distinguished from higher order fringes viewing the model with white light.

At an unloaded (free) boundary, one of the two principal stresses is zero. The fringe order therefore gives a direct indication of the magnitude of the stress at a free boundary. At a loaded boundary it is difficult to determine the fringe order because of contact stresses. However, it is possible to find both σ_1 and σ_2 if the load is known as well as the directions of σ_1 and σ_2.

It is not possible to draw a direct distinction between tensile and compressive stresses. At a free boundary, the sign of the tangential stress may be obtained by observing the effect of a concentrated force. If the force is compressive and the tangential stress is tensile, the fringe order will increase and the fringe will travel inwards.

The sensitivity of the photoelastic method would appear to be limited to the stress difference corresponding to one fringe order. It may be improved by using polarizer and analyser first crossed and then parallel. Half order fringes are then determined.

Fractional fringe orders may be found by the Tardy method. At first, with the quarter-wave plates removed, rotate polarizer and analyser (crossed) until light at observed point is extinguished. The isoclinic for observed point is thus found. The quarter-wave plates are then inserted, and the analyser is rotated, maintaining all other elements of the photoelastic bench fixed, until light is again extinguished at observed point. If this rotation is of α degrees, then ($\alpha/180$) x 100 is the fringe order (per cent). Other methods may be found in specialist publications.

To count fringes, start from isotropic points or neutral (black in white light) or projecting corners of an unloaded boundary. Alternatively increase load gradually and count their appearance as they travel past a given point. The initial order of fringes in the unloaded model—in general less than unity—must be found first. If it is not clear whether a given fringe is to be counted upwards or downwards, view in white light. The red side points downwards.

(b) *The isoclinics.* To obtain isoclinics, the quarter-wave plates must be removed. Isoclinics and fringes interfere and it is advisable when determining the former to,

(i) Use an insensitive material such as perspex or glass under slight load, in such a way that no fringes will appear. The isotropic points and neutral axis will also be shown. They can be recognized easily since the isotropic points are at the intersection of isoclinics and the position of the neutral axes does not change with the rotation of the polarizer-analyser.

(ii) Use a very sensitive material under high loading and with white light. The isoclinics are the only dark lines on a brightly coloured field.

The isoclinics being the loci of all points where the principal stress directions coincide with the polarizer and analyser axes, it is possible to obtain the *stress trajectories* or *isostatics* by a graphical construction. The isostatics show the flow of force through the model. In a good design they should be smooth and evenly spaced.

All isoclinics must pass through points of concentrated loading and through isotropic points. One isoclinic must coincide with each axis of symmetry. The parameter of an isoclinic intersecting a free boundary is equal to the slope of the boundary at that point.

(c) *Separation of principal stresses.* At a free boundary there is only one principal stress different from zero, while at a loaded boundary the value of the principal stresses may be obtained as described previously. Inside the model, the photoelastic model only gives an indication of $(\sigma_1 - \sigma_2)$. Often it is not necessary to separate the two principal stresses at the interior of the model because the maximum stress is present at the boundary. When the maximum shear or the distortion energy criteria are used for the dimensioning of the design, it is also found that the critical position is at the boundary. It may however be necessary to determine the actual values of σ_1 and σ_2 throughout the whole model. This can be done by the direct determination of $(\sigma_1 - \sigma_2)$, counting the fringe order and using one of several methods for the determination of $(\sigma_1 + \sigma_2)$. To this end, the change in thickness of the model may be determined. Since $\sigma_3 = 0$, $e_3 = -\nu(\sigma_1 + \sigma_2)/E$. Special extensometers, interferometers and strain gauges are used.

The sum $\sigma_1 + \sigma_2$ obeys Laplace's equation at the interior of the model, whilst at the boundary, $\sigma_1 + \sigma_2 = \sigma_1 - \sigma_2$ is known. The solution of Laplace's equation with known boundary conditions may be undertaken by means of suitable analogue methods—membrane, electrolytic tank or conducting paper. Alternatively it may be solved with an electronic computer.

7.3.6 Scaling from model to prototype   The compatibility condition in two-dimensional problems does not depend on Poisson's ratio when no body forces are present and when the body is not multiply connected. If it is multiply connected, the stress distribution is independent of Poisson's ratio only when there are no distributed forces at the boundaries. In all cases, the effect of changes in Poisson's ratio on the stress distribution has been found to be negligible.

When the model is loaded, the maximum stress due to a load of magnitude P, σ_{max}, is proportional to P and inversely proportional to the thickness h and to a characteristic dimension, while it is independent of E and, to a large extent, of ν. We can then write,

$$(\sigma_{max})_{model} = C \left(\frac{P}{ha} \right)_{model}$$

also

$$(\sigma_{max})_{prototype} = C \left(\frac{P}{ha} \right)_{prototype}$$

and

$$(\sigma_{max})_{prototype} = (\sigma_{max})_{model} \times \left(\frac{P}{ha} \right)_{prototype} \bigg/ \left(\frac{P}{ha} \right)_{model}$$

provided that the shape of model and prototype remain the same during the test.

7.3.7 Frozen stress method The direct determination of $(\sigma_1 - \sigma_2)$ is only possible in a two-dimensional model. When it is required to determine the stress distribution in a three-dimensional component, the method used consists of locking-in into the model the stresses caused by the specified load and cutting thin slices out of the model. The slices are then viewed through a polariscope.

When a photoelastic plastic is heated above a certain temperature (about $140°$ for epoxy resins) some of the molecular bonds break down, while stronger or primary bonds remain. The material behaviour is then elastic with a very low value of Young's modulus, and, if the load is maintained, large deformations are induced. Upon cooling down, the weaker secondary bonds are reformed, locking into the model the deformations previously induced. When the load is removed, the elastic spring-back at room temperature is insufficient to cause a significant reduction in the locked-in deformation. It is also found that slicing, machining, or sawing does not alter the fringe pattern provided that these operations are performed at reasonably low temperatures. The procedure to 'freeze' the stress pattern into the model is then (1) to load the model in a furnace with controllable temperature (2) to heat until the temperature required to break down the secondary bonds has been reached. The rate of increase is of about $10°C/hr$; (3) to maintain the model at this critical temperature for a period of between 2 and 4 hr; (4) to cool down very slowly— say $1°C/hr$—in order to minimize thermal stresses. Finally, the load is removed and slices are cut off the model and examined in the polariscope.

When a principal plane is known, e.g., in symmetrical configurations, the first slice is cut to coincide with this plane. On this plane, it is easy to determine the direction of the principal stresses as well as their difference. By cutting a narrow strip off the main slice in the direction of one of the two principal stresses and viewing at right angles, as shown in Fig. 7.14, it is possible to find the difference between one of the principal stresses and the third principal stress. In this way, $(\sigma_1 - \sigma_2)$, $(\sigma_1 - \sigma_3)$, and $(\sigma_2 - \sigma_3)$ are determined.

296

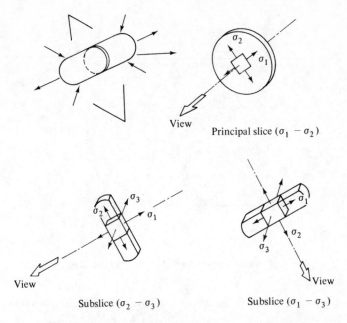

View

Principal slice ($\sigma_1 - \sigma_2$)

View

Subslice ($\sigma_2 - \sigma_3$)

View

Subslice ($\sigma_1 - \sigma_3$)

Fig. 7.14 Slicing when one principal plane is known

Very often it is not possible to predict the orientation of the principal planes. In that case, before proceeding to cutting a general slice, it is usual to cut a surface slice, i.e., one in which one of the faces coincides with a stress-free surface. In a general slice it is always possible to find two secondary stress directions which do not necessarily coincide with the principal stress directions—see Fig. 7.15. In the general slice shown in the figure, the secondary principal stresses are,

$$\sigma'_{1,2} = \frac{\sigma_{xx} + \sigma_{yy}}{2} \pm \left[\left(\frac{\sigma_{xx} - \sigma_{yy}}{2} \right)^2 + \sigma_{xy}^2 \right]^{\frac{1}{2}}$$

when referred to stresses σ_{xx}, σ_{yy}, σ_{xy}. The inclination of σ'_1 with respect to the x-axis is,

$$\tan 2\theta = \frac{2\sigma_{xy}}{\sigma_{xx} - \sigma_{yy}}$$

When the slice is viewed with light incident along the z-axis, it is clear that

$$(\sigma'_1 - \sigma'_2) = [(\sigma_{xx} - \sigma_{yy})^2 + 4\sigma_{xy}^2]^{\frac{1}{2}}$$

can be found as well as the angle θ. From this, it is possible to obtain both $(\sigma_{xx} - \sigma_{yy})$ and σ_{xy}. Cutting first a subslice and then a sub-subslice ($\sigma_{xx} - \sigma_{zz}$)

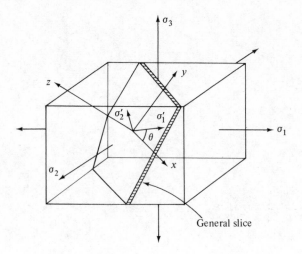

Fig. 7.15 Secondary principal stresses in a general plane

and σ_{xz}, $(\sigma_{yy} - \sigma_{zz})$, and σ_{yz} can then be determined. The effective stress, defined as,

$$\sigma_e = \frac{1}{\sqrt{2}} [(\sigma_{xx} - \sigma_{yy})^2 + (\sigma_{yy} - \sigma_{zz})^2 + (\sigma_{xx} - \sigma_{zz})^2 + 6(\sigma_{xy}^2 + \sigma_{xz}^2 + \sigma_{yz}^2)]^{\frac{1}{2}}$$

can then be obtained.

7.3.8 Birefringent coatings   Normal photoelastic techniques may be used for strain measurement on the surfaces of opaque materials if these are coated with a photoelastic material. The conventional polariscope is then 'folded in half' so that the light after passing through the polarizer, first quarter-wave plate, and coating is reflected back through the coating from the opaque surface and observed through the second quarter-wave plate and polarizer so that the angle between incident and reflected light is small.

Problems

Problems on Chapter 2

2.1 A portion of a beam has unit thickness in the z-direction and depth $2h$ in the y-direction as shown in Fig. P-1, and carries uniform pressure p on its

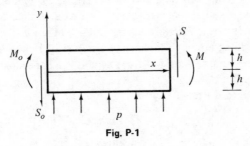

Fig. P-1

lower surface, with shear force S_0 and bending moment M_0 at end $x = 0$. Show that the stress function

$$F = [-M_0 y^3 - S_0 xy(3h^2 - y^2)]/4h^3$$
$$- p[5x^2(2h^3 - 3h^2 y + y^3) + (2h^2 - y^2)y^3]/40h^3$$

gives the required stresses on the upper and lower surfaces, with stresses at other points that can be integrated to give resultants

$$S = S_0 - px \qquad M = M_0 - S_0 x + \tfrac{1}{2}px^2$$

2.2 A semicircle of radius $r = a$ extends from $\theta = 0$ to $\theta = \pi$. On the diameter, the applied stresses are

$$\sigma_{xy} = q \qquad \sigma_y = 0 \qquad (-a < x < 0)$$

$$\sigma_{xy} = \sigma_y = 0 \qquad (0 < x < a)$$

Show that these stresses are given by the stress functions

$$\phi' = -\frac{q}{2\pi} \ln\frac{z}{a} \qquad \psi' = \frac{q}{2\pi}\left(1 + 2\ln\frac{z}{a}\right)$$

299

with no concentrated force or moment acting at the origin, and that in a direction parallel to the diameter, the stress is

$$\sigma_x = \frac{q}{\pi} \left(2 \ln \frac{a}{r} - \sin^2 \theta \right)$$

2.3 An infinite plate has shear stress $\sigma_{xy} = Q$ applied to its outer edges. The stress distribution around a crack in this plate is described by stress function (2.26). Deduce the expression for mean stress

$$\tfrac{1}{2}(\sigma_x + \sigma_y) = -Q \sin 2\beta / (\cosh 2\alpha - \cos 2\beta)$$

and plot contours of this stress on the (x,y) plane.

2.4 For a very large plate carrying some chosen type of external loading and containing a crack in line with the x-axis, find stresses σ_y and σ_{xy} at any point on the x-axis. Integrate these stresses along the x-axis to some large distance x_1 to obtain force resultants. With stresses σ_x and σ_{xy} at points on the y-axis, carry out similar integrations to some large distance y_1. Hence examine the equilibrium of the large rectangle of sides x_1 and y_1. For example, stress function (2.54) may be used.

2.5 The Westergaard stress function

$$Z'' = S \left[1 - \left(\frac{\sin(\pi a/2b)}{\sin(\pi z/2b)} \right)^2 \right]^{-\frac{1}{2}}$$

refers to a series of cracks of length $2a$ and pitch $2b$ situated on the x-axis, with the origin $x = 0$ at the mid-point of one of the cracks. Show that this function represents biaxial tension S at large values of y, with the surfaces of the cracks free from stress, and verify the stress intensity factor

$$K_I = S[2b \tan(\pi a/2b)]^{\frac{1}{2}}$$

2.6 For the problem of a crack in anti-plane strain, use the displacement function (2.59) to calculate the two shear stresses in terms of x and y. Then plot contours on the (x,y) plane of the maximum shear stress, given by

$$\tau_m = (\sigma_{zx}{}^2 + \sigma_{zy}{}^2)^{\frac{1}{2}}$$

Show that these contours become circles centred on the crack tip as the crack tip is closely approached.

Problems on Chapter 3

3.1 A circular plate of uniform thickness $2h$ is simply supported at its outer edge, of radius R, which means that this edge sustains zero bending moment

M_r. Load is applied through the flat end of a rigid cylinder of radius αR, which is pressed upwards against the plate. It is assumed that the cylinder transmits force only at its edge, to give a uniform line distribution of force around a circle concentric with the plate. Hence, when this force is F per unit length, the total load is given by $P = 2\pi\alpha RF$.

Show that bending stresses in the upper surface at radius $r = \rho R$ are given by

$$\sigma_r = \sigma_\theta = \frac{3P}{16\pi h^2}[(1-\nu)(1-\alpha^2) - 2(1+\nu)\ln\alpha] \quad (\rho < \alpha)$$

$$\sigma_r = \frac{3P}{16\pi h^2}\left[(1-\nu)\left(\frac{\alpha^2}{\rho^2} - \alpha^2\right) - 2(1+\nu)\ln\rho\right] \quad (\rho > \alpha)$$

$$\sigma_\theta = \frac{3P}{16\pi h^2}\left[(1-\nu)\left(2 - \alpha^2 - \frac{\alpha^2}{\rho^2}\right) - 2(1+\nu)\ln\rho\right] \quad (\rho > \alpha)$$

3.2   A circular disc of flexural stiffness factor D is simply supported at its outer edge of radius R while a total load P is uniformly distributed within a circle of radius βR. Show that the central deflection is given by

$$w = \frac{PR^2}{64\pi D}\left[\frac{4(3+\nu) - (7+3\nu)\beta^2}{1+\nu} + 4\beta^2 \ln\beta\right]$$

Also show that stresses at the centre are given by

$$\sigma_r = \sigma_\theta = \frac{3P}{32\pi h^2}[4 - (1-\nu)\beta^2 - 4(1+\nu)\ln\beta]$$

3.3   A flat annulus of outside radius R and inside radius γR is simply supported at its circumference and carries a uniform line-distribution of load around its inner edge, the total upward load at this edge being P. Show that hoop stress in the upper surface is given by

$$\sigma_\theta = \frac{3}{8\pi}\frac{P}{h^2}\left[1 - \nu - \frac{(1+\nu)(1+\rho^2)}{\rho^2(1-\gamma^2)}\gamma^2 \ln\gamma - (1+\nu)\ln\rho\right]$$

where ρ is the radius ratio r/R, and hence deduce that the maximum value occurs at the inner edge.

3.4   A solid disc of outside radius R has a line distribution of load $F = F_2 \cos 2\theta$ applied to its rim. Show that deflection is given by

$$w = \frac{F_2 R^3}{12(3+\nu)(1-\nu) D}[6\rho^2 - (1-\nu)\rho^4]\cos 2\theta$$

where ρ is the radius ratio r/R. For a value $\nu = 0\cdot3$, compare the amplitude at the rim with that indicated by the numerical factor in Table 3.1.

With the same loading applied to the rim of a hole of radius R in an infinite plate, show that the deflection amplitude at the rim of the hole is $0.097F_2R^3/D$.

3.5 A flat rectangular plate ABCD has corner A at the origin while sides AB and AD coincide with the x- and y-axes. Show that a deflected shape $w = cxy$ is consistent with the presence of point loads $P = 2(1 - v)Dc$ acting upwards at corners A and C, with equal loads acting downwards at corners B and D, the edges being otherwise stress-free.

If corners A, B, and D are position-fixed while load P is applied to corner C, show that the work done by this load is equal to the gain in strain energy of the plate.

3.6 A spring washer consists of an annulus of inside radius a and outside radius b, with a radial cut along the line $\theta = 0$. Consider a deflected shape $w = c\theta$, where c is a constant. Use equations (3.34) and (3.41) to show that this deflection requires the curved boundaries to be stress-free, with a line-distribution of force along the line $\theta = 0$ and concentrated forces at points $(a,0)$ and $(b,0)$. Show that the force resultant of the line-distribution and point loads passes through the centre of the annulus.

Problems on Chapter 4

4.1 A cantilever of uniform flexural stiffness EI is built-in at end $x = 0$ while the end $x = L$ is free. A single load P acts upwards at point $x = L/2$. Write down six end conditions for this beam and show that they are satisfied by an expression for deflection

$$w = \frac{PL^3}{24EI}[6s^2 - 4s^3 + s^4 - 12a(5s^4 - 6s^5 + 2s^6)]$$

where $s = x/L$, and a is a constant which may be given any value. Use the variational equation for a beam to determine this constant. Hence find the end deflection $w(L)$ and compare it with the value $w(L) = 0.104\,PL^3/EI$ found from solutions of the equilibrium equation.

(*Ans.* $a = 29/640$, $w(L) = 0.102PL^3/EI$.)

4.2 A pin-ended beam of length L and uniform flexural stiffness EI carries an end load Q, and also a transverse load P applied at mid-span. First assume a deflected shape given by $w = a(5 - 24s^2 + 16s^4)$, where a is a constant and $s = x/L$, distance x being measured from the mid-span point. Note that this expression gives zero deflection and moment at the two ends, but predicts incorrect shear forces. By applying the variational method, determine the constants α and β in the expression for transverse stiffness

$$\frac{PL^3}{EIw(0)} = \alpha\left(1 - \frac{QL^2}{\beta EI}\right)$$

Repeat using the deflected shape $w = a \cos \pi s$.

(*Ans.* $\alpha = 49 \cdot 15, \beta = 9 \cdot 88; \alpha = 48 \cdot 70, \beta = 9 \cdot 87$.)

4.3 A flat uniform diaphragm of outside radius R is clamped at its outer edge. Its deflection is assumed to follow the relation

$$w = a(1 - 2\rho^2 + \rho^4) + b(\rho^2 - 2\rho^3 + \rho^4)$$

in which ρ is radius ratio r/R and a and b are constants. Using the variational equation (4.30), find the central deflection produced by various kinds of axisymmetrical loading. In particular, for a central point load P, obtain a value $w = (105/1728\pi)PR^2/D$ by the variational method, and also the value $w = (1/16\pi)PR^2/D$ from the exact solution of this problem, as given in chapter 3.

4.4 A bar has a uniform square section of side a. With coordinates drawn from the centre of the square parallel to its sides, take the torsional stress function

$$\phi = c(a^2 - 4x^2)(a^2 - 4y^2)$$

Verify that this is zero on the boundary of the section and use a variational method to find the constant c. If torque T produces an angle of twist θ over length L, with a shear modulus G, use the stress function to find an approximate value for constant k in the stiffness relationship $TL/G\theta = ka^4$. Compare this with the correct value $k = 0 \cdot 1406$.

(*Ans.* $c = 5/32a^2, k = 5/36$.)

Problems on Chapter 5

5.1 Show that in a shaft of uniform cross-section, the torque T is given by

$$T = 2 \int \phi dA$$

where ϕ is the stress function defined in section 5.1.1, the integration being extended over the whole area A of the cross-section. Referring now to Fig. 5.7, show that the maximum shear stress τ in a shaft of square section of side a is given by $\tau = kT/a^3$ where factor $k = 4 \cdot 81$.

5.2 Follow through the derivation of equation (5.14) governing the stress distribution in a shaft of circular cross-section and variable diameter under torsion, as indicated below.

Assuming that the deformation is such that plane sections, normal to the shaft centre line remain plane and that the distance of any given point to the centre line remains unchanged, all strains are zero except,

$$e_{r\theta} = \frac{1}{2}\left(\frac{\partial v}{\partial r} - \frac{v}{r}\right) \qquad e_{\theta z} = \frac{1}{2}\frac{\partial v}{\partial z}$$

where v is the displacement along the tangential (hoop) direction. The equilibrium equations are reduced to

$$\frac{\partial \sigma_{r\theta}}{\partial r} + \frac{\partial \sigma_{z\theta}}{\partial z} + 2\frac{\sigma_{r\theta}}{r} = 0$$

Defining the stress function of section 5.2.2, this equation is satisfied. Eliminate v and its derivatives from the expression of the shear strains and express these in terms of the stress function ϕ.

5.3 Defining the stress function as in section 5.2.2, draw lines of constant ϕ for the shaft shown in Fig. P-2.

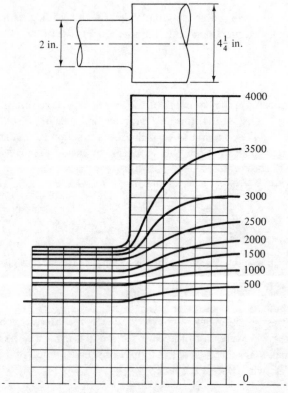

Fig. P-2

5.4 The centrifugal stresses produced in a turbine wheel were determined by R. Guernsey (*Proc. Soc. Exp. Stress Analysis*, Vol. 18, No. 1, p. 1) using the photoelastic model shown in Fig. P-3(a). Compare the stresses, measured on the wheel midplane by Guernsey with those that may be predicted by applying the method described in chapter 5. Note that the experimental values plotted in Fig. P-3(b) have been referred to the hoop stress at the periphery of the wheel; this stress is taken to be equal to unity.

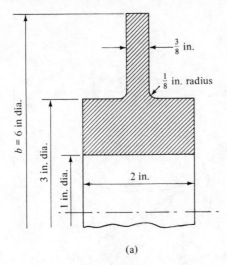

(a)

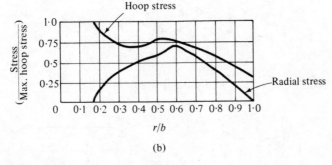

(b)

Fig. P-3

Problems on Chapter 6

6.1 Obtain expressions for the thermal stresses in a ring of rectangular cross-section, inside radius a, outside radius b and height h under linearly varying temperature between T_i (inside) and T_o (outside). The temperature is uniform in the axial direction. Treat the ring (a) as a thin disc, (b) as a thick cylinder. Compare the expressions with the approximate equations derived in chapter 6, Example 2.

6.2 Show that in the general shell of revolution illustrated in Fig. P-4(a), under internal pressure,

$$\frac{\sigma_{\theta\theta}}{R_\theta} + \frac{\sigma_{\phi\phi}}{R_\phi} = \frac{p}{t}$$

where p is the internal pressure and t the shell thickness. Find expressions for the membrane stresses in the conical shell and in the toroidal shell of Fig. P-4(b).

$$\left[\begin{array}{l}
\text{conical shell,} \quad \sigma_{\phi\phi} = \dfrac{pz}{2t}\dfrac{\sin\alpha}{\cos^2\alpha} \qquad \sigma_{\theta\theta} = 2\sigma_{\phi\phi} \\[3ex]
\text{toroidal shell,} \quad \sigma_{\phi\phi} = \dfrac{pr}{2t}\dfrac{2R + r\sin\varphi}{R + r\sin\varphi} \qquad \sigma_{\theta\theta} = \dfrac{pr}{2t}
\end{array}\right]$$

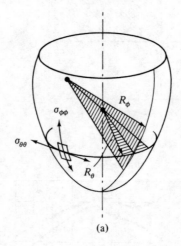

(a)

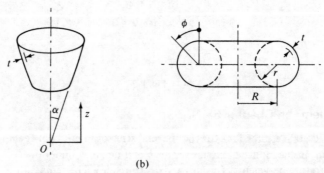

(b)

Fig. P-4

6.3 A thin hemispherical shell of uniform thickness t and radius R is supported on a smooth, horizontal floor and subjected to its own weight, of intensity q

per unit area. Show that the membrane stresses are given by,

$$\sigma_{\phi\phi} = -\frac{qR}{t}\frac{1}{1+\cos\phi} \qquad \sigma_{\theta\theta} = \frac{qR}{t}\left(\frac{1}{1+\cos\phi} - \cos\phi\right)$$

where ϕ is measured from the crown.

6.4 A spherical dome, as shown in Fig. P-5 carries a load w per unit length. Show that the membrane stresses are given by,

$$\sigma_{\phi\phi} = \frac{qR}{t}\frac{\cos\alpha - \cos\phi}{\sin^2\phi} + w\frac{\sin\alpha}{\sin^2\phi}$$

$$\sigma_{\theta\theta} = \frac{qR}{t}\frac{\cos\alpha - \cos\phi - \cos\phi\sin^2\phi}{\sin^2\phi} + w\frac{\sin\alpha}{\sin^2\phi}$$

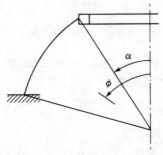

Fig. P-5

6.5 A boiler drum end has the shape of half an ellipsoid of revolution rotated about its minor axis. Show that for uniform pressure p, the membrane stresses are,

$$\sigma_{\phi\phi} = \frac{P}{2b^2 t}(a^4 y^2 + b^4 x^2)^{\frac{1}{2}}$$

$$\sigma_{\theta\theta} = \frac{p(a^4 y^2 + b^4 x^2)^{\frac{1}{2}}}{b^2 t}\left[1 - \frac{a^4 b^2}{2(a^4 y^2 + b^4 x^2)}\right]$$

where a is the semi-major axis and b the semi-minor axis.

6.6 The curves of Fig. P-6 show the variation of the stress concentration factor (S.C.F.) in a spherical shell with a radial branch under internal pressure. The S.C.F., defined as the ratio between the maximum stress near the junction and the membrane stress in the spherical shell, is plotted against the parameter

$$\rho = \frac{d}{D}\left(\frac{D}{2T}\right)^{\frac{1}{2}}$$

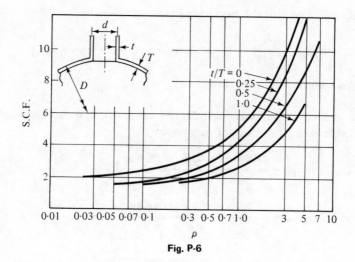

Fig. P-6

for various values of the ratio between the thickness of the branch and that of the shell.

Justify the use of the diagram. (See, for example, M. B. Bickell and C. Ruiz, *Pressure Vessel Analysis and Design.*)

Problems on Chapter 7

7.1   A strain gauge is used to measure the strain at the edge of a circular hole of radius 1 in. on a wide plate. The plate is under uniaxial tension in the y-direction and the gauge is placed along the x-direction, at 90° to y. What is the maximum gauge length if the maximum error due to the length effect is to be limited to 5%? What is the error if a commercially available gauge of $\frac{1}{8}$ in. length is used? Take $\nu = 0.3$.

(*Ans.*   0·019 in., 28·5%.)

7.2   The transversal sensitivity of a strain gauge is 0·04. Determine the resulting error when it is used to measure strains on steel ($\nu = 0.3$) under the conditions, (a) uniaxial strain, (b) biaxial compression $e_{xx} = e_{yy}$, (c) pure shear $e_{xx} = -e_{yy}$. What is the error when two identical gauges are used to find the Poisson's ratio in a uniaxial tension test?

(*Ans.*   −1·2%, −5·2%, 2·8%, 13·9%.)

7.3   A steel bar in uniaxial tension has two strain gauges fixed on its surface, one in the direction of the load and the other in the transversal direction. When the stress in the steel bar is 10 000 lb/in^2, the relative changes in resistance are 0.69×10^{-3} for the longitudinal gauge and -0.195×10^{-3} for the transversal

308

gauge. Find the gauge factor and the transversal sensitivity if $E = 30 \times 10^6$ lb/in^2 and $v = 0.3$ for steel.

(*Ans.* 2·07, 0·019.)

7.4 Calculate the transversal sensitivity of the following two gauges: (A) grid 1 in. long, $\frac{1}{8}$ in. wide, 6 strands. (B) grid $\frac{1}{8}$ in. long, 1 in. wide, 48 strands.
For each gauge, calculate:

(a) The gauge factor, in simple tension, on a steel bar of $v = 0.3$ if the gauge factor for a single strand is 2·05.

(b) The nominal gauge factor for both gauges is given by the manufacturer as equal to 2. Find the true gauge factor when they are used in a steel shaft in torsion (pure shear, $e_{xx} = -e_{yy}$) and when they are fixed on a steel bar in simple tension, at 90° to the direction of the load.

(c) What is the error in the measured strain if the nominal gauge factor is used in the previous questions?

(*Ans.* 0·0208 and 0·1667; 2·0 and 1·66; 1·97 and 1·76; 1·88 and 0·94; errors for gauge A, 0, 1·5%, 6%; for gauge B, 17%, 12%, 53%.)

7.5 If the gauges A and B of the previous question are fixed to a closed cylindrical vessel under internal pressure and the nominal gauge factor is equal to 2, calculate the true gauge factor when the gauges are along the hoop direction and when they are along the longitudinal direction. Take $v = 0.3$.

(*Ans.* Gauge A, 2·02, 2·19; gauge B, 2·18, 3·6.)

7.6 Calculate the angle between the direction of zero strain and the direction of tension in a simple tension test with a steel specimen ($v = 0.3$). If the gauges A, B, of the previous questions are fixed in this direction, calculate the indicated strains in percent of the maximum tensile strain.

(*Ans.* 61·3°, 1·45%, 11·7%.)

7.7 A gauge has a transversal sensitivity 0·03 and $k = 2$ when calibrated on a steel base ($v = 0.3$). Calculate the gauge factor for a single strand of the strain sensing material and the gauge factor when the Poisson's ratio of the base is 0·24 and when it is 0·5, in simple tension.

(*Ans.* 2·02, 2·0, 1·99.)

7.8 Calculate the principal strains and stresses and their direction and the maximum shear stress in a steel plate in the following cases:

(a) Strains measured along direction 0°, 45°, and 90° with rectangular rosette, 250×10^{-6}, -160×10^{-6}, -50×10^{-6} respectively.

(b) Strains measured along direction $0°$, $120°$, and $240°$ with delta rosette, 250×10^{-6}, 250×10^{-6}, -200×10^{-6}.

(*Ans.* 400×10^{-6}, -200×10^{-6}, $30°$.)

7.9 Four strain gauges, connected to a Wheatstone bridge, are fixed to a bar under simple tension to make a load cell. Show the orientation of the gauges in order to obtain maximum sensitivity and to cancel any bending components that may appear due to slight eccentricity in loading.

7.10 A Wheatstone bridge consists of two 1000 ohm resistors and two strain gauges (one active, one dummy for temperature compensation) both of 100 ohm, $k = 2\cdot0$. The maximum current through the strain gauges is of 25 mA and the bridge is to be used to measure strains of less than 10 000 microstrain following the deflexion method. It is connected to a millivoltmeter with an impedance of 10 000 ohm and a sensitivity of $0\cdot1$ m scale deflexion/mV. Determine

(a) the bridge voltage for maximum sensitivity,
(b) the sensitivity.
(c) the deflexion for 10 000 microstrain,
(d) the departure from idealized open-circuit behaviour caused by the millivoltmeter.

(*Ans.* 5 V; $2\cdot37 \times 10^{-3}$ mm microstrain; $23\cdot7$ mm; $5\cdot2\%$.)

7.11 One strain gauge is connected to form a potentiometer circuit with a ballast resistor of 2000 ohm. The gauge resistance is 1000 ohm and $k = 2\cdot0$. The maximum current through the gauge is 15 mA and a cathode-ray oscilloscope of sensitivity 10 mm/mV is connected to the gauge terminals. The strain gauge is fixed to a vibrating beam and subjected to a cyclic strain amplitude of 1000 microstrain. Determine

(a) the maximum input voltage to the potentiometer,
(b) the cathode-ray oscilloscope trace deflexion.

(*Ans.* 45 V, 300 mm.)

7.12 Materials used in photoelasticity have the following fringe values f (lb/in^2/in per fringe) and breaking strength (lb/in^2),

CR39	$f = 100$	UTS = 3600
Perspex	$f = 500$	UTS = 6000
Glass	$f = 800$	UTS = 10 000

Determine the number of fringes that can be shown by a model $\frac{1}{4}$ in. thick of these materials and estimate the highest advisable number of fringes.

(*Ans.* 9, 3, 3; 6, 2, 3.)

7.13 Assuming that the above fringe values have been obtained with mercury light ($\lambda = 5 \cdot 461 \times 10^{-7}$ m) determine the corresponding values when sodium light is used instead ($\lambda = 5 \cdot 893 \times 10^{-7}$ m).

(*Ans.* × 1·08.)

7.14 A model of CR39, $\frac{1}{4}$ in. thick shows six fringes in sodium light and under a load of 20 lb. Determine the stress under a load of 4000 lb in a similar steel component 1 in. thick and twelve times larger in the plane of the plate than the model.

(*Ans.* 10 800 lb/in^2.)

7.15 A photoelastic model in the shape of a beam, 0·625 in. deep and 0·125 in. thick is stressed in pure bending. When a bending moment 58·5 lb/in. is applied, the ninth fringe just appears at the outer edge. Sketch the fringe distribution and calculate the fringe values of model and material.

(*Ans.* 100 lb/in per fringe; 800 lb/in per fringe.)

7.16 A photoelastic model in the shape of a long tension strip, 0·789 in. wide 0·136 in. thick, has a central hole 0·256 in. diameter and is subjected to a uniform tensile stress σ_0 away from the hole. A total force of 183 lb applied at the ends of the strip produces a fringe of order 9 at a point A on the edge of the hole. If the material fringe value f is 86 lb/in per fringe, calculate the stress at A, the ratio (σ_A/σ_0), and the stress concentration factor (σ_A/σ_0), where σ_m is the mean stress over the reduced cross-section at the hole.

(*Ans.* 3·33, 2·25, 5700 lb/in, 1710 lb/in.)

7.17 A photoelastic beam model 0·9 in. high, 0·25 in. thick is freely supported at the ends of a span of 3 in. and loaded by a concentrated force of 100 lb acting at the mid-span on the upper edge. Sketch an approximate stress pattern of the mid-portion of the beam. If the load produces 12 fringes at the most highly stressed point P of the bottom edge, calculate the fringe value of the material and its stress-optic coefficient in Brewsters. The light used is monochromatic of $\lambda = 5 \cdot 461 \times 10^{-7}$ m. What fringe value would you expect on the neutral axis of bending just off the midplane of the model?

(*Ans.* 46 lb/in per fringe, 6·76 B, 3·62.)

7.18 When $\sigma_1 > 0$ and $\sigma_2 < 0$, show the plane upon which the maximum shear stress acts. Show the position of this plane when $\sigma_1 > 0$ and $\sigma_2 > 0$.

7.19   Given a fringe order of 5, a model thickness of 0·250 in., a material fringe value of 90 lb/in^2/in. per fringe and an isoclinic parameter of 20° defining the angle between σ_1 and the x-axes, determine the shear stress σ_{xy}.

(*Ans.* 584·5 lb/in^2.)

7.20   For the determination of the fractional fringe order at a certain point, a compensator is to be made of CR39 (f = 94 lb/in^2/in. per fringe, maximum load 3000 lb/in^2) with a constant width of 0·5 in. and tapering thickness or with a constant thickness of 0·5 in. and tapering width. Calculate the required tension for 14 fringes to appear in each compensator. Indicate whether the tension axis of the compensator should be placed parallel with or normal to the edge of the model.

(*Ans.* 658 lb; normal.)

References

Benham, P. P. and Hoyle, R. D.: *Thermal Stress,* Pitman & Co., London (1964).

Bickell, M. B. and Ruiz, C.: *Pressure Vessel Design and Analysis,* Macmillan & Co., London (1967).

Bijlaard, P. P. and Dohrmann, R. J.: 'Thermal Stress Analysis of Irregular Shapes', *Trans. A.S.M.E.* Ser. B, **83**, 467 (1961).

Cauchy, A. L. (see Love, A. E. H., p. 8).

Dugdale, D. S.: *Elements of Elasticity,* Pergamon Press, Oxford (1968).

Eshelby, J. D.: *Fracture Toughness,* Iron and Steel Institute Publication 121, p. 36, London (1968).

Flügge, W.: *Stresses in Shells,* Springer-Verlag, Berlin (1960).

Frocht, M. M.: *Photoelasticity,* Vol. 1, p. 232, John Wiley & Sons, New York (1941).

Guernsey, R.: 'Photoelastic Study of Centrifugal Stresses in a Single Wheel and Hub', *Proc. Soc. Exp. Stress Analysis.* **18**, No. 1, 1 (1961).

Hertz, H. (see Timoshenko, S. and Goodier, J. N., p. 382).

Heywood, R. B.: *Designing against Fatigue,* Chapman & Hall Ltd., London (1962).

Hill, R.: *Theory of Plasticity,* Clarendon Press, Oxford (1950).

Inglis, C. E.: 'Stresses in a Plate due to the Presence of Cracks and Sharp Corners', *Trans. Inst. Naval Architects,* **55**, 219 (1913).

Kirchhoff, G. (see Timoshenko, S. and Woinowski-Krieger, S., p. 83).

Lagrange, J. L. (see Timoshenko, S. and Woinowski-Krieger, S., p. 82).

Lamé, G. (see Love, A. E. H., p. 142).

Love, A. E. H.: *Theory of Elasticity,* 4th ed., Dover Publications, New York (1944).

Lowell, H. H.: National Aeronautics and Space Administration, Technical Report No. 32, 1959.

Muskhelishvili, N. I.: *Some Basic Problems on the Mathematical Theory of Elasticity,* P. Noordhoff Ltd., Gröningen (1953).

Nadai, A.: *Theory of Flow and Fracture of Solids,* Vol. 1, p. 103, McGraw-Hill Book Company, Inc., New York (1950).

Novozhilov, V. V.: *The Theory of Thin Shells,* P. Noordhoff, Ltd., Gröningen (1959).

Parker, E. R.: *Brittle Behavior of Engineering Structures,* John Wiley & Sons, Inc., New York (1957).

Prescott, J.: *Applied Elasticity,* Dover Publications, Inc., New York (1961).

Rayleigh, Lord: *Theory of Sound,* Vol. 1, p. 359, Dover Publications, Inc., New York (1945).

Relton, F. E.: *Applied Bessel Functions,* p. 143, Blackie & Son Ltd., London (1946).

Sokolnikoff, I. S.: *Mathematical Theory of Elasticity,* 2nd ed., p. 25, McGraw-Hill Book Company, Inc., New York (1956).

Southwell, R. V.: *Relaxation Methods in Engineering Science,* Oxford University Press, London (1956).

Southwell, R. V.: *Theory of Elasticity,* 2nd ed., p. 15, Oxford University Press, London (1941).

Stokes, G. G. (see Love, A. E. H., p. 12).

Timoshenko, S. and Goodier, J. N.: *Theory of Elasticity,* McGraw-Hill Book Company, Inc., New York (1951).

Timoshenko, S. and Woinowski-Krieger, S.: *Theory of Plates and Shells,* 2nd ed., McGraw-Hill Book Company, Inc., New York (1959).

Westergaard, H. M.: 'Bearing Pressures and Cracks', *J. Appl. Mech.,* **6,** A-49 (1939).

Index

Printed by William Clowes & Sons Limited, London, Colchester and Beccles